Advances in Simulation

Volume 1

Editors:

Paul A. Luker
Bernd Schmidt

A. Sydow S. G. Tzafestas R. Vichnevetsky (Editors)

Systems Analysis and Simulation I

Theory and Foundations

Proceedings of the International Symposium
held in Berlin, September 12-16, 1988

With 110 Figures

Springer-Verlag New York Berlin Heidelberg
London Paris Tokyo Hong Kong

Editors:

Achim Sydow
Akademie der Wissenschaften der DDR
Zentralinstitut für Kybernetik und Informationsprozesse
Berlin, GDR

Spyros G. Tzafestas
National Technical University of Athens
Computer Engineering Division
Athens, Greece

Robert Vichnevetsky
Rutgers University
Department of Computer Science
New Brunswick, NJ, USA

Series Editors:

Paul A. Luker
California State University
Department of Computer Science
Chico, CA, USA

Bernd Schmidt
Universität Erlangen-Nürnberg
Institut für Informatik
Erlangen, FRG

Licensed edition for
Springer-Verlag NewYork Berlin Heidelberg London Paris Tokyo Hong Kong

With exclusive distribution rights for all nonsocialistic countries:
Springer-Verlag NewYork Berlin Heidelberg London Paris Tokyo Hong Kong

With exclusive distribution rights for all socialistic countries:
Akademie-Verlag Berlin

ISBN 0-387-97091-6 Springer-Verlag NewYork Berlin Heidelberg
ISBN 3-540-97091-6 Springer-Verlag Berlin Heidelberg NewYork

Library of Congress Cataloging-in-Publication Data
Systems analysis and simulation: proceedings of the international symposium held in Berlin, September 12-16, 1988 /
A. Sydow, S.G. Tzafestas, R. Vichnevetsky (editors).   (Advances in simulation; v. 1-2)
Papers from the 3rd International Symposium for Systems Analysis and Simulation, sponsored by the Central Institute
of Cybernetics and Information Processes of the Academy of Sciences of the GDR, and others.
Includes bibliographical references.   Contents: [1] Theory and foundations – [2] Applications.
ISBN 0-387-97091-6 (v. 1: alk. paper). – ISBN 0-387-97093 (v. 2: alk. paper)
1. System analysis – Congresses. 2. Simulation methods – Congresses. I. Sydow, Achim, II. Tzafestas, S.G. 1939 -.
III. Vichnevetsky, Robert. IV. International Symposium on Systems Analysis and Simulation (3rd: 1988 : Berlin,
Germany) V. Akademie der Wissenschaften der DDR. Zentralinstitut für Kybernetik und Informationsprozesse.
VI. Series.   T57.6.S9769   1989   003–dc20 89-19669 CIP

Offsetprinting: VEB Druckerei »Thomas Müntzer«, Bad Langensalza
Binding: Lüderitz & Bauer, Berlin
2849/3020-543210 Printed on acid-free paper

Preface

The present volume contains the papers which were accepted for presentation at the 3rd International Symposium for Systems Analysis and Simulation held in Berlin (GDR), September 12–16, 1988.

It is already a tradition to meet a broad international community of experts in systems analysis, modelling and simulation at this symposium. This fact shows the requirements for a forum of presentation and discussion of new developments and applications of modelling and simulation in systems analysis.

To realize the great interest in this field one has to take into consideration the developed role of computer simulation as a powerful tool of problem solving. More and more areas in sciences and production have been investigated by mathematical models and computer simulation. Biological sciences and social sciences are even by now influenced by this trend.

The model use on the computer has been very much improved in decision support systems. Parallel simulation will provide drastic shortening of computing time. Parallel simulation and model based decision support systems are brought in the focus of international activities.

Numerical mathematics, systems theory and control sciences provide with algorithms supporting the modelling process itself based on simulation or analytic methods. Such simulation systems equipped with tools for modelling and graphics for representing results are real model support systems.

A new important impact comes from artificial intelligence by knowledge processing. Expert systems may help decision making in case of missing mathematical models. Expert systems may also support teaching and using simulation systems.

New application areas are investigated. Complex systems with multicriteria control problems are in the scope of the symposium as well as problems of qualitative analysis of small scale nonlinear systems. Applications in engineering sciences, economy and management, natural sciences and social sciences are examined but also mixed problems from different areas.

The state of computer technique and programming environment set up efficient conditions for simulations. Personal computer are even used for simulation more and more.

The symposium reflects the state of the art and trends in systems analysis, modelling and simulation.

The 3rd International Symposium is organized by the Central Institute of Cybernetics and Information Processes of the Academy of Sciences of the GDR (ZKI) with cosponsorship of the

— International Association for Mathematics and Computers in Simulation (IMACS),
— International Federation of Automatic Control (IFAC),
— International Institute for Applied Systems Analysis Laxenburg (IIASA),
— Scientific Society of Measurement and Automation (WGMA) in the Chamber of Technology (KdT) of the GDR,
— Mathematical Society (MG) of the GDR.

The papers included in these proceedings were not formally refereed. The authors themselves are fully responsible.
The international Program Committee consisted of:

W. Ameling (FRG), P. Borne (France), L. Dekker (The Netherlands), S. Deng (PRC), A. A. Dorodnicyn (USSR), K. H. Fasol (FRG), W. Findeisen (Poland), O. I. Franksen (Denmark), V. Hamata (Czechoslovakia), C. Hu (PRC), A. Javor (Hungary), K. Kabes (Czechoslovakia), V. V. Kalashnikov (USSR), V. Kempe (GDR), E. J. H. Kerckhoffs (The Netherlands), R. Klötzler (GDR), R. Kulikovski (Poland), A. Kurzhansky (Austria, USSR), N. Levan (USA), A. H. Levis (USA), T. I. Ören (Canada), M. Peschel (GDR), P. D. Dieu (Vietnam), F. Pichler (Austria), K. Reinisch (GDR), W. Schirmer (GDR), B. Schmidt (FRG), V. V. Solodovnikov (USSR), F. Stanciulescu (Romania), J. M. Svirezhev (USSR), M. Thoma (FRG), I. Troch (Austria), S. G. Tzafestas (Greece), G. C. Vansteenkiste (Belgium), R. Vichnevetsky (USA), A. Sydow (GDR).

Many thanks should be given to the members of this committee for the very helpful cooperation.
Special thanks are said to Prof. Dr. V. Kempe, Director of the ZKI, for his great support in preparing and performing the symposium. Furthermore, great gratitude is to express to

Prof. Dr. R. Vichnevetsky (USA), IMACS-President,
Prof. Dr. B. Tamm (USSR), IFAC-President,
Prof. Dr. R. H. Pry (USA), IIASA-Director,
Prof. Dr. W. Richter (GDR). Chairman of WGMA,
Prof. Dr. R. Klötzler (GDR), Chairman of MG,

for help and encouragement.

A lot of the hard preparation work was done by the Department for Systems Analysis and Simulation of the Central Institute of Cybernetics and Information Processes. The editor expresses his thanks to all colleagues and friends who were very much engaged in the research work and in the preparation. First of all I would like to name Dr. P. Rudolph and Dr. A. Wittmüß who helped to prepare the proceedings. Furthermore I thank these colleagues and Dr. K. Bellmann, Dr. W. Jansen, Dr. E. Matthäus, Dr. R. Straubel and all the other colleagues for engaged cooperation for years in developing this research area. Mrs. Ch. Fröhlich and Mrs. J. Obretenov should be named for speedy service in preparing the manuscript. Mrs. S. Böttcher made an excellent job as organizer.
Last not least I would give my thanks to the publishers, especially Mrs. R. Helle and Mrs. G. Reiher, for their assistance and cooperation.

Finally I would like to express my expectation also on behalf of the coeditors Prof. Dr. S. G. Tzafestas and Prof. Dr. R. Vichnevetsky that also the third symposium will be a contribution to the further development in systems analysis, modelling and simulation as well as a place for cooperation and communication like the first both.

April 1988

Achim Sydow
On behalf of the editors

Table of Contents

1.5. Multiobjective Optimization

2. Simulation Techniques

2.1. Simulation of Discrete Systems

2.2. Simulation of Continuous Systems

2.3. Parallel Simulation

List of Authors

A Language to Describe and to Simulate Digital Systems

Garte, Dieter; Haufe, Juergen; Ruelke, Steffen [1]

1. Introduction

On basis of mixed-level simulator KOSIM /3/,/5/ a software is represented to allow a description and simulation from network with logical elements and designer defined functional units. The used description language abbreviated HB consists of a structural part for describing network of elements and units and of a functional part for describing functions of used elements and units. The network description makes use of NBS84 language /4/ and the KOSIM input language. Both were extended by BOOLEAN equations that by means of precompiler are transformed in the format of convential NBS84 or KOSIM input language as the description dimension of BOOLEAN equations is smaller than the description of the above convential net description.

The elements are standard functions of a library. The designer defined functions are described either in the procedural language F77 for the algorithmical level or in the known hardware description language DDL - Digital Description Language /2/ - for the register transfer level. F77 only allows to describe sequential algorithm, whereas DDL can describe parallel processes, too. Moreover, you can use a DDL subset to describe a set of BOOLEAN equations. The F77 units are integrated part of KOSIM simulator. The DDL part and the BOOLEAN equations are new. As further functional part the instructional set process language ISPS is investigated to be used /1/ in this system in order to integrate it in future in equal way as F77 or DDL. According to catalogue of state-of-the-art CAD tools for the design of VLSI IC's /6/ there is moreover the architectural level represented by the PMS language. This language is not yet included in the HB language.

For the purpose of structural and functional simulation of digital systems, a particular simulator of each DDL (BOOLEAN equations or F77) designer defined function is generated and linked to simulator KOSIM. The complete system is illustrated in fig. 1.

The verification of every DDL unit is performed separately because it is inefficient within the structural and functional simulator and because the internal facilities of DDL unit are not directly visible. For that reason, the generated particular DDL simulator is linked to a simulator frame developed for it. The designer view for this simulator was developed analogously to the designer view of KOSIM simulator. The DDL simulator includes time controlled data input like the KOSIM simulator.

The HBD-system runs on the computer K1840 from ROBOTRON and applies operating system SVP1800.

HB language description

The HB description language is based on the NBS84 and KOSIM network description language extended for designer defined functions. It consist of a structural part for describing network of elements and several designer defined units and a functional part for describing these elements and these designer defined units included the pin functions of the circuit. Within the network each element or unit is featured by its name, its type and its terminals. Optionally you can indicate parameters for instance time delay for the outputs. The standard delay is one time unit. All these informations have to meet the type declaration in the functional part. The terminals can be busses, too.

1) Central Institute of Cybernetics and Information Processes, Kurstrasse 33, Berlin, 1080, GDR

Here is an example (fig. 2): a controlled 8 bit circle counter.
It consists of 2 elements: AND1 of type AND and OR1 of type OR and 2
shifter register units SR1 and SR2 of type SR.

First, there is a connection description for AND1, OR1, SR1 and
SR2 with the name of element or unit, connection list and type of
element or unit. Moreover, the structural description of the
functions for the pins INP, TOR and RING by a table controlled
generator INPUT of type TABSD and for the pin TAKT by the clock
generator LPER is following. The types TABSD and LPER are standard
types of the KOSIM-library.

Second, the element declarations are following beginning with
'E: ' for the types AND, OR, TABSD and LPER and the unit declaration
for the type SR beginning with 'P: ', latter are continued with the
language type the functional description 'DDL'. In the case of
BOOLEAN equations units is continued with 'BOOLE', of F77 units with
'F77'. The syntax of all unit terminals is equal to that of the
elements. Input, output and bidirectional terminals are following in
turn. Another modified DDL description – a 4 bit counter description
– is shown in fig. 3.

BOOLEAN equations and register transfers can be declared outside
automata (global declaration) or within the automata (local
declaration). Moreover, they can be declared within the automaton
outside the state declaration or within the state declaration.

Accordingly, the header of SR declaration in HB language of this
example is equivalent to
        <SY> SR:
        <TI> TKT.
        <TE> EIN, TOR, OUT.
in DDL language.

3. DDL model and simulation

The register transfer language DDL is based on the multiautomaton
model (see fig. 4). For the simulation this model is transformed in
to a single automaton model (see fig. 5) which consists of a set of
BOOLEAN equations and a set of register transfers. The set of BOOLEAN
equations is solved by means of static simulation. Incorrect cycles
in the BOOLEAN equations are indicated. Features the separate DDL
simulator are:
– The separate DDL simulator frame is based on the developed DDL
debugger. Its commands are formally identically with the debugger
commands of the used operating system SVP 1800. Its results shown a
compromise between the system debugger results and the KOSIM simulator
results.
– Refering to its defined terminals there are only inputs, outputs and
internal terminals but no bidirectional terminals. The latters are
necessary to link DDL units in a circuit network. Therefore, they
have to be considered. It depends on automaton state whether the
bidirectional terminal operates like an input or like an output. As
an input the terminal is on the right side of a register transfers or
connections otherwise as output on the left hand side. In dependence
on given internal conditions the bidirectional terminal in the output
mode is set to lower impedance otherwise to higher impedance.
– The implication of the DDL system clock (<TI> clock.) is shown in
fig. 6 (see also part 4.). Especially if no system clock is defined
the DDL unit is started by an event of its inputs and works
autonomously as long as no internal state changes occur This
interpretation is based on the process model from ISPS.
– In order to reduce the description predefined operators are
developed as usual in KOSIM and ISPS. In the HB system there is a
library of these predefined operators. These operators can be called
as macros.

4. HB model and simulation

The HB simulator is based on the event oriented simulator KOSIM also. The event oriented call of the DDL unit within KOSIM simulator separated in a BOOLEAN equation call and in a register transfer call accordingly fig. 5 is shown in fig. 6. Here for the evaluation of the combinatoric part from fig. 5 the BOOLEAN call is present and for the evaluation of the sequence part the register transfer call.

The visibility of the internal DDL facilities is obtained by the developed DDL debugger (see also part 3.).

5. Design strategy and simulation dialogue

The HB system can be used by the top down design (see fig. 7) started for example on the register transfer level and stopped on the logic gate level. On this occasion the DDL units are transformed into a network of logical standard gate functions.

The HB dialogue commands are deduced from this strategy and support it.

6. Examples

The following examples characterize the application volume of HB language:
- counters, shifters
- asynchronous automata,
- automata net, coupled automata,
- processors, multi processor systems and cellular automata.

7. References

/1/ Barbacci, M. R.; Siewioreck, D. P.:
"The design and analysis of instruction set processes" McGraw Hill, 1982
/2/ Dietmeyer, Donald L.:
"Logic Design of Digital Systems" Allyn and Bacon, Inc., 470 Atlantic Avenue, Boston 1971
/3/ Donath, U.; Schwarz, P.; Trappe, P.:
"Dynamische Logiksimulation auf Bit- und Wortniveau" 19. Fachkolloquium Informationstechnik Dresden 1986
/4/ Issel, W. et al.:
"NBS-84: A structural description language for VLSI design", Circuit theory and design 85, Proceedings of the 1985 European Conference, p. 62-5
/5/ Schwarz, P.:
"A program for the mixed-level simulation of digital integrated circuits" Proc. ECCTD'85, Prag 1985, p. 133-136
/6/ Siewioreck, D. P. et al.:
"Proposal for resaerch on DEMETER — a design methodology and environment" Research Report No. CMUCAD-P3-5, Jan. 1983, CMU

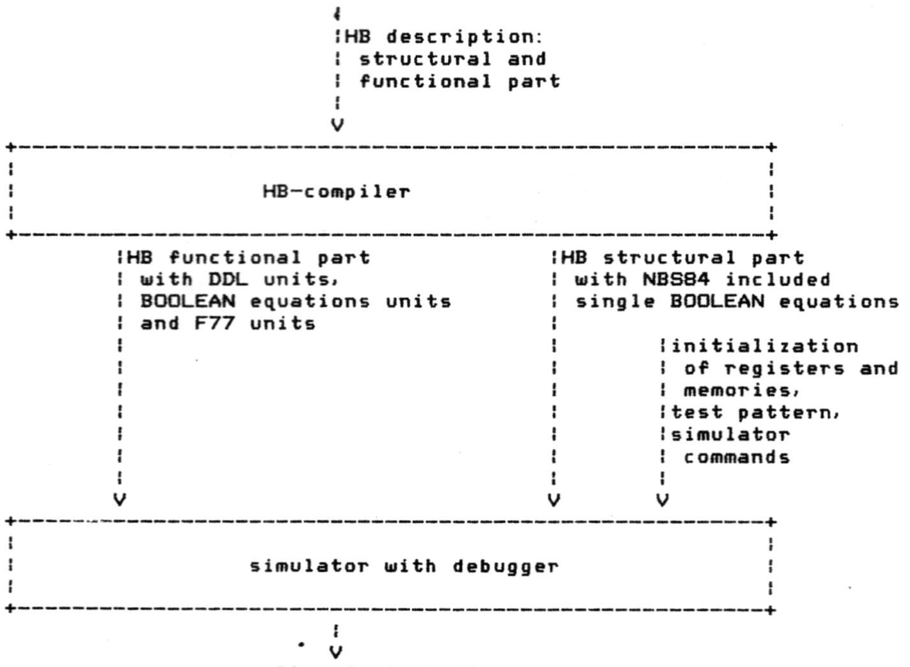

Fig. 1:   View of the system.

```
! CIRCLE COUNTER:
! 2 DDL SHIFT REGISTERS INCLUDED STANDARD AND
! INPUT ELEMENTS WITH KOSIM INPUT LANGUAGE LINKED
!
! TKT      >------------------------*--------------.
!                                    !             !
! TOR      >------------------*--------------.     !
!                             !      !SR_1 !       !SR_2
!                   OR_1      ! +---------+ !   +---------+
!                   +---+   '-!       ! '---!         !OUT2
! INP      >---------!   !   ! SR 'DDL'!   ! SR 'DDL'!--*->
!          +---+     !OR !----! (R1..R4)!-----! (R1..R4)!  !
! RING     >-!   ! .-!   !EIN1+---------+ OUT1+---------+  !
!            !AND!-' +---+                                 !
!          .-!   !' V1                                     !
!          !  +---+                                        !
!          ! AND_1                                         !
!           '-------------------------------------------------'
! CIRCUIT HEADER:
S:        RING/L
**LOGIK
! CONNECTION DESCRIPTION:
          AND_1     RING - OUT2 - V1            =AND
          OR_1      INP  - V1   - EIN1          =OR
          SR_1      TKT  - EIN1 - TOR - OUT1    =SR
          SR_2      TKT  - OUT1 - TOR - OUT2    =SR
! PIN DESCRIPTION
          INPUT     INP  - TOR  - RING          =TABSD
          TAKTGEN TKT                           =LPER
! DESCRIPTION OF STANDARD AND PIN ELEMENTS
E:        AND/L   E: *    A: 1 W: TO1=0, T10=0
E:        OR/L    E: *    A: 1 W: TO1=0, T10=0
E:        TABSD/L          A: * W: RCN=0, FILE='Q01'
E:        LPER/L  A: 1 W: S1='0', S2='1', TST=3, TIM=5, TPR=10
##
! DESCRIPTION OF SHIFT REGISTER IN DDL
P:        SR/S    DDL     E: TKT TAKT, EIN, TOR      A: OUT
          $ 'TAKT' IS KEY WORD FOR SYSTEMCLOCK <TI>          $
          <RE> R1, R2, R3, R4.         $ REGISTER DECLARATION $
          <BO> OUT=R4.                 $ OUTPUT CONNECTION     $
                                       $ SHIFT OPERATION       $
              !TOR!    R1<-EIN.,
                       R2<-R1,
                       R3<-R2,
                       R4<-R3
##

Fig. 2: HB description of a circle counter
```

```
P:        ZT/S      DDL E: TAKT TAKT,GO,INC,STOP,DIR          A: A[4]
          $ZT       SYSTEM NAME <SY>                                      $
          $TAKT     CLOCK <TI>                                            $
          $GO       COUNTER START WITH GO=1 <TE>                          $
          $START    COUNTER INITIALIZATION  <TE>                          $
          $INCR     COUNTER INCREMENT       <TE>                          $
          $STOP     COUNTER STOP VALUE      <TE>                          $
          $DIR      COUNTER DIRECTION: O FORWARD,   1 RETURN <TE>         $
          $A[4]     COUNTER OUTPUT <TE>                                   $
<TE>      H1[4].              $ SUPPLEMENT ARRAY                          $
<RE>      INC[4],ST[4],D,X1[4].
                              $ INCR, STOP, DIR, COUNTER REGISTER         $
<BO>      A=X1.
$ EQUIVALENCE OPERATOR FOR COUNTER STOP                                  $
<OP>      EQ(EEE1,EEE2)
          <TE>      EEE1[4],EEE2[4],HHH[4].
          <BO>      HHH=EEE1*EEE2 + ^EEE1*^EEE2,
                    EQ=*/HHH. .
$ ANTIVALENCE OPERATOR FOR ALU                                           $
<OP>      ANT3[4](E1,E2,E3)
          <TE>      E1[4],E2[4],E3[4].
          <BO>      ANT3=E1*E2*E3 + ^E1*^E2*E3 + ^E1*E2*^E3
                    + E1*^E2*^E3..
$ ARITHMETIC LOGICAL UNIT: ALU                                           $
<OP>      ALU[4](X,Y,CIN)
          <TE> X[4],Y[4],CIN, C[4]=CX'CC[3].
          <BO>      C=X*Y +CC'CIN*(X+Y),
                    ALU=ANT3(X,Y,CC'CIN)..
$ COUNTER AUTOMATON                                                      $
<AU>      STAU[2]:          $ COUNTER STATE REGISTER                      $
          TAKT:            $ AUTOMATON CLOCK                              $
          <ST>             $ STATE DECLARATION                            $
                    S1[0]:   GO:       $WAIT AND START STATE              $
                             X1<-START,        ST<-STOP,         D<-DIR,
                             ¦DIR¦ INC<-^INCR; INC<-INCR.,       ->S2.
                    S2[1]:   X1<-H1, H1=ALU(X1,INC,D),$COUNT-STATE        $
                             ¦EQ(H1,ST)¦              ->S1;      ->S2
                             .             $ END OF ¦EQ¦                  $
                             .             $ END OF S2                    $
                             .             $ END OF <ST>                  $
                             .             $ END OF <AU>                  $
##                                         $ END OF <SY>                  $
```

Fig. 3: Modified DDL description of a 4 bit counter

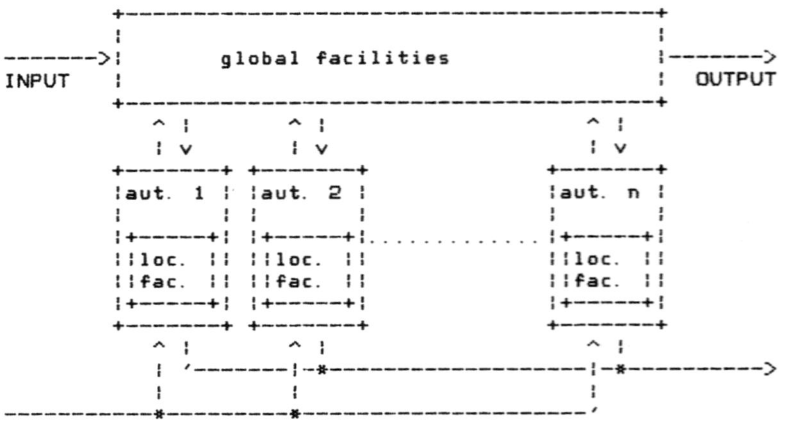

Fig. 4: DDL model

```
DDL description: model of several automata
                          !
                          V
+-----------------------------------------------+
!                                               !
! DDL translator: reduction of automata         !
!                                               !
+-----------------------------------------------+
                 !
                 V
descritpion of the result: model of only one automaton

        -> model DDL simulator:

E(t)       +-------------------------------+
o------->! combinatoric K                  !              A(t)
           !                               !------->*------>o
   .-->! A(t)=K(E(t),R(t))                 !        !
   !       +---------------------------+   !        !
   !                                        !       !
R(t)!      +---------------------------+    !       !
   !       ! sequence S                !    !       !
   '---!                               !<------'
           ! R(t+1)=S(R(t))            !
           +---------------------------+
```

Fig. 5: Principle of the DDL simulator solution with the clock 't'

contents of the DDL unit	clocked	unclocked
BOOLEAN equation	by clock state 'O' and '1' one BOOLEAN call	only one BOOLEAN call
register transfer	DDL separatly: by clock state 1 one register transfer call	only one register transfer call
automaton	KOSIM: by clock slope 1 to O one register transfer call	register transfer calls as long as no internal state changes occur

Fig. 6: Event oriented BOOLEAN and register transfer calls of the DDL
        units within separate DDL simulator and within KOSIM simulator

Design strategy / Top down

```
        (  1.         1 DDL unit
        (
HB      (             <------------------------------->
        (                                                 } increasing
        (  2.         Net several DDL-subunits,           } separation
        (             logical gates and registers         } in control
        (                                                 } and data
        (                                                 } path
        (             <------------------------------->

NBS       3.          Net of logical gates and registers
                      without DDL subunits
```

Fig. 7: Design strategy with HB language

The State Space Approach to the Analysis of Discrete Event Dynamic Systems

Jozef V ö r ö s [1]

INTRODUCTION

There are many complex dynamic systems, evaluation of which is governed not by diferential or difference equations, but by interactions of discrete events. Such systems are known as discrete event dynamic systems /DEDS/. A flexible manufacturing system is a typical example of DEDS, where the completion of a task represents the discrete event.

The behaviour of real processes, which can be considered as DEDS, is generally given by sequences of discrete activities /operations/ carried out with objects passing through the process. Neither activities nor objects can be described and studied involving only continuous variables and traditionally used differential/difference equations. The analysis and synthesis of DEDS require dealing with more and different types of knowledge and information. Models of DEDS, in particular, have to reflect the discontinuous nature of discrete events and the dynamics.

Recently a new approach has been introduced to the analysis of DEDS. The proposed object-oriented method uses so-called nonhomogeneous /NH/ elements for the description of DEDS attributes and their behaviour. This enables computer simulation of a great variety of production processes on different levels of abstraction / 1 /.

The paper deals with an object-oriented approach to the analysis of DEDS from the state space representation point of view. The objects passing through a discrete process are associated with NH-vectors describing them as a complex of relevant information concerning their attributes. Defining algebraic structures and special superstructures on the sets of NH-elements enables to introduce proper mappings which can be used for representation of basic discrete activities in a broad class of DEDS. An extension of n-ary transformation / 2 / provides framework for description of the discrete operations encountered in flexible production systems.

NONHOMOGENEOUS ELEMENTS

Let A_i, i=1,..,n, be mutually disjoint sets of elements /numbers, words, patterns, and so on/. If an ordered n_i-tuple of elements from A_i exists, it will be called a partial vector defined on the set A_i. In the same way a partial matrix is defined on A_i. The partial elements provide the basis for the definition of new mathematical notions being of polyalgebraic nature.

An ordered k-tuple of partial vectors from different A_i, i.e. $x^T = /x_1,\ldots,x_k/$, if $1 < k \leqslant n$, is called a nonhomogeneous vector of $/n_1,\ldots,n_k/$-type. The totoal vector dimension N is equal to the sum of partial vectors dimensions. The set of NH-vectors is assigned as X_N. Similarly, an ordered m-tuple $P = /p_1,\ldots,p_m/$ of N-dimensional NH-vectors p_j is called /N,m/-dimensional nonhomogeneous matrix. The matrix P is

[1] VUNAR – Tool Research Institute, Nove Zamky, CZECHOSLOVAKIA

said to be row-nonhomogeneous, and its transpose is column-nonhomogeneous. The matrix set with the elements of the same type and dimension is assigned as $M_{N,m}$.

The definitions of NH-elements imply that generally different mathematical rules /algebras/ are valid for individual partial components within the globally considered NH-element and cross-operations are inadmissible. In spite of the fact that the NH-elements represent very complex entities, they can be treated as standard terms and on the sets of NH-elements we may define proper mappings and relations.

Relations between NH-elements can be discussed in the same way as those of any set. In accordance with analogical situation arising in the case of ordinary vectors, let us consider the simple transformation of NH-vectors which can be written

$$y = P \cdot x$$

where P is a block-diagonal NH-matrix of corresponding type and dimension. This transformation is nonhomogeneous, too, and consists of individual partial transformations. Then the so-called neutral element e from X_N is such that for any P

$$P \cdot e = e$$

and e is called the empty element — empty NH-vector. The empty NH-matrix E is then such that for any NH-vector x

$$E \cdot x = e \ .$$

It will be useful to introduce the unit NH-matrix L, which realizes the identity transformation, i.e.

$$L \cdot x = x \ .$$

Naturally, both E and L are diagonal NH-matrices of corresponding type.

Given the set of NH-elements we can introduce a special structure on it. Let X_N be a nonempty set of NH-vectors. We define the hierarchy of binded sets by

$$S_N^1 = X_N$$

$$S_N^{i+1} = S_N^i \ I \ X_N$$

i=1,2,..., where the symbol "I" denotes the so-called binding sign. The union S_N of all the sets S_N^i will be called the superstructure defined on X_N. Equally the superstructure $G_{N,m}$ on the set $M_{N,m}$ can be defined by means of the sets given as

$$G_{N,m}^1 = M_{N,m}$$

$$G_{N,m}^{i+1} = G_{N,m}^i \ I \ M_{N,m} \ .$$

Elements of the superstructures are called binded terms and the binding sign represents a concatenation-like operation that connects the elements in a fixed order. However, it is easy to prove that this operation is an inner binary operation defined on the superstructure. The superstructure has the properties of grupoid with respect to the binding operation and since only one neutral element with respect to an inner operation is allowed by definition, it must fulfil

$$e \ I \ e \ I \ ... \ I \ e = e$$
$$E \ I \ E \ I \ ... \ I \ E = E \ .$$

Now assigning as ϵ and $\mathcal{E}$ the binded terms on the left-hand sides of preceding equations, the property of neutral element implies that for any $\xi \in S_N$ we have

$$\xi \, I \, \epsilon = \epsilon \, I \, \xi = \xi .$$

We can define the N-ary NH-transformation as the mapping from X_N into S_N by means of NH-matrix binded term as follows:

$$(P_1 \, I \, P_2 \, I \, \dots \, I \, P_n) \cdot x = y_1 \, I \, y_2 \, I \, \dots \, I \, y_n$$

where $y_i = P_i \cdot x$. This transformation can be generalized as the mapping from S_N into S_N, so the general transformation of binded NH-vectors may be written as

$$\Pi \cdot \xi = \sigma_1 \, I \, \sigma_2 \, I \, \dots \, I \, \sigma_n = \sigma$$

where $\xi = x_1 \, I \, \dots \, I \, x_n$ and $\Pi = \Pi_1 \, I \, \dots \, I \, \Pi_n$ are binded terms of NH-vectorss and NH-matrices, and σ_i is realized as n-ary transformation $\sigma_i = \Pi_i \cdot x_i$.

In the case of general NH-transformation the neutral elements of superstructures will act as follows:

$$\Pi \cdot \epsilon = \epsilon$$
$$\mathcal{E} \cdot \xi = \mathcal{E}$$

and defining the special NH-matrix binded term $\Lambda = L \, I \, L \, I \, \dots \, I \, L$ we have

$$\Lambda \cdot \xi = \xi .$$

On the set of NH-elements the composition operation can be introduced, too. Let x and s are NH-vectors, the the compound NH-vector is given by the composition

$$x(s) = \begin{bmatrix} x \\ s \end{bmatrix} \qquad \text{or} \qquad x = \begin{bmatrix} x_o \\ x_i \end{bmatrix}$$

where x_o is the outer NH-element and x_i is the inner element of compound NH-vector. The transformation of compound vector consists of two parallel simple transformations

$$\begin{bmatrix} P \\ Q \end{bmatrix} \cdot \begin{bmatrix} x \\ s \end{bmatrix} = \begin{bmatrix} P \cdot x \\ Q \cdot s \end{bmatrix} \qquad \text{or} \qquad P \cdot x = \begin{bmatrix} P_o \cdot x_o \\ P_i \cdot x_i \end{bmatrix}$$

We can observe the following property of the empty NH-vector appearing in a compound vector

$$\begin{bmatrix} x \\ e \end{bmatrix} = x \qquad \text{and} \qquad \begin{bmatrix} e \\ x \end{bmatrix} = x .$$

The compound elements can be generalized and defined on the superstructures, too.

DISCRETE EVENT STATE SPACE MODELS

As the proposed method of DEDS analysis is object-oriented, the first step is to declare a proper object representation. An object is assumed to consist of elementary objects-segments. We identify the segment with a proper NH-vector, where its partial vectors correspond to encountered object attributes. Now the object can be associated with a binded term of NH-vectors. A discrete operation on the object may be seen as the general transformation of binded NH-vectors, which is represented by a binded term of NH-matrices.

As the DEDS is given by a sequence of operations performed on objects, a set of objects/segments reflects the recent situations-states during the process. Hence it is reasonable to introduce the state space model of DEDS. As the notion of state is connected with the time flow, we define discrete time units as the differences between the beginnings of corresponding couples of discrete operations. These units are generally nonequal, but it plays no role, because we are interested in the process state after its observable change, irrespective of its duration.

The state space model is based on the notions of input-i, output-o, state-s and time-t, and is uniquely determined by state and output mappings as follows:

$$s(t) = f(s,i,t)$$
$$o(t) = g(s,i,t) \ .$$

Interpretation of this model depends on the application of variables included. It will be appropriate to associate the DEDS outputs and states with objects. The inputs will depend on the type of process and may be omitted in the case of seemingly autonomous systems / 2 /. Naturally, there can be more inputs, states and outputs in the DEDS models.

The kernel of DEDS analysis is now to determine the model variables and mappings. A possible approach is based on the idea of splitting the processed object into a fix part /not changing/ and a variable part /still being processed/. Associating these parts with states we receive a feasible framework for modelling and simulation of technological, assembly and disassembly processes.

The state space model of assembly process consists of two state and one output mappings, as that of machining one / 2 /. Using the before mentioned notions and transformations, the model equations are:

$$A(t) = A(t-1) \ I \left[\begin{matrix} D(t) \\ (L \ I \ \epsilon).B(t-1) \end{matrix} \right]$$

$$B(t) = (E \ I \wedge).B(t-1)$$

$$Y(t) = A(t) \ I \ B(t)$$

where A represents the fix part, B is the variable part of assembled object /machine unit/, D is the input detail and Y is the output assembly. This is proper for the shaft-type assembly processes. For the housing-type assembly the model differs only in the state A, where the compound term has the inner and outer components interchanged. The initial conditions of A and B result from the analysis of given process and concern the base or support parts of machine units.

A disassembly process is the reverse of assembly one, but the corresponding state space model is a little more complex:

$$A(t) = A(t-1) \ I \left[\begin{bmatrix} \epsilon \\ \wedge \end{bmatrix} . \left((L \ I \ \epsilon).B(t-1) \right) \right]$$

$$B(t) = (E \ I \wedge).B(t-1)$$
$$Y(t) = A(t) \ I \ B(t)$$

$$C(t) = \begin{bmatrix} \wedge \\ \epsilon \end{bmatrix} . \left((L \ I \ \epsilon).B(t-1) \right)$$

where the second output represents the disassembled details. This model is proper

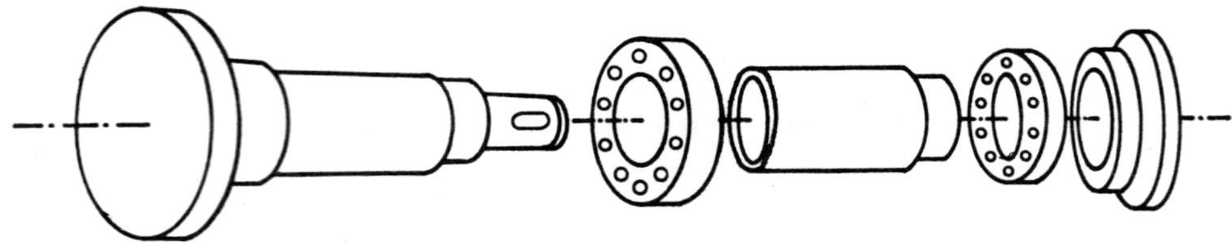

Fig. 1. Example of assembly process

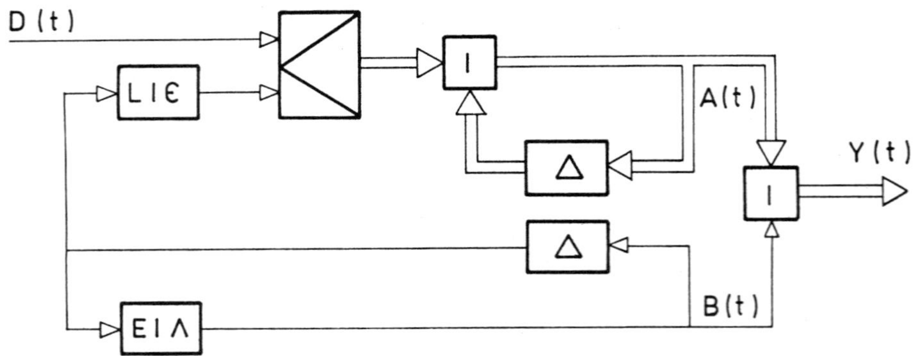

Fig. 2. Flow diagram of assembly system

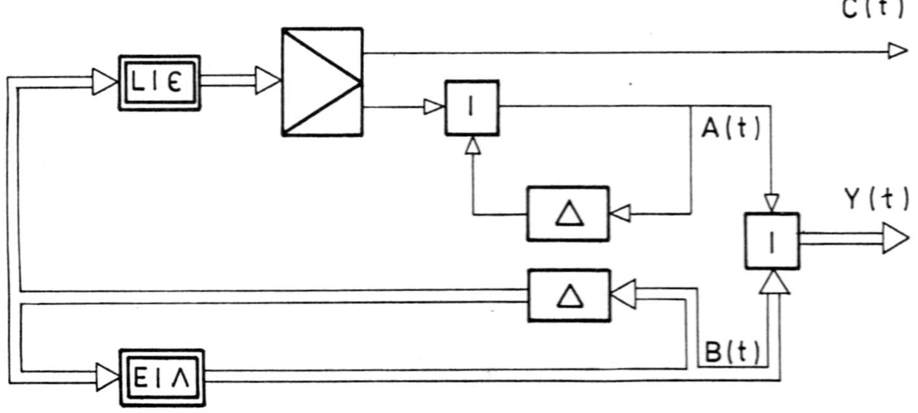

Fig. 3. Flow diagram of disassembly system

for the shaft-type dissassembly, while the model for the housing-type has reverse
compound operators in the first state and second output equations.

CONCLUSION

The proposed method of DEDS analysis is object-oriented and based on the use
of nonhomogeneous elements representations of discrete objects and operations. Hence the modelling and simulation of DEDS can be performed on the chosen level of abstraction, that appears only in the change of type and dimension of NH-elements.
The presented state space models for assembly and disassembly processes have shown
that even such complex processes can be treated within common framework and described by means of proper mathematical tools.

REFERENCES

/ 1 /   Vörös, J.: Discrete event dynamic systems analysis using nonhomogeneous ele-
        ments representation. 7th IFAC Conf. Digital Computer Application to Process
        Control, Wien, 1985.
/ 2 /   Vörös, J.: On modelling and simulation of discrete event dynamic systems.
        IFAC Symp. Simulation of Control Systems, Wien, 1986.

Catastrophe, Chaos, Synergetics and Thermodynamics.
A Unified Approach via Information
of Deterministic Maps

Guy Jumarie*

ABSTRACT. Shannon information theory contains in itself all the elements which we need to define the entropy of deterministic patterns; without using probability,but in a way which is nevertheless fully consistent with its framework. Shannon entropy, Renyi entropy and entropies of family of deterministic maps are so obtained. We can then define thermodynamic entropy of deterministic maps, therefore the thermodynamic meaning of the Liapunov exponent so useful in the analysis of chaotic dynamics. The relation between the se entropies of maps and synergetics is exhibited, and one examines the types of results one may so expect when applies the model to the catastrophe theory.

1. INTRODUCTION

Many authors have criticized the use of information as a basic tool to analyze gene ral systems, and if one believes Rosen [9] the time and effort spent on the analysis of this concept "must surely ranks as one of the most unprofitable investments in modern scientific history". More objectively, Thom [11] pointed out that an information theory should be rather topological to deal with qualitative features, and Haken [4] remarked that information theory, at least in its present form, is unsuitable to analyze systems which are driven far from their equilibrium position.

As a matter of fact, we must distinguish between the information of mathematicians which is nothing else but a set of preliminary axioms followed by mathematical deductions [1] and the information of system scientists, which is more or less rigorously defined , but which, loosely speaking, would be information in the customary sense of everybody (that is to say someone who is ignorant of Shannon theory!)

Despite that the use of statistical entropy in thermodynamics is beginning to be controversial (see for instance Guy [2]), we believe that a good suitable extension of information theory would provide a possible approach to defining thermodynamics of gene- ral systems, subject to the condition that it satisfy requirements which can be summari - zed as follows. *First, it should deal with both symbols and meanings that is to say syn- tax and semantics; second, it should be able to deal with the information involved by pa- tterns and forms;and third, it should provide a model of information which does not nece- ssarily refer to probability.*

Recently [6] we derived models of entropies without probability for deterministic maps; and our purpose herein is to generalize these results, and then to show how, to a large extent, they satisfy the above conditions,so that they could provide a framework for a unified approach to general systems. Mainly, we shall derive a thermodynamic mea- ning for deterministic chaos, a thermodynamic meaning for the catastrophe theory and we shall show how this entropy of deterministic pattern is related with synergetics.

---

* Department of Mathematics and Computer Science; Université du Québec à Montréal; P.O. Box 8888, St A; Montréal, QUE; H3C 3P8; Canada

For the sake of length, we shall deal with continuous maps only, but all the results so obtained can be generalized easily to discrete maps bu using the concept of *total discrete entropy* or *complete discrete entropy* which we introduced recently [7] to
obtain a unified approach to Shannon entropy and Boltzmann entropy.

2. SHANNON ENTROPY OF DETERMINISTIC MAPS

2.1 The General Approach

Definition 2.1 Assume that $X \in R^n$ is a random variable and let $\underline{P}$ denote the set of
its admissible probability density functions $p(x)$. Then the entropy $H(f(.)/\underline{P})$ of the $R^n \to$
R^n deterministic map $f(.)$ given $\underline{P}$ is defined by the expression

$$H(f(.)/\underline{P}) := \max_{p(x) \in \underline{P}} \int_{R^n} p(x) \ln|f'(x)| dx \qquad (2.1)$$

where $|f'(x)|$ is the Jacobian determinant of $f(x)$. ∎

Derivation. (i) Consider two random vectors X and Y; it is well known that the conditional entropy

$$H(Y/X) := \int p(x) H(Y/x) \, dx \qquad (2.2)$$

satisfies the inequality

$$H(Y/X) \leq H(Y) \qquad (2.3)$$

so that one can write

$$H(Y) = \max_{X} H(Y/X) \qquad (2.4)$$

(ii) This being so, given two random variables X and Y related by the equation $Y = f(X)$, their Shannon entropies satisfy the relation

$$H(Y) = H(X) + \int p(x) \ln|f'(x)| dx \qquad (2.5)$$

(iii) Analogously with the equation

$$H(X,Y) = H(X) + H(Y/X) \qquad (2.6)$$

we ascribe Equation (2.5) the following meaning, that is

$$H(f(.)/X) = \int p(x) \ln|f'(x)| dx \qquad (2.7)$$

(iv) We now apply the remark (i) above and we so obtain the expression (2.1). ∎

2.2 Shannon Entropy of Degree c of Continuous Maps

As a special case, assume that the admissible probability densities $p(x)$'s are defined by the given value of their Shannon entropies, namely

$$- \int p(x) \ln p(x) \, dx = C = \text{constant} \qquad (2.8)$$

then one can derive the entropy, *call it Shannon entropy of degree c*, in the form

$$H_c(f(.);\Omega) = \int_\Omega |f'(x)|^c \ln|f'(x)| dx \, / \, \int_\Omega |f'(x)| dx \qquad (2.9)$$

where Ω is the range of variation of X, and $c \in R$ is a parameter which depends upon the
constant C, say $c = c(C)$. ∎

Remark. The condition (2.8) is quite meaningful. Indeed, Equation (2.6) will yield

$$H(X,Y) = H(X) + H(Y)$$

if and only if we maximize $H(Y/X)$ given a fixed value of $H(X)$. ∎

3. RENYI ENTROPY OF DETERMINISTIC MAPS

Definition 3.1. Let $f: R^n \to R^n$ denote a continuously differentiable map with the Ja-
cobian determinant $|f'(x)|$. Its *generalized Renyi entropy* on the domain Ω is defined by

the expression

$$H^R_s(f(.);u,\Omega) \;=\; \frac{1}{1-s} \; \ln \; \frac{\int_\Omega (u + |f'(x)|^{1-s})^{\frac{s}{1-s}} \, |f'(x)|^{1-s} \, dx}{\int_\Omega (u + |f'(x)|^{1-s})^{\frac{s}{1-s}} \, dx} \tag{3.1}$$

with $s \geq 0$ and $u \geq 0$.

In the special case where $u = 0$, one then has the expression

$$H^R_s(f(.);\Omega) \;=\; \frac{1}{s-1} \ln \; [\int_\Omega |f'(x)|^s dx \;/\; \int_\Omega |f'(x)| dx \tag{3.2}$$

which is referred to as the *Renyi entropy of order s of f(.) on the domain* Ω. ∎

Derivation. In order to obtain this definition, we shall proceed as follows.

(i) If $Y = f(X)$, then one has

$$H^R_s(Y) \;=\; \frac{1}{1-s} \ln \int p^s(x) |f'(x)|^{1-s} \, dx \tag{3.3}$$

(ii) Next, if X_1 and X_2 are two random variables with the respective   Renyi entropies $H^R_s(X_1)$ and $H^R_s(X_2)$, then $H^R_s(X_1,X_2)$ is maximum when and only when X_1 and X_2 are inde - pendent, in which case one has

$$H^R_s(X_2) \;=\; H^R_s(X_1,X_2) \;-\; H^R_s(X_1) \tag{3.4}$$

(iii) We shall consider $H^R_s(Y)$ in Equation (3.3) as being the entropy $H^R_s(X,f(.))$ of the pair $(X,f(.))$, and according to (3.4), we shall write

$$H^R_s(f(.)) \;=\; \max_p \; \frac{1}{1-s} \ln [\int p^s(x)|f'(x)|^{1-s} dx \;/\; \int p^s(x) \, dx] \tag{3.5}$$

subject to the condition

$$H^R_s(X) \;=\; C \;=\; \text{constant} \tag{3.6}$$

Applying the Lagrange multiplier technique for instance (don't forget the additional condition $\int p(x) \, dx = 1!$) yields the result. ∎

4. SHANNON ENTROPY OF $R^n \to R$ MAPS

Definition 4.1. Let $g: R^n \to R$, $x \to g(x)$, $x^T := (x_1,x_2,\ldots,x_n)$ denote a continuously differentiable map; its Shannon entropy is defined by the expression

$$H(g(.);\Omega) := H_c(g(.);\Omega) := \frac{\int_\Omega |g^{(n)}_{1,2,\ldots,n}(x)|^c \ln|g^{(n)}_{1,2,\ldots,n}(x)| \, dx}{\int_\Omega |g^{(n)}_{1,2\ldots,n}(x)|^c dx} \tag{4.1}$$

with the notation $g^{(n)}_{1,2\ldots,n}(x) := \partial g(x) / \partial x_1 \partial x_2 \ldots \partial x_n$. ∎

Derivation. Assume that $X \in R$ is a random variable with the probability density $p(x)$, and let $F(x)$ denote its cumulative distribution, $p(x) =: dF(x)/dx$, then one has the equality

$$(1-s)\frac{d}{ds} H^R_s(X) \;-\; H^R_s(X) \;=\; H_s(F(.),R) \tag{4.2}$$

where the right-side term is exactly the Shannon entropy of order s of the map $F(.)$. We then extend this remark to the multi-variate distribution $F(x_1,x_2,x_3,\ldots,x_n)$ therefore we obtain the expression (4.1). ∎

Remark. For $g(x,y) = h(x)r(y)$, $x \in R$, $y \in R$, defined on the rectangle $\Omega_1 \times \Omega_2$, one has expectfully the relation $H_c(h(.)r(.); \Omega_1 \times \Omega_2) = H_c(h(.);\Omega_1) + H_c(r(.);\Omega_2)$. ∎

5. RENYI ENTROPY OF $R^n \to R$ MAPS

Definition 5.1. The Renyi entropy of order s of the differentiable $R^n \to R$ map $g(.)$ is

$$H_s^R(g(.);\Omega) := - \frac{1}{1-s} \ln \frac{\int_\Omega |g_{1,2,\dots,n}^{(n)}(x)|^s dx}{\int_\Omega |g_{1,2,\dots,n}^{(n)}(x)| sx} \qquad \blacksquare \qquad (5.1)$$

Derivation. The rationale of this derivation is exactly similar to that one which led us to the definition of Renyi entropy of $R^n \to R^n$ maps, see equation (3.2). $\blacksquare$

6. SHANNON ENTROPY OF DISTRIBUTED MAPS

Definition 6.1. Let $f: \Omega \times D \to R^n$, $\Omega \subset R^n$, $\Omega \subset R^n$, $D \subset R^m$, $(x,z) \to f(x,z)$ denote a map continuous w.r.t. z and continuously differentiable w.r.t. x, with the Jacobian deter minand $|f'(x,z)|$; then as a direct consequence of the definition of Shannon entropy for random variables, its Shannon entropy of degree $(b,c(z))$ on the domain $\Omega \times D$ is given by the expression

$$H_{b,c(z)}(f_z(.);\Omega,D) := \frac{\int_D e^{bH_{c(z)}(f(.,z);\Omega)} H_{c(z)}(f(.,z);\Omega) dz}{\int_D e^{bH_{c(z)}(f(,.z);\Omega)} dz} \qquad (6.1)$$

where $H_{c(z)}(f(.,z);\Omega)$ is expressed by the equation (2.9), and b with $c(z)$, $b \geq 0$, $c(z) \geq 0$, denote two parameters which describe how the curves defined by the maps are observed.$\blacksquare$

Indication on the derivation. The basic remark is that, because of their defini - tions themselves, the respective effects of x and z are considered at different levels of observation: x is the variable, and z is a distributed parameter. For a given fixed z, the Shannon entropy $H_{c(z)}(f(.,z);\Omega)$ of the map $f(.,z)$ is well defined, and the problem is then to define the aggregate value of this family of entropies. To this end, we use a ran domization with respect to z. $\blacksquare$

7. THERMODYNAMIC ENTROPY OF DETERMINISTIC MAPS

The Shannon entropy of degree 1 of the one-dimensional map $f(.)$ is

$$H_1(f(.);\Omega) = \int_\Omega |f'(x)| \ln f'(x) \, dx \Big/ \int_\Omega |f'(x)| dx \qquad (7.1)$$

In the special case where $f(x)$ is the cumulative distribution function $F(x)$ of a random variable X, then one has

$$H_1(F(.);R) = - H(X) \qquad (7.2)$$

therefore the

Definition 7.1. The identification (7.2) together with the Boltzmann equation su- ggest to consider the Shannon entropy of order one $H_1(f(.);\Omega)$ as being the thermodynamic entropy of the map $f(.)$. $\blacksquare$

8. THERMODYNAMIC ENTROPY AND LIAPUNOV EXPONENT

Consider the dynamical equation

$$\dot{x}(t) = - V_x(x) \, , \quad x(0) = x_o \in R \qquad (8.1)$$

where $V(x)$ denotes the potential function of the system, and $V_x(x)$ holds for $dV(x)/dx$.

It is customary to measure the chaos of the trajectory $x(t)$ by means of the Liapu- nov exponent λ with respect to the invariant (or natural) measure $\rho(x)$, (see for instan - ce Schuster [10]), the determination of which is not a simple task.

So our concern is the following one: *how can we measure the chaos generated by x(t) priorly to any experiment, and by using the dynamical equation (8.1) only ?*

A straightforward suggestion is to use the natural entropy

$$H(x(.);t_1,t_2) = \int_{t_1}^{t_2} |\dot{x}(t)| \ln|\dot{x}(t)| dt \; / \; \int_{t_1}^{t_2} |\dot{x}(t)| dt \qquad (8.2)$$

and a simple calculation yields

$$H(x(.),t_1,t_2) = \int_a^b \varepsilon(x) \ln|V_x(x)| dx \; / \; \int_a^b \varepsilon(x) dx \qquad (8.3)$$

with

$$\varepsilon(x) := - \operatorname{sgn} V_x(x) \; ; \quad a := x(t_1) \; ; \quad b := x(t_2) \qquad (8.4)$$

Assume that $\varepsilon(x) \geq 0$ for $a \leq x \leq b$, then one has

$$H(x(.);t_1,t_2) = \frac{1}{b-a} \int_a^b \ln|V_x(x)| dx \qquad (8.5)$$

in other words, *the thermodynamic entropy of the trajectory x(t) is equal to the Liapunov exponent of V(x) with respect to the uniform probability density.* ■

9. SYNERGETICS AND ENTROPY OF MAPS

9.1 A Few Prerequisites

Haken [3][4] coined the term of *synergetics* to refer to the cooperation of subsys tems which results in spatial, temporal or functional structures on macrospic scales (in our ten yers ago work on the same topic we used the term of general system) and he sugges ted that the so-called slaving principle would govern the processes of self-organization irrespective of the nature of the subsystems.

For instance, loosely speaking, consider the system

$$\dot{x}(t) = \alpha x - xs \; , \quad \alpha > 0 \qquad (9.1)$$
$$\dot{s}(t) = - \beta s + x^2, \quad \beta > 0 \qquad (9.2)$$

then, under large mathematical conditions, s will be slaved by x in the sense   that   it will be possible to express it as a function $s = f(x)$ of x only. x is then referred to as the order parameter of the system.

9.2 Relation With Entropy of Maps

The question is then whether one could not address this problem on a thermodynamic standpoint by using entropy of maps. All the problem would be to define the thermodynamic entropy of a vector.

Close to the equilibrium position (0,0) of the system, the dynamics of the devia- (η,ξ) is

$$\dot{\eta} = \alpha\eta - x\xi \qquad (9.3)$$
$$\dot{\xi} = 2x\eta - \beta\xi \qquad (9.4)$$

and analogously with the equation (8.2), one may define the thermodynamic entropy $S(\eta,\xi)$ by the expression

$$S(\eta,\xi) = \iint \varepsilon(\eta,\xi)\ln|\eta\xi(- \alpha\beta + 2x^2)| d\eta d\xi \; / \; \iint \varepsilon(\eta,\xi) d\eta d\xi \qquad (9.5)$$

and for small (η,ξ) one would have

$$S(\eta,\xi) - \ln(\eta\xi) \cong \ln|\alpha\beta - 2x^2| \qquad (9.6)$$

This relation (9.6) could translate the fact that x is the driving variable of the system on the thermodynamic standpoint.

More generally, one could consider the entropy of the second variation of the state to get more insight in the dynamics of the system.

10. STATIC INSTABILITIES AND ENTROPY OF MAPS

10.1 Entropic Distance Between Maps

(i) According to Equation (6.1), the Shannon entropy of degree $(b,c(z))$ of the map $f_z(.)$ is the mathematical expectation

$$H_{b,c(z)}(f_z(.);\Omega,D) = E\{H_{c(z)}(f(.,z);\Omega)\} \qquad (10.1)$$

with respect to the probability density referred to as $p_f(z)$.

(ii) A parameter of interest is the variance

$$\sigma_f^2 := \int_D p_f(z)H_{b,c(z)}^2(f(.);\Omega,D)dz - (\int_D p_f(z)H_{b,c(z)}(f(.);\Omega,D)dz)^2 \qquad (10.1)$$

which characterizes how uniformly distributed is $H_{c(z)}(f(.,z);\Omega)$ around its mean value.

Entropic Distance. Given a map $l(.)$, we shall identify it with the map $f_z(.)$ whenever the following condition is satisfied, that is

$$(H_{b,c(z)}(f_z(.);\Omega,D) - H_c(l(.);\Omega)^2 \leq k\,\sigma_f^2$$

where k denotes a positive constant the value of which is chosen via practical considerations. ∎

10.3 Application to Potential Functions

These results apply directly to the potential functions $V(x,u)$ of the catastrophe theory, and one can then envisage the exhaustive study of the latter in terms of entropy of deterministic maps.

11. CONCLUDING REMARK

The present theory of entropy of deterministic maps provides new basic approaches to fuzzy sets, probabilistic sets, pattern recognition, logical inference, combination of evidences, approximate reasoning, and so on.

REFERENCES

[1]   ACZEL, J; DAROCZY, Z; *On Measures of Information and Their Characterizations*, Academic Press, New York, 1975

[2]   GUY, A. G.; The scientific revolution of 1987. Abdication of Boltzmann's Entropy, *Applied Physics Communication*, Vol 7, No 3, 217-235, 1987

[3]   HAKEN, H.; *Synergetics*, Springer Verlag, New York, Berlin, 1978

[4]   HAKEN, H.; *Advanced Synergetics*, Springer Verlag, New York, Berlin, 1983

[5]   JUMARIE, G.; *Subjectivity, Information, Systems. Introduction to a Theory of Relativistic Cybernetics*, Gordon and Breach, New York, London, 1986

[6]   JUMARIE, G.; New results on the information theory of patterns and forms, *J. Systems Analysis, Modelling and Simulation*, Vol 4, No 6, 483-520, 1987

[7]   JUMARIE, G.; A Minkowskian theory of observation. Application to uncertainty and fuzziness, *Fuzzy Sets and Systems*, Vol 24, No 2, 231-254, 1987

[8]   JUMARIE, G.; *Relative Information. Theories and Applications*, Springer Verlag, New York, London (to appear)

[9]   ROSEN, R.; On information and complexity, in *Complexity, Language and Life: Mathematical Approaches; Biomathematics*, Vol 16, 174-196, 1986

[10] SCHUSTER, H.G.; *Deterministic Chaos*, Physik-Verlag, Weinhein 1984

[11] THOM, R.; *Structural Stability and Morphogenesis*, translated by D.H. Fowler, Benjamin, New York, 1975

A Route to Chaos

Wolfgang Metzler [1)]

Abstract. The route to chaos of the coupled logistic map is studied. We observe a flip bifurcation from a stable fixed point to a stable period-2 orbit which is followed by a Hopf bifurcation, quasiperiodic behaviour and periodic orbits. At the end of the route the iteration scheme tends to a fascinating strange attractor looking like the Eiffel tower.

1. INTRODUCTION

Chaotic behaviour of simple dynamical systems is today widely believed to model temporarily irregular phenomena in many fields of science [4, 8]. Driven by graphic computers, chaotic dynamical systems also deliver to us a large variety of new fantastic insights into mathematical structures (Julia sets, strange attractors; cf.[2,12,14,15,17]). This paper provides an example of a plane dynamical system where computer graphics have opened a wide field of interesting mathematical questions (stability, bifurcations and chaos) and helped to formulate hypotheses and to prove theorems.

2. THE MAP

It is well known that the iteration scheme

$$x_{k+1} = x_k + h\,x_k(1 - x_k) \,, \qquad h > 0 \,, \tag{1}$$

which is Euler's method for solving the logistic equation $\dot{x} = x\,(1 - x)$, may be transformed into

$$u_{k+1} = r\,u_k(1 - u_k) \,, \tag{2}$$

where $u_k = (h / (1 + h))x_k$ and $r = 1 + h$.

The latter iteration scheme is one of the most classical examples in the study of complex behaviour and chaos of dynamical systems associated with interval maps [3,5,11,13].

In this paper we study the coupled logistic map

$$u_{k+1} = r\,u_k(1 - u_k) + (r - 1)\,v_k \tag{3a}$$

$$v_{k+1} = r\,v_k(1 - v_k) + (r - 1)\,u_k, \qquad r > 1. \tag{3b}$$

This map represents a system of two logistic models coupled symmetrically by an additive term. It can be regarded as a simple model of two coupled systems each of which exhibits a period-doubling bifurcation route to chaos.

Using the above transformation, that is $u_k = (h / (1 + h))x_k$, $v_k = (h / (1 + h))y_k$, we can write

$$x_{k+1} = x_k + h(x_k - x_k^2 + y_k) \tag{4a}$$

$$y_{k+1} = y_k + h(y_k - y_k^2 + x_k), \qquad h > 0. \tag{4b}$$

[1)] Department of Mathematics, University of Kassel, P.O. Box 101380, D-3500 Kassel, F.R. Germany.

Relative to (4), $M = \{(x,y) \in \mathbb{R}^2 \mid x = y\}$ is an invariant manifold. Reduction of (4) onto M results in the one-dimensional iteration scheme

$$x_{k+1} = x_k + h(2x_k - x_k^2) \tag{5}$$

which may be transformed into (2) by $u_k = (h/(1+2h))x_k$ and $r = 1 + 2h$.

The dynamics of the coupled system (4) will be discussed in this paper. It is based on an unpublished note [1] which has been the reason for a computergraphics film [16]. Related couplings of nonlinear oscillators have been studied by Hogg and Hubermann [9] as well as Waller and Kapral [18]. Kaneko [10] investigated the transition to chaos of a coupled logistic map with a linear coupling term.

3. PERIOD-DOUBLING

Considering eq. (4) we find, as the parameter h is varied, that there are two fixed points given by

$$E_1 = (0,0) \; , \quad E_2 = (2.2). \tag{6}$$

Linearizing (4) in the neighbourhood of E_1, we obtain the eigenvalue equation

$$\left| DF_h(0,0) - \lambda I \right| = \begin{vmatrix} 1+h-\lambda & h \\ h & 1+h-\lambda \end{vmatrix} = 0 \; , \tag{7}$$

where DF_h denotes the Jacobian matrix of the first partial derivatives of the function

$$F_h(x,y) = \left(x + h(x - x^2 + y), \, y + h(y - y^2 + x) \right) \tag{8}$$

defined by (4). Because of its roots $\lambda_1^{(1)} = 1 + 2h$ and $\lambda_2^{(1)} = 1$ the fixed point E_1 is unstable for $h > 0$. For E_2 we obtain the eigenvalue equation

$$\left| DF_h(2,2) - \lambda I \right| = \begin{vmatrix} 1-3h-\lambda & h \\ h & 1-3h-\lambda \end{vmatrix} = 0 \tag{9}$$

with the eigenvalues $\lambda_1^{(2)} = 1 - 2h$ and $\lambda_2^{(2)} = 1 - 4h$. Obviously E_2 is a sink only if $0 < h < 0.5$. For $h > 0.5$ the period-1 cycle E_2 is unstable. There is a critical situation at $h = 0.5$.

If $h > 0.5$, then

$$0_h := \left\{ \left(\frac{1}{h} + \frac{1}{h}\sqrt{2h-1}, \frac{1}{h} - \frac{1}{h}\sqrt{2h-1} \, , \, \frac{1}{h} - \frac{1}{h}\sqrt{2h-1}, \frac{1}{h} + \frac{1}{h}\sqrt{2h-1} \right) \right\} \tag{10}$$

is a hyperbolic period-2 orbit. It is born in the equilibrium point $E_2 = (2,2)$ and moves with increasing h to $E_1 = (0,0)$ on a circle, which is given by $(x-1)^2 + (y-1)^2 = 2$. In particular we have $0_{0.5} = \{(2,2)\}$ and $0_\infty = \{(0,0)\}$.

The first part of this statement on 0_h follows by inserting $x_k = \frac{1}{h} + \frac{1}{h}\sqrt{2h-1}$, $y_k = \frac{1}{h} - \frac{1}{h}\sqrt{2h-1}$ into eq. (4). Then we get

$$x_{k+1} = \frac{1}{h} - \frac{1}{h}\sqrt{2h-1} \; , \quad y_{k+1} = \frac{1}{h} + \frac{1}{h}\sqrt{2h-1}\cdot \tag{11}$$

and

$$x_{k+2} = y_{k+1} = x_k \text{ as well as } y_{k+2} = x_{k+1} = y_k \; . \tag{12}$$

Furthermore, using (12), eq. (4) can easily be shown to be equivalent to

$$2hy_k + 2hx_k - hy_k^2 - hx_k^2 = 0 \tag{13}$$

and this proves the second part of the statement.

Moreover, for $0.5 < h < 0.6$ O_h is a stable orbit. If $h > 0.6$, O_h is unstable. This can be proved straightforward looking at

$$F_1(x,y) = x + h(x - x^2 + y) \tag{14a}$$

$$F_2(x,y) = y + h(y - y^2 + x) \tag{14b}$$

and

$$\tilde{F}_1(x,y) := F_1(F_1(x,y), F_2(x,y)) \tag{15a}$$

$$\tilde{F}_2(x,y) := F_2(F_1(x,y), F_2(x,y)) , \tag{15b}$$

respectively, for $\bar{x} = \frac{1}{h} + \frac{1}{h}\sqrt{2h-1}$, $\bar{y} = \frac{1}{h} - \frac{1}{h}\sqrt{2h-1}$.

A lengthy calculation yields

$$\frac{\partial}{\partial x}\tilde{F}_1(\bar{x},\bar{y}) = 5 - 10h + 2h^2 = \frac{\partial}{\partial y}\tilde{F}_2(\bar{x},\bar{y}) \tag{16}$$

and

$$\frac{\partial}{\partial y}\tilde{F}_1(\bar{x},\bar{y}) = 4h\sqrt{2h-1} - 2h + 2h^2, \tag{17a}$$

$$\frac{\partial}{\partial x}\tilde{F}_2(\bar{x},\bar{y}) = -4h\sqrt{2h-1} - 2h + 2h^2. \tag{17b}$$

This leads to the characteristic equation

$$\begin{vmatrix} 5 - 10h + 2h^2 - \lambda & 4h\sqrt{2h-1} - 2h + 2h^2 \\ -4h\sqrt{2h-1} - 2h + 2h^2 & 5 - 10h + 2h^2 - \lambda \end{vmatrix} = 0, \tag{18}$$

equivalent to $\lambda^2 + (-10 + 20h - 4h^2)\lambda + (25 - 100h + 100h^2) = 0$.

The eigenvalues of (18) are

$$\lambda_{1,2} = (5 - 10h + 2h^2) \pm \sqrt{20h^2 - 40h^3 + 4h^4} . \tag{19}$$

Analyzing the roots of the radicant $20h^2 - 40h^3 + 4h^4$, one can easily show that

$$|\lambda_i| < 1 \quad \text{if} \quad 0.5 < h < 0.6, \tag{20a}$$

$$|\lambda_i| = 1 \quad \text{if} \quad h = 0.6, \tag{20b}$$

and

$$|\lambda_i| > 1 \quad \text{if} \quad h > 0.6. \tag{21}$$

The considerations concerning $(\bar{y}, \bar{x})$ are analogous. Thus (20) and (21) prove the above stability statements about the period-2 orbit O_h.

4. THE ROUTE TO CHAOS

Now we investigate the system's behaviour in the range $0.6 \le h \le 0.684$. As shown in the section before, there exists a stable period-2 orbit for $0.5 < h < 0.6$. First we consider the bifurcation at $h = 0.6$. We know that the one parameter family $\tilde{F}_h := F_h \circ F_h$ satisfies

$\tilde{F}_h(\bar{x},\bar{y}) = (\bar{x},\bar{y})$ for $h > 0.5$ with $\bar{x} = \frac{1}{h} + \frac{1}{h}\sqrt{2h-1}$ and $\bar{y} = \frac{1}{h} - \frac{1}{h}\sqrt{2h-1}$. Its Jacobian matrix $D\tilde{F}_h(\bar{x},\bar{y})$ has two non-real eigenvalues λ_1 and $\overline{\lambda_1}$ satisfying $|\lambda_1| < 1$ for $h < 0.6$ and $|\lambda_1| > 1$ for $h > 0.6$. Also we have

$$\left. \frac{d\,|\lambda_1(h)|}{dh} \right|_{h=0.6} = 10 > 0. \tag{22}$$

These three properties of the map (4) and further numerical studies (cf.[14]) indicate a Hopf bifurcation at (see [7], p. 162) $h = 0.6$ Accordingly for $(\bar{y},\bar{x})$.

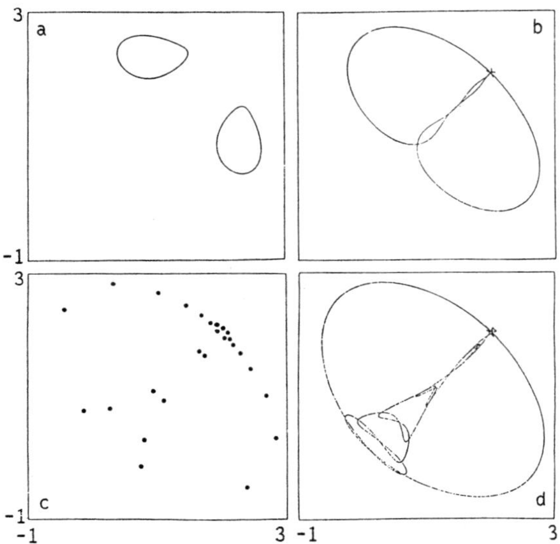

Fig. 1. Route to chaos (after [15]):
    (a) Stable loops ($h = 0.614$),   (b) Overlapping loops ($h = 0.66$),
    (c) Period-26 attractor ($h = 0.678$), (d) "Eiffel tower" ($h = 0.684$).

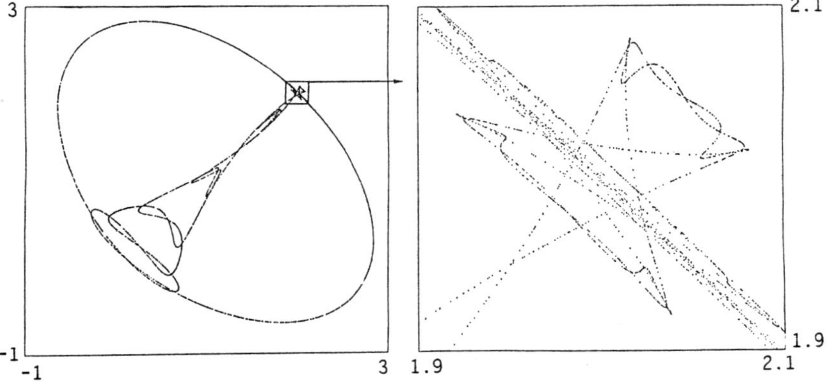

Fig. 2. The attractor of the map (4) at $h = 0.684$ with a blow up of the squared region $[1.9,2.1] \times [1.9,2.1]$.

Next we outline some computational results about the map (4) in the range $0.6 \leq h \leq 0.684$. These calculations have been performed on a graphic computer with double precision arithmetic. First we iterate the map 10.000 times to avoid transients. For $h > 0.6$ the period-2 orbit blossoms out into two stable loops (Fig. 1a). The loops grow by turn with various periods, overlap themselves (Fig. 1b) by turn with other periods (Fig. 1c) and finally reach the structure of the Eiffel tower (Fig. 1d).

A blow up of the Eiffel tower's topstar (Fig. 2b) delivers a fractal structure which indicates sensitive dependence on the initial conditions. Indeed, one computes a positive characteristic exponent (cf. [6]) for $h \gtrsim 0.651$, as shown in Fig. 3. Increasing the values of h, the chaotic attractor of the map (4) continuously changes its shape, interrupted by different periodic orbits neglected in Fig. 3. The attractor's final shape at $h = 0.684$ is shown in Fig. 2. It is followed by escape which can be observed for $h \gtrsim 0.686$.

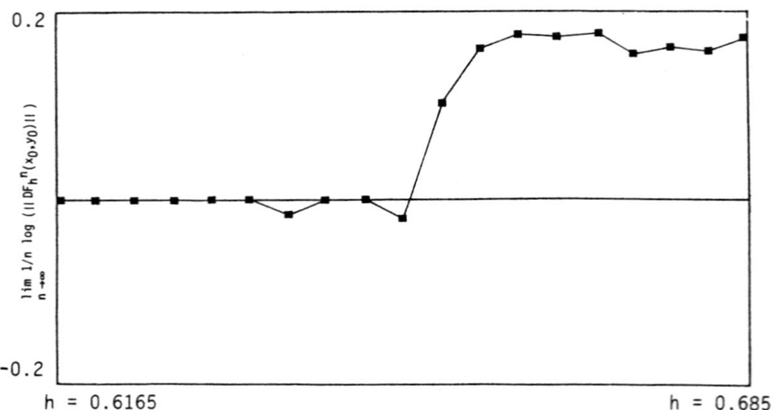

Fig. 3. Spectrum of the characteristic exponents $\lim_{n\to\infty} 1/n \log \left\| DF_h^n(x_0,y_0) \right\|$ for $0.6165 \leq h \leq 0.685$ and $(x_0,y_0) = (0.4,0.5)$.

REFERENCES

[1] Beau, W., W. Metzler and A. Überla: The Route to Chaos of Two Coupled Logistic Maps. Preprint (1986).

[2] Beau, W., W.H. Hehl, W. Metzler: Computerbilder zur Analyse chaoserzeugender Abbildungen. Informatik Forsch. Entw. 2 (1987), 122-130.

[3] Collet, P. and J.-P. Eckmann: Iterated Maps on the Interval as Dynamical Systems. A. Jaffe and D. Ruelle (eds.). Birkhäuser, Basel/Boston/Stuttgart 1980.

[4] Cvitanović, P. (ed.): Universality in Chaos. Adam Hilger Ltd., Bristol 1983.

[5] Feigenbaum, M.: The Universal Metric Properties of Nonlinear Transformations. J. Stat. Phys. 21 (1979), 669-706.

[6] Feit, S.D.: Characteristic Exponents and Strange Attractors. Commun. math. Phys. 61 (1978), 249.

[7] Guckenheimer, J., P. Holmes: Nonlinear Oscillations, Dynamical Systems and Birfurcations of Vector Fields. Springer, New York/Berlin/Heidelberg/Tokyo 1986².

[8] Haken, H. (ed.): Evolution of Order and Chaos in Physics, Chemistry, and Biology. Springer, Berlin 1982.

[9] Hogg, T. and B.A. Huberman: Generic Behavior of Coupled Oscillators, Phys. Rev. A 29 (1984), 275.

[10] Kaneko, K.: Transition from Torus to Chaos Accompanied by Frequency Lockings with
     Symmetry Breaking. Prog. Theor. Phys. $\underline{69}$ (1983), 1427.

[11] Li, T.Y. and J.A. Yorke: Period Three Implies Chaos. Amer. Math. Monthly (1975),
     958-992.

[12] Mandelbrot, M.S.: The Fractal Geometry of Nature. Freeman, San Francisco 1982.

[13] May, R.B.: Simple Mathematical Models with very Complicated Dynamics. Nature $\underline{261}$
     (1976), 459-467.

[14] Metzler, W., W. Beau, W. Frees, A. Überla: Symmetry and Self-similarity with
     Coupled Logistic Maps. Z. Naturforsch. $\underline{42a}$ (1987), 310-318.

[15] Metzler, W.: Chaos und Fraktale bei zwei gekoppelten nichtlinearen Modelloszilla-
     toren. PdN Physik $\underline{7/36}$ (1987), 23-29.

[16] Metzler, W., W. Beau, A. Überla: A Route to Chaos. Computergraphics Film.
     Inst. f.d. Wiss. Film, $\underline{C\ 1641}$, Göttingen 1987.

[17] Peitgen, H.O., P.H. Richter: The Beauty of Fractals. Springer, Berlin/Heidelberg/
     New York/Tokyo 1986.

[18] Waller, H. and Kapral, R.: Spatial and Temporal Structure in Systems of Coupled
     Nonlinear Oscillators. Phys. Rev. $\underline{A\ 30}$ (1984), 2047.

Bifurcations of Two-Dimensional Tori and Chaos in Dissipative Systems

V.S.Anishchenko, T.E.Vadivasova, M.A.Safonova [*)]

Transition to dynamical chaos in different distributed and multi-dimensional systems is often preceded by a quasiperiodic motion bifurcations. In simplest case, chaos arises via distruction of two-dimensional torus (T_2) [1] . This communication represents the results of computer and physical experiments on the investigation of torus distruction regularities, mechanisms of appearance of quasiattractors (CA_1) and their characteristics in different flow and diskrete systems. The following systems realising regim of quasiperiodic oscillations were investigated: driven generator with inertial nonlineority, two coupled generators and discrete system of coupled Feigenbaum maps.

Numerical simulation was carried out with help of computer programs permitting to calculate the lines of limit cycle bifurcations on the parameter plane, various dynamical and statistical characteristics of oscillation regimes: Poincare section, probability distribution density, power spectra, Lapunov characteristic oxponents. Random noise generator was added in numerical scheme for fluctuation excitation simulation.

Universal character of following bifurcation mechanisms of transition T CA_1 predicted by torus distruction theorem was confirmed [2,3]

1. Loss of smoothness and break-down of ergodic torus with soft appearance of torus-attractor. Smoothness loss effect can be preceded by the ergodic torus period-doubling.

2. Loss of smoothness of torus with resonance structure bn it, its distruction on the resonance cycle stability loss line and appearance of chaos within looking region through Feigenbaum sequence of period-doubling or through emergence of new torus and its following break-down.

3. Hard appearance of torus-attractor via the saddle-node bifurcation on unsmooth torus or after its distruction.

4. Torus with resonance structure breaks also on the line of homoclinic tangency of resonance saddle cycle manifolds. However, appearing hyperbolic subset of trajectorics is nonattracting. Numerical experiments permitted to establish the role of natural fluctuations in situation under consideration. Noise influence on system dynamics in the regime of stable limit cycle with homoclinic structure in its neighbourhood leads to chaos arising.

For example let us regard the results of numerical analysis of discrete system [3]

$$x_{n+1} = 1 - \alpha x_n^2 + \gamma (y_n - x_n), \quad y_{n+1} = 1 - \alpha y_n^2 + \gamma (x_n - y_n), \quad (1)$$

[*)] Saratov State University, Departament of Physics, SU-410601,Saratov, USSR

which simulates dynamics of two coupled generators [2] .

Bifurcational diagram on parameter plane in neighbourhood of phase locking with winding number $\theta = 2{:}5$ is represented in Fig. 1.

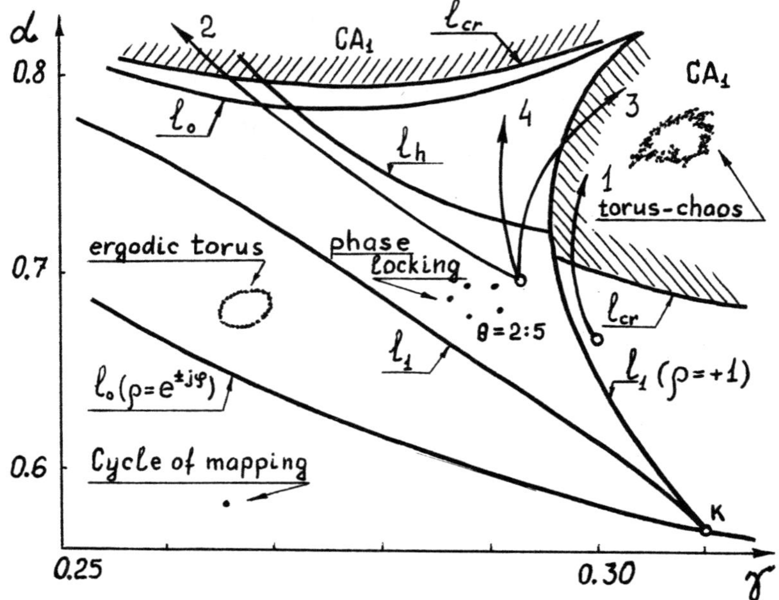

Fig. 1. l_0 – the Hopfe bifurcation line; l_1 – the saddle-node bifurcation line; l_h – the invariant manyfold homoclinic tangency line; l_{cr} – the boundary of CA_1 – region; ρ – multiplicator of limit cycle.

The digits on Fig. 1. indicate the motion directions on parameter plane where the bifurcational mechanisms mentioned above are realised.

Regularities of transition "torus-chaos" predicted by one-dimensional circle mapping theory are regarded for flow systems. Quantitative relations in oscillation power spectrum at the T_2-distruction moment are investigated in physical and computer experiments. Universal regularity in frequency distribution of spectral lines proved for model circle mapping are confirmed. Good agreement with one-dimensional theory results is obtained for fractal dimension of the set of irrational values of winding number near the torus break-down line.

References

[1] Anishchenko V.S.: Dynamical Chaos-Basic Concepts. Teubner-Verlag, Leipzig 1987.

[2] Afraimovich V.S., Shilnikov L.P.: Invariant two-dimensional tori, their distruction and stochastisity. Methods of qualitative theory of differential equations. University Gorky, 1983, 3-25.

[3] Anishchenko V.S.: Distruction of quasiperiodic oscillations and chaos in dissipative systems. Journal of Technical Physics, 56, (1986) 2, 225-237.

Shape and Dimension of Certain Hyperbolic Invariant Sets

H.G. Bothe

In many cases the evolution of a system whose state in each moment
is determined by a point in a phase space P can be described by a
dynamical system on P, i.e. by a family $\{\varphi^t\}_{t\in T}$ of mappings
$\varphi^t : P \longrightarrow P$ depending on a time parameter t belonging to a set T
which consists either of all real numbers, all non-negative real
numbers, all integers or all non-negative integers. These mappings
satisfy

$$\varphi^o = id, \quad \varphi^{s+t} = \varphi^s \circ \varphi^t$$

where $\circ$ denotes the composition of mappings. If at the beginning
(i.e. at time 0) the system has the state corresponding to the point
p_o in P, then its state at time t corresponds to the point $\varphi^t(p_o)$
and the whole evolution of the system is described by the **orbit**
$\{\varphi^t(p_o)\}_{t\in T}$ starting at p_o.

If a dynamical system $\{\varphi^t\}_{t\in T}$ on a space P is given then it
suggests itself to look for invariant sets Λ in P, i.e. for sets
satisfying $\varphi^t(\Lambda) = \Lambda$ $(t\in T)$. If Λ is an attractor, i.e. if for each
point p in P sufficiently close to Λ we have $\varphi^t(p) \longrightarrow \Lambda$ if
$t \longrightarrow \infty$, and if φ^t is sufficiently mixing on Λ , then for $t \longrightarrow \infty$
the orbits starting near Λ are approximately determined by the shape
of Λ . If Λ is not an attractor then orbits possibly may stay for a
long time in the vicinity of Λ , and the shape of Λ approximately
determines the evolution during this period.

Now it has turned out (by mathematical reasoning and by experiments)
that such invariant sets often have a special geometric structure
which is related to Cantor sets or to certain kinds of fractals. It
is the aim of this note to suggest how some mathematical ideas can
help to explain why the shape of invariant sets, though quite unfamiliar to "classically minded mathematiciens", seems to be strongly
restricted. The reasoning is, roughly speaking, as follows:

If a special shape of an invariant set can be observed, then it must
be stable with respect to small perturbation; i.e. all dynamical
systems which differ sufficiently little from the given one must have
invariant sets with the same structure. This implies (using topological
properties of the space of all dynamical systems on P) that the number
of essentially different observable shapes for invariant sets is at

Karl-Weierstraß-Institut für Mathematik
Akademie der Wissenschaften d. DDR
Mohrenstr. 39
Berlin
1086
DDR

most countable; i.e. all these shapes can be arranged in a sequence
$S_1, S_2, \ldots$. Therefore the number of stable shapes is not "too high",
and there is some hope that a satisfactory description of these shapes
might be possible.

To get a precise mathematical framework one has to formulate assumptions under which invariant sets have a stable shape. By a well known
theory of S. Smale it is reasonable to consider hyperbolic invariant
sets which are stable and not too special.

Let us describe roughly how in some cases the local structure of
these sets can be described: Start with a finite graph G. Then according to a fixed law replace each vertex of G by a smaller graph
where these smaller graphs are connected along the edges of G. This
process

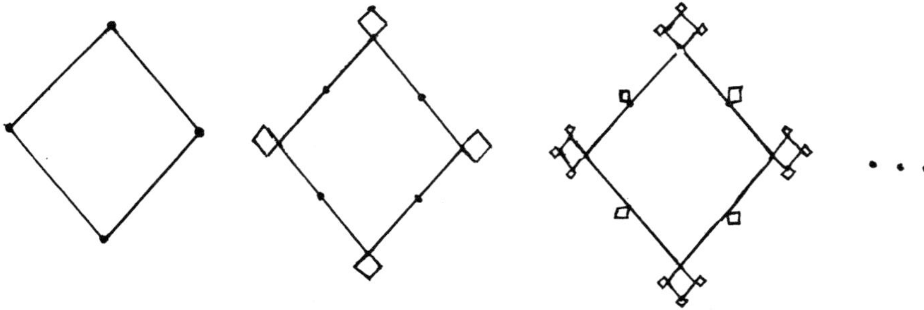

has to be repeated again and again. The limit of this construction
(which may have a higher dimension than 1) is a typical structure
which hyperbolic invariant sets may have locally.

The following remark about the dimension of hyperbolic invariant
sets concerns the classical Hausdorff dimension which in our cases
coincides with the capacity dimension. Compared with some other
dimensions which were introduced for invariant sets Hausdorff dimension is difficult to handle, but from a geometric point of view it
should play a central role, since for the sets considered here it
has the following property: If a hyperbolic set Λ in an n-dimensional
space P (n is the number of parameters necessary to fix the state
of the system) has Hausdorff dimension m, then after neglecting
n-(2m+1) parameters (i.e. after a projection to a (2m+1)-dimensional
subspace of P) we can expect to get a true picture of Λ . (This is a
result due to R. Mañe and F. Takens). Here we make the following
remark: The Hausdorff dimension of a set Λ is a local invariant which
may have different values at different points of Λ , and it is natural to ask how much the local Hausdorff dimension of a hyperbolic

invariant set Λ may vary. There is no complete answer to this question, but if Λ is a special attractor (an expanding attractor), then, once more very roughly speaking, we can say that the ratio of inner distortion in Λ under the dynamics and the rate of attraction of Λ gives a bound for the variation of the local Hausdorff dimension.

Since one can not expect that invariant sets which appear in modelling of evolution processes are in general hyperbolic, some words about the justification of our restriction seem to be necessary. Let us mention two reasons why mathematicians consider hyperbolic invariant sets: 1) In connection with hyperbolic sets we are able to understand and drescribe phenomena with mathematical precision which with some modifications are observed very often for more general invariant sets. 2) A precise description of an essentially larger class of invariant sets seems to be impossible at least now.

CANDYS/QA – A Software System for Qualitative Analysis of the Behaviour of the Solutions of Nonlinear Dynamical Systems

W. Jansen, U. Feudel[1]

Many processes in physics, chemistry, and biology are described by models of ordinary differential equations or difference equations. The investigation of the behaviour of specific solutions of such models in dependence on the model parameters is the main task of qualitative analysis. In general, nonlinear dynamical systems cannot be solved analytically, therefore a lot of numerical methods have been developed by several authors to study bifurcation phenomena in ordinary differential equations [1,2,3]. The present paper is concerned with the description of a software system for the numerical analysis of the qualitative behaviour of evolution equations.

Three classes of dynamical systems are considered:

a) autonomous differential equations

$$\frac{dX_i}{dt} = f_i(X_1 \ldots X_n, \vec{p}) \qquad i = 1 \ldots n$$

b) iterations of the form

$$X_i(k+1) = f_i(X_1(k) \ldots X_n(k), \vec{p}) \qquad i = 1 \ldots n$$

c) periodically forced differential equations

$$\frac{dX_i}{dt} = f_i(X_1 \ldots X_n, t, \vec{p}) \qquad X_i(t+T) = X_i(t) \qquad i = 1 \ldots n$$

The qualitative analysis of dynamical systems deals with properties which are invariant with respect to nonlinear, differentiable transformations of the coordinate system. Transformations of the time are not included.

The basic idea of qualitative analysis is the investigation of time-invariant sets M in the state space $\mathbb{R}^n$ which are mapped onto themselves via evolution equations for all times. For the above mentioned system classes two different types of invariant sets are important in practice and therefore taken into account:

- steady-states (fixed points) for autonomous differential equations
- periodic solutions (cycles) for all classes

In the case of autonomous systems the period length is an invariant. Fixed points of iterations are assumed to be cycles of period 1.

---

[1] Academy of Sciences, Central Institute for Cybernetics and Information Processes, Kurstrasse 33, Berlin, 1086 , GDR

Using Poincare's method of first return maps the search for cycles has been transformed into an algebraic problem. For this reason both, the steady-states and the periodic solutions are computed by standard algorithms for solving nonlinear equations. The program includes Monte-Carlo-Search, search by evolution strategies, homotopy methods and a regularized Newton-method. In order to find a suitable starting point for these methods, the simulation of the trajectories of the nonlinear dynamical system is possible. As the result of one or a combination of these techniques the fixed points or one point of the cycle, respectively, is obtained.

For applications not only the existence and location of invariant sets is an important question but also the stability of the calculated solution. Linearizing the system around the solution found, the stability is determined by analyzing the magnitude of the eigenvalues of the corresponding Jacobian. Using these algorithms mentioned above stable as well as unstable steady-states and cycles have been evaluated. Varying the model parameters the behaviour of the solutions of the dynamical system changes, the location of the fixed points and cycles shifts in dependence on the parameter changes. At certain values in parameter space, so-called turning and bifurcation points, the following phenomena may occur: Stable solutions loose their stability and unstable solutions become stable (turning and bifurcation points), the multiplicity of steady-states or cycles increases or decreases (bifurcation points), periodic solutions branch off a stationary solution (Hopf bifurcation) or cycles with a multiple of the period arise (period doubling points). Various algorithms for the numerical determination of bifurcation and turning points are known from the literature [4]. They all use different extensions of the originally nonlinear equations system. The implemented methods base on a minimally extended system of nonlinear equations [5,6] which allow an easy calculation of the branching solutions.

The solution branches are computed by a predictor corrector continuation technique with adaptive step size control. During the continuation of the solution branch the special points are detected and calculated. The computation of these special points and the corresponding solution branches between them leads to the construction of solution diagrams (or bifurcation diagrams) which reflect the parameter dependent behaviour of the time-invariant sets of the model equations. The bifurcation diagram will be generated by the program as a graph in the parameter-state space where the nodes denote the interesting special points as well as the temporary end points and where the edges represent the joining branches. In order to start further research at arbitrary node of this graph it is possible to jump from one node to any other, thereby the results of this node is used as starting information. If one chooses a node representing a special point then the program automatically switches to another working mode. Instead of varying only one parameter as before now two

parameters are changed and the continuation technique is used to
construct a branch of such special points in the two-parameter space.

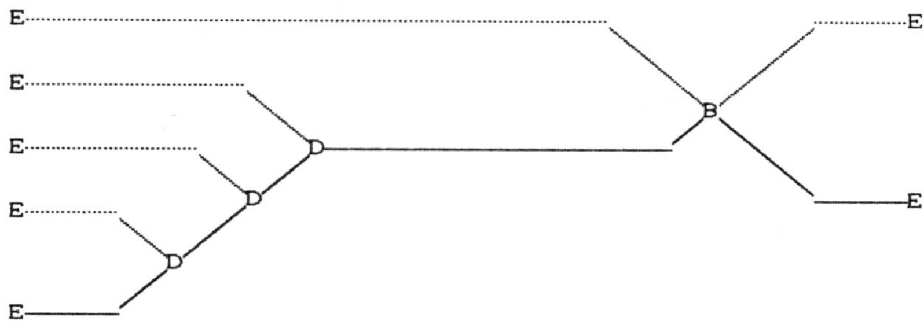

Fig. 1: Bifurcation diagram for cycles of the well known Lorenz-system
        (Parameters: r ∈ [334,500], σ = 16, b = 4).
        Bifurcation (B), period doubling (D), and end points (E)
        solid lines: stable cycles, dotted lines: unstable cycles

The software system is written in the programming language MODULA-2 and
is implemented on 32-bit minicomputers CM 52. The man-machine interface
is fully interactive. The only programming activity required from the
user is the preparation of the model's right hand side f(X,p). The
program generates some graphical output, e. g. the bifurcation diagram,
and a so-called restart file that enables the continuation of the
investigation at later computer sessions. The package integrates a
collection of modules, each of them contains the complete algorithm for
one computational method. Therefore it is easy to replace the current
implementation of some algorithms or to include additional techniques to
solve the problems described above. The extension of the numerical
analysis to other classes of dynamical systems is also possible in this
way.
A first application is the investigation of a water quality model. This
is a periodically forced ordinary differential equation system
consisting of 8 state variables.

References:

[1] Chibnik, A.I., E.E. Schnol : Programs for qualitative analysis of
    differential equations (in Russian). Preprint 1982.
[2] Doedel, E.J., J.P. Kernevez : Software for Continuation Problems in
    Ordinary Differential Equations with Application. Preprint 1985.
[3] Kubiček, M., M. Marek : Computational Methods in Bifurcation
    Theory and Dissipative Structures. Springer-Verlag, New York 1983.
[4] Mittelmann, H.D., H. Weber : In: Bifurcation Problems and their
    Numerical Solution (Eds.: Mittelmann, H.D., H. Weber) Birkhaeuser-
    Verlag, Stuttgart 1980.
[5] Poenisch, G. : Beitr. Numer. Math. 9(1981)147.
[6] Poenisch, G. : Computing 35(1985)277.

Qualitative Behaviour of Ordinary Differential Equation Models Describing Forest Growth Under Air Pollution

W. Metzler and H. Krieger [1]

Abstract. Two related two-dimensional differential equation models of forest growth are considered which concentrate on the interactions between photosynthesis and the development of fine roots. Dependent on a bifurcation parameter which represents pollution stress, the qualitative behaviour of either ODE model undergoes a typical change from subcritical survival to a supercritical dying-off of the model forest.

1. INTRODUCTION

Since the early 80th forest declines in Europe were recognized as a very severe problem. Numerous attempts have been made to describe this phenomenon with the help of models consisting of a system of differential equations. The approach to model a tree as a system of interacting living compartments connected to the environment, leads to complex forest ecosystem models [1,6,8]. Although numerical solution of these models shows reliable results, they are lacking the possibility to recognize essential processes by applying mathematical tools.

Other scientists concentrate on modelling compartments of the forest growth process like photosynthesis, height growth, allocation of assimilate, light competition etc. (e.g. [3,10]). Treating forest growth as a whole on a more general level, a third method introduces compact models neglecting incidental effects. Some of these are working with annual representation of forest growth principles (e.g. [7]). Therefore, short-time influences like water stress etc. are excluded. Other compact models were derived from complex forest ecosystem models with the help of numerical analysis and using some simplifying assumptions [2,5]. For these models a mathematical analysis of the qualitative behaviour of the correspondent differential equation systems is possible.

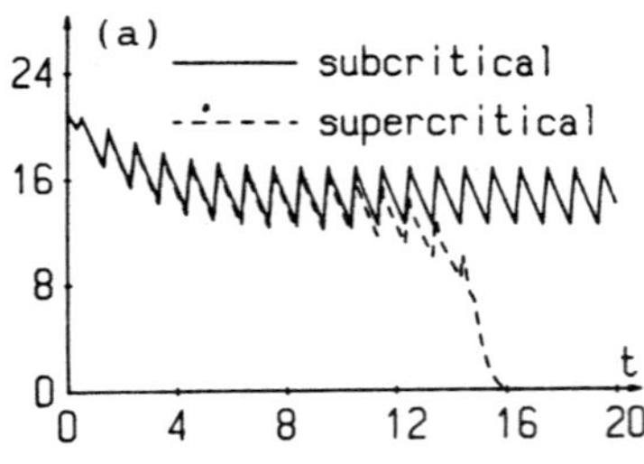

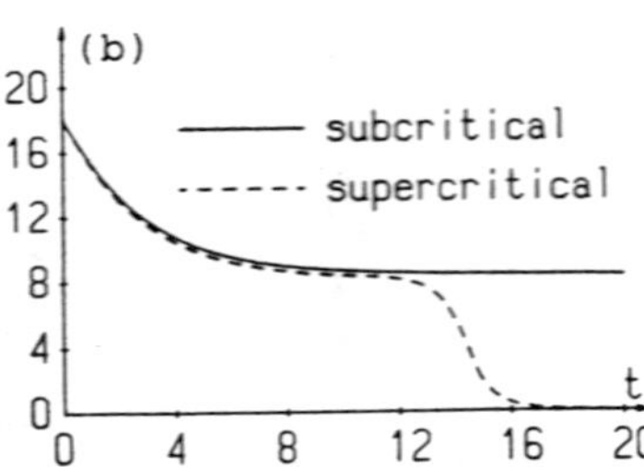

Fig. 1. Subcritical and supercritical growth mode of the state variable leaves (1):
    (a) for the original simulation model [1],
    (b) for the autonomous leaf-root model (3.2) below ($\lambda = 0.28$ and $\lambda = 0.29$).

Bossel et al. [1] have proposed a discrete simulation model which concentrates on the interactions between photosynthesis and the development of fine roots, both influenced by air pollution and acid precipitation. The heart of this simulation model is a four-dimensional tree model which is completely documented in [4]. Dependent on the intensity

[1] Department of Mathematics, University of Kassel, P.O. Box 101380, D-3500 Kassel, F.R. Germany.

of pollution stress, simulation results in [2,5] exhibit a bifurcation of solutions into (*normal*,) *subcritical* and *supercritical* growth of forest trees, the latter mode being terminated by a sudden dieback of all state variables (cf. Fig. 1(a)).

The following differential equation models of sect. 2 and 3 have been derived from [1] by reducing the number of state variable and averaging time-dependencies (cf. [4]).

2. STABILITY DEPENDING ON THRESHOLD BEHAVIOUR

In this section, we consider an autonomous parameter dependent ODE system with two state variable l and a representing the actual amounts of leaves and assimilate. Its solutions qualitatively reproduce the two characteristical modes of forest growth (subcritical - supercritical) discussed in the previous section.

As a basic modelling tool a threshold function [9, p.48] is used given by

$$c(a,l) := \frac{y_3 a}{y_1 + y_2 l + y_3 a} \tag{2.1}$$

(for parameters y_i see Table 1 below). It compares the actual amount $a = a(t)$ of assimilate with the total demand for the growth of roots, leaves and biomass, i.e. in terms of the state variable leaves (see Table 1)

$$y_1 + y_2 l, \tag{2.2}$$

by using the smooth limiter (or threshold response [9]) function (cf. sect. 3)

$$L(x) = \frac{y_3 x}{1 + y_3 x} \qquad (y_3 > 0, \text{ fixed}). \tag{2.3}$$

(2.3) maps the positive real axis $[0, \infty)$ onto $[0, 1)$. Inserting the proportion

$$x = \frac{a}{y_1 + y_2 l} \; , \tag{2.4}$$

(2.3) yields (2.1). Now, correcting the total demand (2.2) of assimilate by (2.1) results in the first system equation for the assimilate:

$$\dot{a} = y_4 l - (y_1 + y_2 l) \frac{y_3 a}{y_1 + y_2 l + y_3 a}$$
$$= (x_4 a + x_5 l + x_6 a l + x_7 l^2)(y_1 + y_2 l + y_3 a)^{-1} \; , \tag{2.5a}$$

where $y_4 l$ expresses the net photosynthesis rate (cf. Table 1). Analogously [5] for the second state variable leaves we obtain

$$\dot{l} = (x_1 a + x_2 l + x_3 l^2)(y_1 + y_2 l + y_3 a)^{-1} \; . \tag{2.5b}$$

Typical solution curves of (2.5) turn out to be very similar to those of Fig. 1(a) (see [4], p.172f.). The system tends to exponential growth in the subcritical case, but a sudden change to supercritical behaviour is observed, when the pollution parameter ψ increases. Before analyzing this matter of fact, for the sake of completeness we compile all system parameters, constants and coefficients of (2.5) in Table 1.

Via two pollution functions $\phi_1(\psi)$ and $\phi_2(\psi)$, the solutions of (2.5) depend on a pollution parameter $\psi \in [0, 1.5]$ (df. [4], p.168), i.e., $l = l(t, \psi)$ and $a = a(t, \psi)$. The first quadrant $D = \mathbb{R}_0^+ \times \mathbb{R}_0^+$ is an invariant set for the flow of (2.5). In D, (2.5) has two fixed points: $(0,0)$ is asymptotically stable for all values of $\psi \in [0, 1.5]$, and the nontrivial

constants: $c_1 = 7[a]$: average lifespan of needles

$c_2 = 16[kg\ ASSI/(kg\ LEAF \cdot a)]$: optimal net photosynthesis rate

$c_3 = 241[kg\ H_2O/(kg\ ASSI)]$: transpiration coefficient

$c_4 = 2000[kg\ H_2O/(kg\ ROOT \cdot a)]$: water supply coefficient

$c_5 = 6[t\ ASSI/(a \cdot ha)]$: growth of wooden biomass

pollution functions:

$\phi_1(\psi) = 0.5^2(0.6 - 0.4\psi)$: efficiency of needles

$\phi_2(\psi) = \psi(1+\psi)^{-1}$: proportion of damaged needles

$\psi \in [0,1.5]$: pollution parameter

coefficients: $c_2 c_3 c_4^{-1}\phi_1 l + (c_1^{-1}c_5 0.06 + c_1^{-1}l + \phi_2 l) + (1/6 + c_5) =: y_1 + y_2 l$.

$$y_4 = (1 - c_3 c_4^{-1})c_2\phi_1 - c_1/6, \qquad y_5 = \phi_2 + c_1^{-1}.$$

$$x_1 = 0.06 y_3 c_1^{-1}c_5, \quad x_2 = -y_1 y_5, \quad x_3 = -y_2 y_5,$$

$$x_4 = -y_1 y_3, \quad x_5 = y_1 y_4, \quad x_6 = y_3(y_4 - y_2), \quad x_7 = y_2 y_4.$$

Table 1. System parameters of (2.5), y_3 free for choice (cf. (2.3))

equilibrium $(l_1, a_1) = (l_1(\psi), a_1(\psi))$ in the interior of D is an unstable saddle point for all parameter values ψ with $(l_1(\psi), a_1(\psi)) \in D$. After [5] there exists ψ_0, $0 < \psi_0 < 1.5$, with

$$\lim_{\psi \nearrow \psi_0} l_1(\psi) = \lim_{\psi \nearrow \psi_0} a_1(\psi) = \infty. \tag{2.6}$$

Fig. 2 shows a decomposition of D into subsets defined by the coordinates $l_1(\psi)$ and $a_1(\psi)$ which are invariant under the flow of (2.5). This can easily be checked by considering the signs of $\dot{l} = dl/dt$ and $\dot{a}$. Obviously, the invariant set E in Fig. 2 is contained in the basin of (0,0). Thus, regarding (2.6), the more the pollution parameter ψ increases the larger the initial values of leaves and assimilate must be, to guarantee survival of the model tree. This behaviour exclusively depends on ψ, variations of the free parameter y_3 (as well as c_2 and c_5) only change the position of (l_1, a_1).

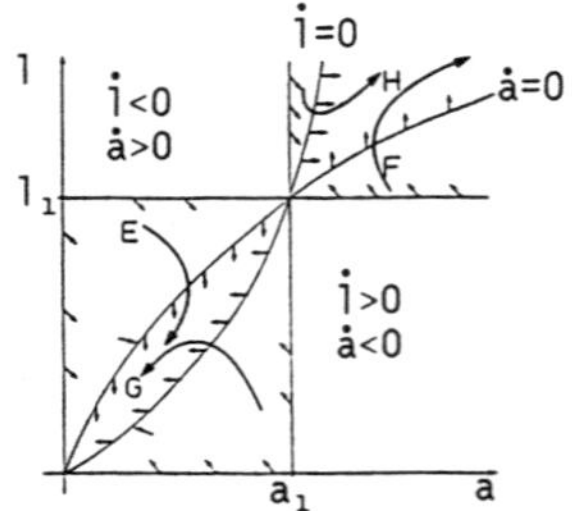

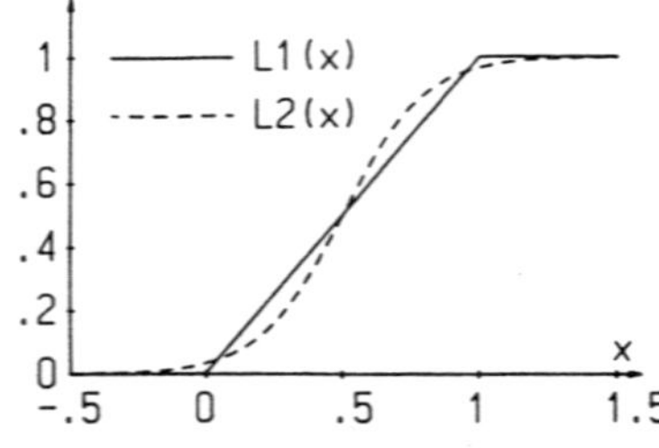

Fig. 2. Decomposition of the phase space of (2.5) for fixed ψ: E is attracted by the origin (after [5]).

Fig. 3. Limiter functions L_1 (3.1) and L_2 ((3.2), $a=0.5$, $b=0.3$) for $x \in [-0.5, 1.5]$.

3. THE LIMITER-FUNCTION CONCEPT

We will focuss now on another basic principle of forest ecosystems modelling turning out to be responsible for the distinct behavioural modes mentioned above. It is often useful to compare the resource demand of a specific internal process (e.g. assimilate demand for growth) with the appropriate supply (photosynthetic production). If the demand can be matched, the process will work unaffected. In case of insufficient supply the activity of the process must be diminished proportional to the supply demand ratio, while it doesn't make sense to increase the activity if the demand is oversupplied. Therefore, the supply demand ratio is restricted artificially to values lower than 1 before using it to control the process. Additionally, the ratio is often restricted to positive values.

The simplest approach to realize this kind of restriction is the *limiter-function*

$$L_1(x) := \begin{cases} 0 & \text{if } x < 0 \\ x & \text{if } x \in [0,1] \\ 1 & \text{if } x > 1 \end{cases} \quad . \tag{3.1}$$

As shown in Fig 3, it is also possible to apply the more smooth function

$$L_2(x) := 1/(1 + \exp(-2 \cdot (x-a)/b)) \tag{3.2}$$

with real parameters a,b. KRIEGER [2] shows that the complex forest ecosystem model described in [1] can be reduced to a two-dimensional, parameter-dependent model (3.3) containing only one simple-type limiter-function and showing qualitatively the same behavioural modes as the complex version (cf. Fig. 1(a),(b))

$$\frac{dl}{dr} = \frac{-c_1 \cdot \Phi_1 \cdot (c_2 + \Phi_2) \cdot l^2 + c_3 \cdot (c_4 \cdot \Phi_1 \cdot L_1(r \cdot (c_1 \cdot \Phi_1 \cdot l)^{-1}) - c_5 - c_2 - \Phi_2) \cdot l - c_3 \cdot c_6 \cdot r}{c_3 + c_1 \cdot \Phi_1 \cdot l}$$

$$\frac{dr}{dt} = \frac{c_1 \cdot \Phi_1 \cdot (c_4 \cdot \Phi_1 \cdot L_1(r \cdot (c_1 \cdot \Phi_1 \cdot l)^{-1}) - c_5) \cdot l^2 - c_1 \cdot \Phi_1 \cdot (c_6 + 1) \cdot l \cdot r - c_3 \cdot r}{c_3 + c_1 \cdot \Phi_1 \cdot l} \quad . \tag{3.3}$$

Here l and r denote leaf respectively root biomass, and L_1 is defined in (3.1). $c_1, \ldots, c_6$ are positive constants ($c_1 = 0.773$, $c_2 = 1/7$, $c_3 = 3$, $c_4 = 6$, $c_5 = 1.5$, $c_6 = 2$), and the functions $\Phi_1(\lambda) := 3/5 - 2/5 \cdot \lambda$, $\Phi_2(\lambda) := \lambda/(\lambda + 1)$ again introduce the pollution parameter λ. Although the right-hand side of (3.3) is not differentiable, local existence and uniqueness of solutions can be proved [2]. If treated as a whole, for (3.3) it is not possible to derive conditions explaining the parameter-dependent change of behaviour. Thus, each of the cases in (3.1) was seperately applied to (3.3). The resulting three branches are of the form

$$\frac{dl}{dt} = \frac{\alpha_1 \cdot l^2 + \alpha_2 \cdot l + k \cdot r}{c_3 + \alpha_5 \cdot l} \qquad \frac{dr}{dt} = \frac{\alpha_3 \cdot l^2 + \alpha_4 \cdot l \cdot r - c_3 \cdot r}{c_3 + \alpha_5 \cdot l} \tag{3.4}$$

with parameter-dependent functions $\alpha_1, \ldots, \alpha_5$ and a constant k that are different for each of the cases. (3.4) has two equilibria: the origin $(0,0)$ and another non-trivial equilibrium (l_1, r_1). Within the interesting parameter interval $[0.28, 0.29]$ (the change of behaviour was numerically found at $\lambda^* \approx 0.2837$, cf. [2]) none of the three systems' equilibria change stability. On the contrary, the location of (l_1, r_1) in the phase plane plays the major role to describe the qualitative behaviour of (3.3). The cases in (3.1) are represented by subsets of the phase plane separated by the straight lines $r = \alpha_5 \cdot l$ ($x = 1$) and $r = 0$ ($x = 0$).

From the chosen initial value $(l_0, r_0) = (18,7)$, where $x = 1$ applies, the trajectory of (3.4) is attracted by the equlibrium (l_1, r_1) which is asymptotically stable in that case [2]. For $\lambda < \lambda^*$, (l_1, r_1) is located in the same sector. Thus, the case $x = 1$ applies for all $t > t_0$. Increasing λ moves (l_1, r_1) on the opposite side of the line $r = \alpha_5 \cdot l$, and the trajectory, again starting at (l_0, r_0), also enters that sector. Now the case $0 < x < 1$ applies: the limiter-function (3.1) switches to another branch of (3.4) for which $(0,0)$ is the only stable equilibrium. The origin attracts the trajectory.

4. OUTLOOK

Mathematical analysis of the properties resulting from a nonlinear ODE-model turn out to be very difficult for complex, high-dimensional systems. To get compact models allowing

to apply mathematical tools, we use numerical analysis to identify the essential processes. At present, we focuss on two basic problems. On the one hand, we develop simple
models on a more general level neglecting short-time effects. As outlined in [7], the
partitioning of assimilate to leaf, root and stem growth, strongly influences forest
productivity. Assuming trees to optimize growth to survive among others, we will try to
apply optimal control theory to find (in that sense) optimal partitioning patterns.

On the other hand, models referred to in this paper ([1],[2],[5]) are condensed to their
essential mathematical structure. The results will be linear coupled ODE-models depending on more than one parameter.

REFERENCES

[1] Bossel, H., W. Metzler, H. Schäfer (eds.): Dynamik des Waldsterbens - Mathematisches
     Modell und Computersimulation. Fachberichte Simulation Bd.4, Springer, Berlin/
     Heidelberg/New York/Tokyo 1985, 265 p.

[2] Krieger, H.: Analysis of a Parameter Dependent Tree Model Introducing the Limiter-
     Function Concept (in preparation).

[3] Mäkelä, A.: Implication of the Pipe Model Theory on Dry Matter Partitioning and
     Height Growth in Trees. J. Theor. Biol. 123 (1986), 103-120.

[4] Metzler, W.: Dynamische Systeme in der Ökologie. Teubner, Stuttgart 1987, 210 p.

[5] Metzler, W., D. Gockert: Dynamical Simulation of Air Polluted Forests. Proceedings
     IASTED Sixth International Symposium on Modelling, Identification and Control.
     Acta Press, Anaheim 1987, 445-449.

[6] Mohren, G.M.J.: Simulation of Forest Growth Applied to Douglas Fir Stands in the
     Netherlands. PhD Thesis Wageningen, Netherlands, 1987, 184 p.

[7] Linder, S., R.E. McMurtrie, J.J. Landsberg: Growth of Eucalyptus: A Mathematical
     Model Applied to Eucalyptus Globulus. In: Tigerstedt P.M.A., Puttonen, P.,
     Koski, V. (eds.): Crop Physiology of Forest Trees. Helsinki (University Press)
     1985, 117-126.

[8] Pastor, J., W.M. Post: Development of a Linked Forest Productivity-Soil Process
     Model. Oak Ridge National Laboratory, Env. Sci. Div. Publ. No. 2455. Oak Ridge,
     Tennessee 1985, 162 p.

[9] Thornley, J.H.M.: Mathematical Models in Plant Physiology. Academic Press, London
     1976.

[10] Valentine, H.T.: A Carbon-Balance Model of Stand Growth: A Derivation Employing
     the Pipe-Model Theory and the Self-Thinning Rule. IIASA Working Paper No. WP-87-56
     1986, 17 p.

The Analysis of Decision-Making Systems

László CSERNY[§]

The *decision-making system* of an organization is understood as the totality of decisions, of decision-points, of decision-makers/abbreviated by DM in the following/ and relationships among them.

1.THE ELEMENTARY DECISION-MAKING SYSTEM

1.1.The model of the system

Any state of a /decision-making/ system can be given in the form of an n-dimensional vector $\underline{s}$ in a state space S. The input and output variables, as well as the objective variables of the system are regarded as the state variables of the system i.e.

$$\underline{s} = (s_1, s_2, \ldots, s_n) = (\underline{x}, \underline{y}, \underline{z}) = (\underline{x}, \underline{y}, F(\underline{x}, \underline{y})) \in S = S_1 x S_2 x \ldots x S_n \qquad (1)$$

where $\underline{x}, \underline{y}, \underline{z}$ are the vectors of the input, the output and the objective variables,

$\underline{z} = F(\underline{x}, \underline{y})$ the goal function of the system. $\qquad (2)$

A simplified model of an elementary decision-making system can be seen in figure 1 with reference to [4]. Referring to the picture the input and the output vectors of the system are the following:

$$\underline{x} = (\underline{x}_i, \underline{x}_v) \qquad \underline{y} = (\underline{y}_i, \underline{y}_v) \qquad (3)$$

where $\underline{x}_i, \underline{y}_i$ information input and output vectors respectively,

$\underline{x}_v, \underline{y}_v$ non-information input and output vectors respectively.

The DM's inputs and outputs can be given by the following vectors:

$$\underline{x}_D = (\underline{v}, \underline{t}) \qquad \underline{y}_D = (\underline{u}, \underline{w}) \qquad (4)$$

where $\underline{v} = (\underline{x}, \underline{y})$ information about the inputs and the outputs of the system

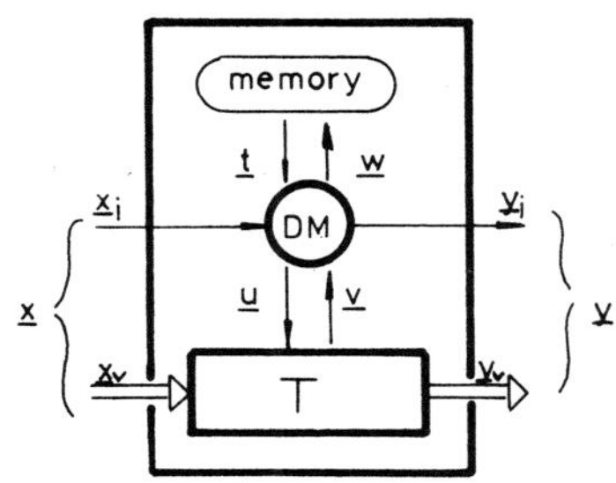

Figure 1

§ author: lecturer, Ybl Miklós College for Building Industry
H-1442 Budapest,70. P.O.Box 117. HUNGARY

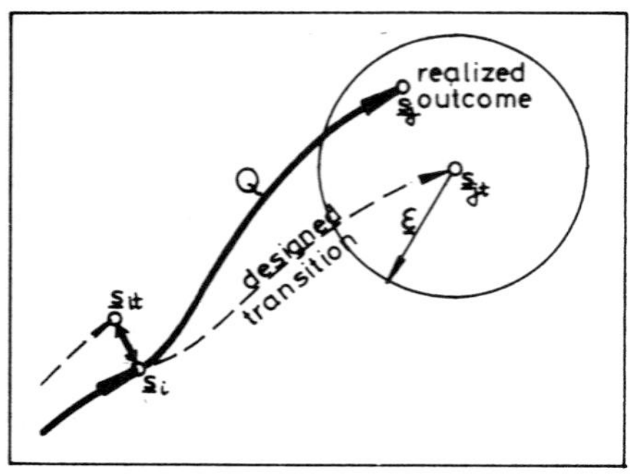

Figure 2

Finally using the formulas (1) and (2) the definition of the system can be given in the form below:

$$S = X \times Y \times Z \tag{5}$$

and

$$F: X \times Y \; ----\!> Z \tag{6}$$

$$T: X \; ----\!> Y \tag{7}$$

where X,Y and Z are the sets of inputs, of outputs and objectives,

    F   the goal function/operator/ of the system that might be the decision function at the same time,

    T   the transformation function/operator/ of the system.

1.2. The decision-making process

The process, the functioning of the model goes through the following stages[3,5]:

a.) *The determination of the present state, the decision situation*

The deviation of the present state $\underline{s}_i$ from the designed state $\underline{s}_{it}$ /Fig. 2/ leads to the development of a decision situation. Such decision constraint is going to develop if

$$\Delta\underline{s} = |\underline{s}_i - \underline{s}_{it}| \sim |\underline{z}_i - \underline{z}_{it}| > \epsilon \tag{8}$$

i.e. if the deviation from the goal state is larger than a prescribed value ϵ.

The decision situation, the decision problem can be investigated   in   the decision space D, which is a subspace of the space S at the same time. The selection of the decision variables,   the considered number of   the variables/the dimension of D/ highly depends on the DM's intention. It can be written

$$D = D_1 \times D_2 \times \ldots \times D_p \qquad \text{and} \qquad \dim D \leqq \dim S \qquad (9)$$

Thus the relation (8) corresponds to the relation in D given below:

$$|\underline{d}_i - \underline{d}_{it}| > \epsilon \qquad (10)$$

b.)*The decision problem and its solving procedure*

The DM should order his problems to be solved according to their importance so that he should choose the most important one of them

$$D = \{\underline{d}_{i1}, \underline{d}_{i2}, \ldots, \underline{d}_{ij}, \ldots \mid \underline{d}_{ij} \prec \underline{d}_{ik}, \quad \text{if} \quad j<k\} \qquad (11)$$

The DM drops some of the problems being unimportant for him whilst others are unsolvable for him.

To solve a *decision problem*
1.)it should be assigned the *decision variables*, the *decision space* D spanned by them according to (9), as well as the goal function operator F. Usually the task is to minimize F i.e.

$$\underline{z} = F(\underline{x},\underline{y}) \quad \text{-----}> \quad \min \qquad (12)$$

If it gives the solution of the problem then F will be the decision function of the problem at the same time.
2.)it should be determined the *possible alternatives of acting* or the *possible goals* taking into consideration of the objective of the system[3].
   - If the set of alternatives of acting A is given then $\underline{x}_{jt}, \underline{y}_{jt}$ and $\underline{z}_{jt}$ can be determined too.
   Here
$$A = \{a_{ij} \mid a_{ij}=(\underline{s}_i,\underline{s}_j), \ \underline{s}_j=Q\underline{s}_i, \quad \underline{s}_i,\underline{s}_j \in S\} \subseteq S \times S \qquad (13)$$

   and Q is the transition operator in the state $\underline{s}_i$.
   - If the goal state $\underline{d}_{jt}$ or rather $\underline{z}_{jt}$ is known then by using the goal function operator F, it could be determined those values of $\underline{x}_{jt}$ and $\underline{y}_{jt}$ which enable the equation

$$\underline{z}_{jt} = F(\underline{x}_{jt},\underline{y}_{jt}) \qquad (14)$$

   and the alternative of acting realizing the transition can be assigned too.
3.)In the greater part of the problems the decision situation is *stochastic-like*, i.e. some probability distribution can be assigned to each of the alternatives presuming a /subjective/ probability field (Ω,B,P). For a given ϵ the risk-event $\omega_e \in \Omega$ is the following

$$(\, | \underline{s}_j - \underline{s}_{jt} | \, > \in \, | a_{ij}) \tag{15}$$

where $a_{ij} \in A$ is the alternative to be realized. Thus *the risk* of an a_{ij} is the conditional probability

$$p_j = P(a_{ij}, \omega_e) = P(\, | \underline{s}_j - \underline{s}_{jt} | > \in \, | a_{ij}) \tag{16}$$

i.e.

$$P: \, A \times \Omega \, \dashrightarrow \, V = [0,1] \subseteq R^{+} \tag{17}$$

c.)*The decision, the decision strategy*

After having solved the problem and considered the goal and the risk of the alternatives, the DM should make a /subjective/ ordering on the alternatives and should choose only and only one alternative of them. This would be his decision.
If λ is an ordering relation and N is the set of the natural numbers then

$$\lambda: \, Z \times (A \times \Omega) \, \dashrightarrow \, N \tag{18}$$

and then the set of optimal alternatives is

$$A_o = \{ a_{ij} \, | a_{ij} \in A \quad \text{and} \quad \lambda(Z,A,\Omega) = \min \} \tag{19}$$

Hence the decision function is

$$\gamma: \, A_o \, \dashrightarrow \, A^{*} = \{ a^{*} \} \tag{20}$$

The DM's successive decisions γ_i can be considered as a *strategy* of short-term decisions[2] i.e.

$$\varphi = \{ \gamma_1, \gamma_2, \ldots, \gamma_i, \ldots \} \in \Phi \tag{21}$$

Let the objective function operators considered above the short- and the long-term decision strategies be signed by G and G_o respectively. Then if

$$\gamma_o: \, A_o \, \dashrightarrow \, A_o^{*} = \{ a_o^{*} \} \tag{22}$$

is a long-term decision and

$$\varphi_o = \{ \gamma_o \} \in \Phi_o \tag{23}$$

then the DM's purposes are to minimize

$$G_o(\varphi_o) \, \dashrightarrow \, \min \quad \text{and} \quad G(\varphi | \varphi_o) \, \dashrightarrow \, \min \quad \text{respectively.} \tag{24}$$

2.ORGANIZATIONAL DECISION-MAKING SYSTEM

The organizational decision-making systems which consist of more or less
elementary decision-making systems are very complicated systems[5,6] and
the investigation of them is a very difficult thing.

2.1.The decomposition of the systems

The simplified model of the decision-making system is in Figure 3 Without
the decomposition of the process itself.

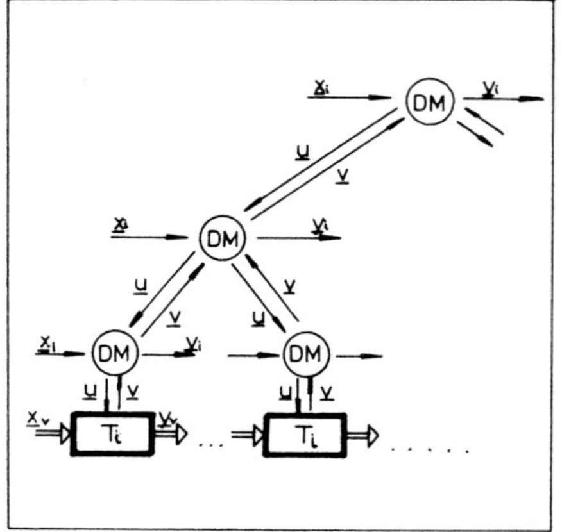

Figure 3.

The inputs and the outputs of the
different level DM's subsystems are the
following similarly to the formula (3):

$$\underline{x}=(\underline{x}_{ik},\underline{x}_{io},\underline{x}_{il},\underline{x}_v)\quad \underline{y}=(\underline{y}_{ik},\underline{y}_{io},\underline{y}_{il},\underline{y}_v)$$

$$(25)$$

where $\underline{x}_{ik},\underline{y}_{ik}$ upper level origin information,

$\underline{x}_{io},\underline{y}_{io}$ information from the DM's
level,

$\underline{x}_{il},\underline{y}_{il}$ lower level information,

$\underline{x}_v,\ \underline{y}_v$ non-information type input and output.

2.2.The organizational decision problem

The *organizational decision problem* is understood as a decision problem to
be solved in a system of several DMs that cooperate for common goals and
objectives.

Let $\varphi_o \in \mathbb{O}_o$ be the upper level DM's strategy and $H_i: \mathbb{O}_o \longrightarrow \mathbb{O}_i$ the set of
strategies of the i-th DM which may be used in the case of $\varphi_o \in \mathbb{O}_o$. Then the
set of the *collective strategies* of the group is

$$\mathbb{O} = \{\underline{\varphi}\,|\,\underline{\varphi} = (\varphi_1,\varphi_2,\ldots,\varphi_i,\ldots,\varphi_t),\ \varphi_i \in \mathbb{O}_i,\ 1 \leq i \leq t\} \tag{26}$$

i.e.

$$\mathbb{O} = \mathbb{O}_1 \times \mathbb{O}_2 \times \ldots \times \mathbb{O}_t \tag{27}$$

Furthermore let the DMs' objective function operator be signed G_i, for $\forall i$.

a.)If all of the members of the group use cooperative, e.g.*Pareto-optimal
strategies* then they only choose their own strategies from the
following set of rational strategies[1]:

$$R(\varphi_o) = \{\underline{\varphi}^* | G_i(\varphi_i^* | \varphi_o) \leq G_i(\varphi_i | \varphi_o) \quad \text{and at least for one } i$$

$$G_i(\varphi_i^* | \varphi_o) < G_i(\varphi_i | \varphi_o), \quad \varphi_i^*, \varphi_i \in \mathbb{O}_i \text{ for } \forall i\} \tag{28}$$

In this case the upper level DM's *Stackelberg-strategy* is that $\varphi_o^* \in \mathbb{O}_o$ that

$$G_o(\varphi_o^* | \underline{\varphi}^*) \leq G_o(\varphi_o | \underline{\varphi})$$

and
$$\varphi_o^*, \varphi_o \in \mathbb{O}_o \qquad \underline{\varphi}^* \in R(\varphi_o^*), \qquad \underline{\varphi} \in R(\varphi_o) \tag{29}$$

b.)If the members of the group use non-cooperative strategies/e.g. *Nash-strategy*[7]/ then the set of the rational strategies belonging to $\varphi_o \in \mathbb{O}_o$ is the following:

$$R(\varphi_o) = \{\underline{\varphi}^* | G_i(\varphi_i^* | \varphi_o, \underline{\varphi}_{iP}^*) \leq G_i(\varphi_i | \varphi_o, \underline{\varphi}_{iP}^*) \qquad \text{for } \forall i$$

$$\text{and } \underline{\varphi}^*, \underline{\varphi} \in \mathbb{O}, \qquad \varphi_o \in \mathbb{O}_o\} \tag{30}$$

in (30)
$$\underline{\varphi}_{iP}^* = (\varphi_1^*, \varphi_2^*, \ldots, \varphi_{i-1}^*, \varphi_{i+1}^*, \ldots, \varphi_t^*) \tag{31}$$

Thus the upper level DM's strategy to be chosen is that $\varphi_o^* \in \mathbb{O}_o$ that

$$G_o(\varphi_o^* | \underline{\varphi}^*) \leq G_o(\varphi_o | \underline{\varphi})$$

and
$$\varphi_o^*, \varphi_o \in \mathbb{O}_o, \qquad \underline{\varphi}^* \in R(\varphi_o^*), \qquad \underline{\varphi} \in R(\varphi_o) \tag{32}$$

REFERENCES:

[1] Başar,T.-Cruz,J.B.Jr.: Concepts and Methods in Multi-Person Coordination and Control, in: Tzafestas/ed./, Optimisation and Control of Dynamic Operational Research Models, North-Holland, Amsterdam, 1982, pp.351-396

[2] Cserny L.: The Investigation of Sequential Decision-Making Strategies in Cases of Insuffient Information, 3rd Czechoslovak-Soviet-Hungarian Seminar on Information Theory, Liblice,1980,pp.87-92

[3] Cserny L.: Computer-Aided Modelling of Multi-Level Decision-Making Systems, International Conference on Computer Applications, in Building Design, Architecture and Town Planning /IBA-DAT '82/, Berlin, 1982, pp.98-104

[4] Cserny L.: A vállalat, mint rendszer elemzése/The Analysis of Enterprises as Systems/, Ipargazdaság, XXXV(1), 1983, pp.25-32,/in Hungarian/

[5] Kickert,W.J.M.: Organisation of Decision-Making.A Systems Theoretical Approach, North-Holland, Amsterdam-New York-Oxford, 1980

[6] Mesarovic,M.D.-Macko,D.-Takahara,Y.: Theory of Hierarchical, Multilevel Systems, Academic Press, New York-London, 1970

[7] Nash,J.: Noncooperative Games, Annuals of Mathematics, 54(2), 1951, pp.284-295

Interactive Procedures for Multicriteria Decision Support in Bargaining Problem*

P. Bronisz, L. Krus**

Abstract. Multicriteria bargaining problem is considered. Two interactive procedures supporting solution of the problem are described and discussed.

Key words. Bargaining problem, multicriteria optimization, decision support.

Introduction

The paper deals with a multicriteria bargaining problem. The problem is a generalization of classical formulation proposed by Nash [5], developed by Raiffa [6], Roth [7], Kalai [4] and others. It is assumed that for some reasons — such as practical limitations of the utility theory — the aggregation of player's objectives is impossible. In such a case the classical approach can not be applied. In that place, interactive learning procedures can be constructed which allow the players looking for efficient solutions according to their preferences. Two interactive procedures supporting the multicriteria bargainig problem are considered.

Problem formulation and definitions

Let $N = \{1,2,..,n\}$ be a finite set of players, each player having m objectives. A multicriteria bargaining game is defined as a pair (S,d), where an agreement set S is a subset of $n*m$-dimensional Euclidean space $\mathbb{R}^{n*m}$, and a disagreement point d belongs to S. Any outcome $x \in S$ can be the result of the game if it is specified by unanimous agreement of the players, if there is no such agreement, the disagreement point is the result.

For every point $x = (x_1,..,x_n) \in \mathbb{R}^{n*m}$, $x_i \in \mathbb{R}^m$, let x_{ij} denote the amount of the j-th objective for the i-th player. We assume that each player tries to maximize all his objectives.

We confine our consideration to the class B^m of all multicriteria bargaining games (S,d) satisfying the following conditions:

(i) S is compact and there is $x \in S$ such that $x > d$,

(ii) S is comprehensive, i.e. for $x \in S$ if $d \leq y \leq x$ then $y \in S$.

---

*   supported by Program CPBP 02.15 and IIASA contracted study
**   Systems Research Institute, Polish Academy of Sciences, Newelska 6, 01-447 Warsaw, Poland

(iii)   For any $x \in S$, let $Q(x) = \{i: \ y \geq x, \ y_i > x_i \ $ for some $y \in S\}$. Then
         for any $x \in S$, there exists $y \in S$ such that $y \geq x$, $y_i > x_i$ for
         each $i \in Q(x)$.

Assumption (ii) states that objectives are disposable, i.e. that if the players can reach the outcome x then they can reach any outcome worse than x. Observe that we do not assume convexity of the agreement set. In that place we assume no restrictive condition (iii) which states that the set of Pareto optimal points in S contains no "holes".

The problem consists in looking for nondominated solution in the agreement set according to the preferences of the players.

A point $x^i \in S$ is defined as i-*nondominated*, $i \in N$, if there is no $y \in S$ such that $y_i \geq x^i_i$, $y_i \neq x^i_i$. A point $u \in \mathbb{R}^{n*m}$ is defined as a *utopia point relative to aspirations* (RA utopia point) if for each player $i \in N$, there is an i-nondominated point $x^i \in S$ such that $u_i = x^i_i$.

The i-nondominated point is an outcome which could be achieved by a rational player i if he would have full control of the moves of other players. In the multicriterial case considered here there is a set of such points. That requires each player to investigate his set of i-nondominated points and then to select the most preferable one using multicriteria decision support. The preferable i-nondominated point can be selected by the i-th player using, for example, the achievement function approach proposed by Wierzbicki [10].

Let $U(S,d)$ denote the set of all the RA utopia points u for a bargaining game (S,d) such that $u > d$. A solution for the multi-criteria bargaining problem is a function $f^m : B^m \times \mathbb{R}^{n*m} \ ---> \ \mathbb{R}^{n*m}$ which associates to each $(S,d) \in B^m$ and each RA utopia point $u \in U(S,d)$ a point of S, denoted $f(S,d,u)$.

Interactive procedure based on one-shot solution concept

The following one-shot solution concept is used in the procedure
$$G(S,d,u) = d + h(S,d,u) \ [\ u - d \ ] ,$$
where $(S,d) \in B^m$, $u \in U(S,d)$, and
$$h(S,d,u) = \max \ \{ \ h \in \mathbb{R} : \ d + h(u-d) \in S \ \}.$$

The outcome $G(S,d,u)$ is a unique point of intersection of the line connecting u to d with the boundary of S. Axiomatic characterization of the solution can be found in [1].

The procedure consists of two phases.

In the first one, each player $i \in N$, acts separately. He tests his different i-nondominated points. This phase allows each player to estimate the ranges of his criteria and enables the initial approximation of the most preferable i-nondominated point.

The second phase is played in several rounds. At each round t, $t=1,2,\ldots$, the outcome $G(S,d,u^{t-1})$ is calculated, on the basis of RA-utopia point u^{t-1} generated by i-nondominated points, $i \in N$, selected by players in round $(t-1)$. If the players agree on the outcome $G(S,d,u^{t-1})$, the process ends, otherwise they modify their i-nondominated points. For that, each player $i \in N$ can independently test his outcome for different i-nondominated points assuming actions of the other players and then he selects a new preferable i-nondominated point being component of new RA-utopia u^t point.

The presented scheme can support achievement of an agreement outcome. The scheme based on the presented solution concept seems to be reasonable because the properties resulting from the axioms assure "fairness" of the solution.

Interactive procedure based on limited confidence principle

The solution in this procedure is considered as a result of an iterative process defined by sequence of succesive outcomes d^t, $t=1,2,\ldots,T$ (T can be equal ∞). It is assumed that the process starts at disagreement point $d^o = d$, is progressive $d^t \geq d^{t-1}$, $d^t \in S$ for $t = 1,2,\ldots,T$, and tends to strongly Pareto frontier of the set S: i.e. d^T ($= \lim_{t \to \infty} d^t$ if $T = \infty$) is a strongly Pareto optimal point in S . The limited confidence principle [8], [9], [3] is basis for the process description . The principle says that during the negotiation process the players try to limit conter-players outcomes . According to the principle the acceptable demands are limited by:

$$d^t - d^{t-1} \leq \alpha_{min}^t [u(d^{t-1}) - d^{t-1}]$$

for $t = 1,\ldots,T$, where α_{min}^t is a joint confidence coefficient, $\alpha_{min}^t = \min \{\alpha_1^t,\ldots,\alpha_n^t\}$, and $0 < \alpha_i^t \leq 1$ is a given confidence coefficient of the i-th player at round t.

The procedure consists of a number of rounds. Each round t starts from the current status quo point d^{t-1}, each player $i \in N$ specifies a current confidence coefficient α_i^t (describing his limited confidence in his ability to predict the posible outcomes) and tests different i-nondominated points assuming some moves of other players. For such generated RA-utopia points the procedure calculates the corresponding solution. This information allows the player to specify his the most preferable i-nondominated point at round t. This stage of the procedure is performed independently for all the players. After that on the basis of the confidence coefficients and the most preferable i-nondominated points of the players, the procedure calculates new status quo point d^t at round t according to the formula:

$$d^t = d^{t-1} + \alpha^t [u^t - d^{t-1}].$$

where $\alpha^t = \min \{ \alpha_1^t, \ldots, \alpha_n^t, \alpha_{max}^t \}$ and u^t is RA-utopia point generated by selected i-nondominated points.

It has been shown [2], [3] that assuming rationality of the players the procedure is well-defined and converges to strongly Pareto optimal point in the agreement set.

The presented procedure has been apllied in an experimental system supporting negotiation on joint development program [2].

References

[1]   Bronisz P., L. Krus, "The Raiffa Solution for Multicriterial Bargaining Problems", draft paper, Systems Research Institute, Polish Academy of Sciences, 1986.

[2]   Bronisz P., L. Krus, B. Lopuch, "An Experimental System Supporting Multiobjective Bargaining Problem. A Methodological Guide", in Theory, Software and Testing Examples for Decision Support Systems, ed. A.Lewandowski, A.P.Wierzbicki, IIASA, Laxenburg, 1987 (forthcoming).

[3]   Bronisz P., L. Krus, A. Wierzbicki, "Towards interactive solutions in Bargaining Problem", RP.I.02.1.4, Report of Institute of Automatic Control, Warsaw University of Technology, 1987.

[4]   Kalai E., M. Smorodinsky, "Other Solutions to Nash's Bargaining Problem", Econometrica, Vol. 43,(1975), pp. 513-518.

[5]   Nash J.F., "The Bargaining Problem", Econometrica, Vol. 18,(1950), pp. 155-162.

[6]   Raiffa H., "Arbitration Schemes for Generalized Two-Person Games", Annals of Mathematics Studies, No. 28 1953,pp. 361-387, Princeton.

[7]   Roth A.E., "Axiomatic Model of Bargaining", Lecture Notes in Economics and Mathematical Systems, Vol. 170, Springer-Verlag, Berlin, 1979.

[8]   Fandel G., Optimale Entscheidungen in Organizationen, Springer-Verlag, Heidelberg,1979.

[9]   Fandel G., A.P. Wierzbicki, "A Procedural Selection of Equilibria for Supergames", 1985, (private unpublished communication).

[10] Wierzbicki A.P., "A Mathematical Basis for Satisficing Decision Making", Mathematical Modelling, Vol. 3,(1982), pp. 391-405.

[11] Wierzbicki A.P., "Towards Interactive Procedures in Simulation and Gaming: Implications for Multiperson Decision Support", in Methodology and Software for Interactive Decision Support, Proceedings of International Workshop, Albena, 1987, (forthcoming).

A Procedure for Decision Support Systems Design: Modelling and Simulation Environment

Z. Strezova[*]

Abstract. In the paper a framework for model-based decision support systems design is considered. A procedure for multistage assessment and validation of a set of modelled and simulated alternatives is described. A knowledge-oriented approach to DSS design is discussed. Applications and software environment are presented.

Keywords. Model-based DSS; decision support; model assessment; model validation.

INTRODUCTION

Over the last decade a new information technology has been a challenge for research: the decision support systems technology. Developing in the field of Decision Support Systems (DSS) have included frameworks for DSS design, structures of DSS software, concepts of model management, artificial intelligence techniques used into DSS.

A direction of the DSS movement, attracting a lot of interest, is the development of model-based DSS (MBDSS). Such systems provide decision support models: a new generation of tools in the modelling technology.

This paper focusses on the modelling and simulation environment in DSS design. After pointing out some features of the DSS technology a procedure for developing MBDSS is presented. The procedure incorporates simulation and metamodelling activities for multistage assessment and validation of a set of modelled alternatives. A knowledge-oriented approach to DSS design is discussed. Applications and software environment of specific MBDSS in the policy analysis field are described.

GENERAL ASPECTS OF DECISION SUPPORT SYSTEMS

The features of DSS are linked to the decision making (problem solving) process. Following the popular model of H. Simon, this process comprises three phases: a) intelligence; b) design; c) choice and implementation. In the terms of these phases a given problem can be identified as: a) well structured: one in which the three phases are fully structured; b) unstructured: when the conditions of recognizing the problem can not be defined; c) semistructured: when only one or two of the phases are defined (undefined).

There exist many definitions of DSS. The basis for defining DSS has been migrating from an explicit statement of what a DSS does to some ideas about how the DSS objective can be accomplished, and also from what is to what is not a DSS.

In the early 70's DSS was described as a computer-based system to aid in decision making. In the mid to late 70's the emphasis was rather on the support provided

---
[*] Institute for Information Technologies
Volov st., 2, 1504 Sofia, Bulgaria

by DSS. In the later 70's to early 80's DSS is considered as "an application of available and suitable computer-based technology to improve the effectiveness of managerial decision making in semistructured tasks" /3/.

In DSS development three levels of technology can be identified /4/. The first and most fundamental level is called <u>DSS Tools</u>: hardware and software elements which facilitate the DSS design. The second technological level is called a <u>DSS Generator</u>: a "package" of related hardware and software which provides a set of capabilities to build specific DSS quickly and easily. The third level involves specific <u>DSS Applications</u>, i.e. the hardware/software that allow a specific decision maker or group of them to deal with specific sets of related problems.

An effective MBDSS consists of three subsystems: a data subsystem, a model subsystem and a complex software system for linking the user to each of them (a dialog subsystem). The MBDSS provide optimization and simulation models to help make decisions.

The key capabilities of MBDSS include:
* the ability to create new models quickly and easily
* the ability to access and integrate model "building blocks"
* the ability to manage the model base with management functions analogous to data base management
* the ability to assess, validate and select models.

Following Keen /2/, the possible levels of support provided by DSS are:
a) <u>passive support</u>: tools that the users are comfortable with, without changing the existing modes of operation
b) <u>traditional support</u>: tools for improving the current decision process; selection of alternatives only on the basis of the "what-if" question
c) <u>normative support</u>: tools based on a normative view of how decisions should be made; using the concept of feasibility
d) <u>extended support</u>: an explicit effort to apply analytic and optimization models and methods, to embody limited forms of knowledge (building semi-expert systems), to select a feasible alternative, rather than just evaluate a given set of alternatives.

A PROCEDURE FOR MODEL-BASED DECISION SUPPORT SYSTEMS DESIGN

The procedure developed meets some recent requirements concerning DSS design, mainly generating a set of explicit models of the problem stated, and successive screening and selection of alternatives.

The procedure embodies phases and stages as is shown in Fig. 1. More details about the structure of the procedure and related works can be found in /5/, /6/, /7/.

APPLICATIONS AND SOFTWARE ENVIRONMENT

The procedure for MBDSS design, illustrated in Fig. 1, has been used in two ways: a) as a basis of a general-purpose DSS technology, and b) in developing a number of specific DSS.

A. PROBLEM DEFINITION PHASE
problem formulation
description of the problem environment
B. MODEL DEVELOPMENT PHASE
description of the model environment
design of a general model
SES
generating a set of models {M_1}
C. MODEL ASSESSMENT PHASE
simulation of the set {M_1}
1st validation: empirical tests; selection of {M_2}
SES
2nd validation: sens. analysis I selection of {M_3}
3rd validation: sens. analysis II selection of {M_4}
D. DECISION SUPPORT PHASE
presentation of feasible alternatives
SES
choice of alternative for implementation
to phase A, if necessary
possible combinations

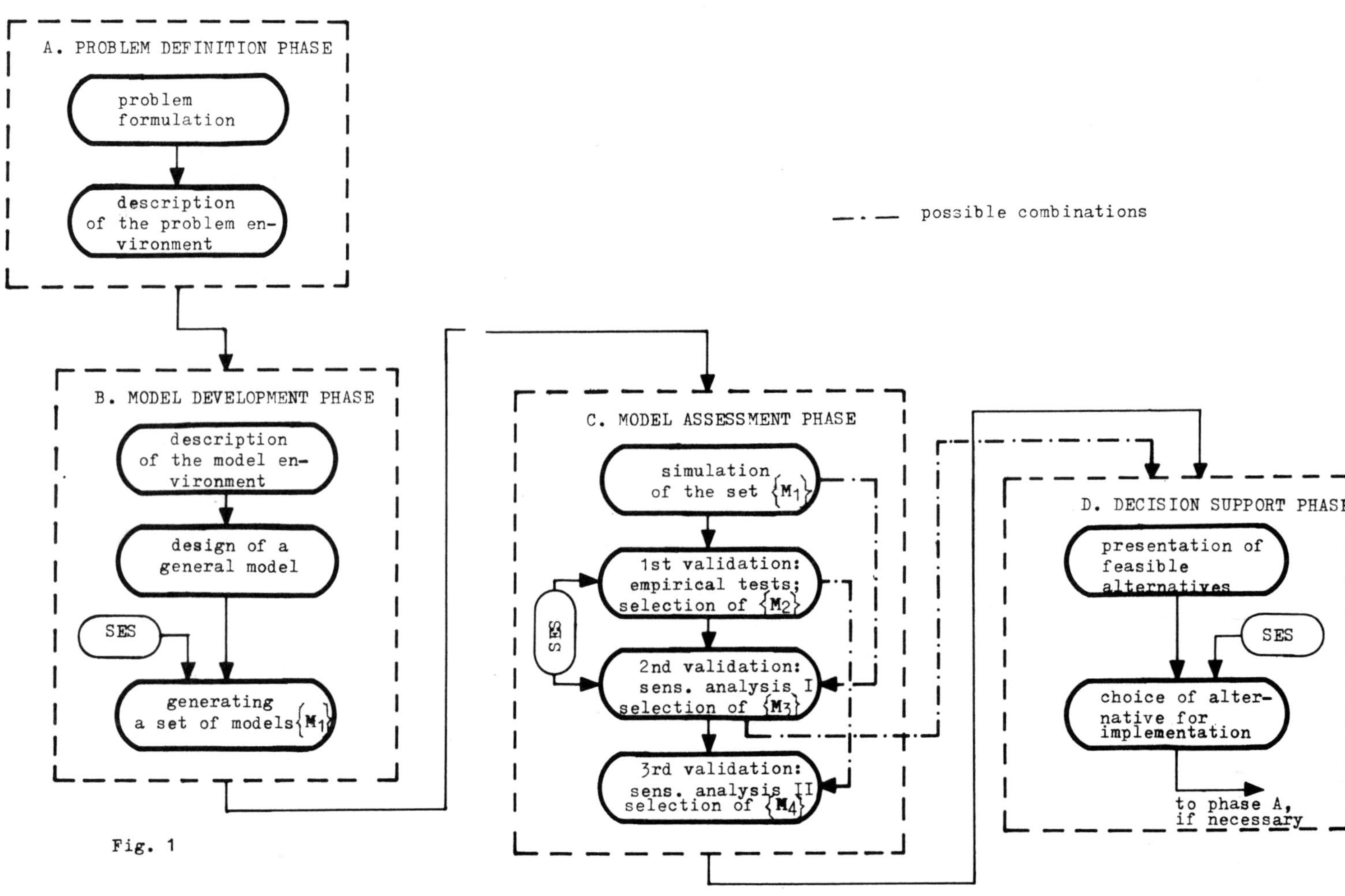

Fig. 1

As for the general technology, it concerns a class of real-world problems in the
management and policy analysis area. The specific DSS applications are concerned
with information and computer resource allocation problems. A number of such problems, related to analysis of regional (local) and sectoral policy of computer and
information resource allocation has been performed.

For supporting all modelling, simulation and validation functions of the entity
of specific DSS a modular software system has been developed. SIMOS I consists of
four standard program packages (discriminant analysis, ILP, GPSS, linear regression analysis), a centralized information base, and a module of special purpose
programs. The information base includes data and information concerning the stated
real-world problem, model description data, simulation results , validation data,
as well as all modelled and simulated alternatives, presented explicitly. Designed
in such a way SIMOS I can be considered as a step toward an integrated modelling
environment.

AI - SUPPORT IN MBDSS DESIGN AND IMPLEMENTATION

Artificial Intelligence (AI) is the subfield of the computer science which has
perhaps the greatest potential with respect to decision making. Expert systems,
incorporated in MBDSS can assist a number of tasks. The so-called "semi-expert"
systems /2/ combine the power of the expert system technology and the knowledge -
already available - about problems whose solution requires extended support, as
this term was described above. The practical realization of such systems will allow to switch from the question "what...if" to "what ought to be" in areas which
are challenge for technologies such as DSS, modelling, simulation and AI.

In Fig. 1 are pointed out the stages of the procedure which can be supported by
a semi-experted system (SES). Some SES components are included in SIMOS I.

CONCLUSION

The procedure for MBDSS, considered briefly above, is consistent with the requirements of the concept of flexibility of decision options and those of the versati-
lity approach /1/. If applied adequately this procedure can provide extended support to ill-defined problems solving.

REFERENCES

1. Bonder, S.: Changing the future of operations research. Opns Res <u>27</u> (1979).
2. Keen, P.G.W.: Decision support systems: the next decade. Decision Support
   Systems, <u>3</u> (1987), 3.
3. Keen, P.G.W. and M.S. Scott Morton: Decision Support Systems: An Organizational Perspective. Addison-Wesley, Reading, MA, 1978.
4. Sprague, R.H. and E.D. Carlson: Building Effective Decision Support Systems.
   Prentice Hall, Englewood Cliffs, N.J., 1982.
5. Strezova, Z.: Technical Reports IR 109, 111, 115, SC MIS, 1981-83. (in Bulg.)
6. Strezova, Z.: An approximate approach to the design of decentralized management systems structures. Proceedings of the 8th IFAC Congress, vol. 7,
   Pergamon Press, 1982.
7. Strezova, Z.: Validation in ill-defined problem modelling: resource allocation policy analysis. Proceedings of the 5th IFAC/IFORS Dynamic Modelling
   Conference. Pergamon Press, 1986.

Knowledge-Based Decision Aid in Textile Technology

Dr.rer.nat Gert Kreiselmeier[*], Dr.-Ing. Roland Seidl[*]

1. DECISION SITUATIONS IN TEXTILES

Textile technology integrates the skill of many special disciplines. Technicians and the operators of textile plants fit the definition of a human expert completely and provide the potential for textile technology to take advantage of knowledge-based computer programs. To capture and widely apply the experience and expertice of this experts is the aim of developing automatic decision aid.

To understand in which respect these computer programms can be useful in the field of textiles suppose what the following tasks have in common?

> diagnosing fabric or machine faults,
> choosing an appropriat machine to produce a desired fabic,
> configuring the many components that make up a clothing

In all cases decision is made in a complex situation from many singulare data and information. This is a tasks that only a few experts do well. Experts are people who can do things other people cannot do because of training and experience. Their reasoning process in each case includes use of judgment and rules of thumb.

As experts in the field of textiles face a more complex and involved world within which to operate, the need for some decision support is becoming urgent and thus computer programs which do this continue to spread out. This is due to the fact that for the considered applications human expertise is rare, in high demand and expensive, often humen expertise is required in remote areas.

Although computers and simulation models have become indespensible tools in many textile processes, there is continued reliance on the human experts ability to identify and synthesize diverse factors, to form judgements, evaluate alternatives and to aid decisions.

---

[*] Forschungsinstitut für Textiltechnologie Karl-Marx-Stadt, DDR

the strategy it follows is called the <u>control</u> <u>strategy</u>. One strategy is to scan through the rules until one is found whose conditions match facts in the data base, the rule is applied, updating the data base and the scanning resumes. This is known as forward-chaining. A different strategy is to select a goal to be achieved and scan the rules to find those whose consequent action can achieve the goal. If there are no rules to establish the new subgoal, the program asks the user for the necessary facts and enters them in the data base. This strategy is known as backward-chaining.

We constructed a backward-chaining rule interpreter in PROLOG which operates over an exchangable rule-base and the basic idea of which is according to Sterling and Shapiro [1]. The essence of the program is, that the rule base is used to examine facts supplied by the user to arrive at a conclusion, that is to establish the truth of a hypothesis. This general problem can be used to classify an objekt, event or situation on the basis of information about its characteristics. This information will be acquired sequentially in dialog with the user.

The system asks questions of the user about "primitive" information, which cannot be found in the knowledge base or derived from other information. The user can respond to such questions in two ways:

  by supplying the relevant information as an yes/no-answer to the
  query, choosing an answer from a menue or typing a numerical value
- ask the system, why this information is needed.

The latter option is useful in order to enable the user to get insight into the systems current intentions. The user can ask "why" in cases that the systems query appears irrelevant, or in cases that answering the query would require additional effort on the part of the user. From the systems explanations the user will judge wether the information the system is asking for is worth the extra effort of obtaining that information. Suppose, for example, the system is asking "are more than 5 knitting-systems required ?". Then the user, not yet knowing the consequences of a yes or no reply, may type "why" repeatly to see what kind of knitting machine the system has choosen as its actuall hyposesis.

Once the system has come up with an answer to the users question, the user may like to see how this conclusion was reached. A proper way of answering such a how-question is to display the subgoals and user-supplied primitive information from which the conclusion was reached.

2. EXPERT SYSTEMS PUT EXPERIENCE IN THE PRODUCTION PROCESS

In recent years, research in the field of artificial intelligence has made many important successes. Among the most significant of these has been the developement of powerful new computer software known as "Expert systems".

An "Expert system" is a computer programm that uses knowledge and inference procedures to solve poblems that are complex enough to require significant human expertise for their solution. The main components of an expert system are:

>   the knowledge base of facts and rules
>   the inference engine for reasoning
>   the explanatory interface(why/how-component)
>   the knowlede acquisiton component.

An expert system differs from more conventional computer programs in several important respects. In a conventional computer program, knowledge pertinent to the problem and means and methods for utilizing this knowledge are all intermixed, so that it is difficult to modify the program. In an expert system there is a clear separation of the general knowledge about the problem and the methods for applying the general knowledge to the problem. The system can be changed by adding or subtracting rules in the knowledge base.

3. ONE POPULAR APPROACH FOR KNOWLEDGE REPRESENTATION

AI researchers have worked out a variety of methods for representing and using knowledge in computers. This is still a matter of debate and active research in AI circles. One popular approach to represent expert knowledge is to use situation-action or production rules, which can be connected to each other to form rule networks.

Once assembled, such networks provide a suitible means for representing knowledge, required by the textile expert to decide in a certain situation. The set of rules is often referred to as the "rule-base" and the part of the program that decides which rule to apply as the "rule interpreter" or inference engine.

The task of the inference engine is to decide which rules to apply,

The diagnosis-shell **WIDIMO** is in practical use on 16-bit PC for the following problems:

- selection of suitable knitting machines on the basis of desired properties of the textile fabric;
- machine fault diagosis on multi shed weaving looms on the basis of information about elemantary defects;
- selection of suitable standards for special textile testing problems;
- interpretation of faults in textile fabrics during finishing processes; and
- predetermination of properties in yarn production.

These problems are complicated enough to justify developement and use of expert systems.

4. DECISION AID FOR DIAGNOSING FAULTS ON WEAVING LOOMS

One major problem in designing an expert system is to find a method for the acquisition and modelling of expert knowledge and for effectively overcoming the "knowledge engineering bottleneck". In this field psychological aspects of knowledge elicitation, interview techniques and analysis of expertise are important. We used interview techniques to extract knowledge from textile experts and formulated und structured this knowledge in knowledge rules about weaving looms. The knowledge has been represented as a hierachically organised search space, as an AND/OR tree. The body of this tree is formed from inter-related rules with the facts occupying terminal positions. This knowledge base is consulted by **WIDIMO** so that advice can be given if a fault occurs. At present the knowledge base contains 90 objects and 150 rules. This fault diagnostic system beats the human textile experts in speed and availability so that a new area of machine maintenance can be put into effect. It also can be effectively used if only less skilled staff is available or if the plant is working with reduced personnel.

REFERENCES

[1] Sterling, L., E. Shapiro: The Art of Prolog. The MIT Press, Cambridge, Mass., London, 1987.

A Multiobjective Decision Support System
for the Top Management (DSS-CAPS)

Dipl.-Ing. Rolf Schmidt,VEB Trafowerk "K.Liebknecht",
        Marketing,1160 Berlin, Wilhelminenhofstr. 83-86
Dr.-Ing. Bernd Koch,AdW der DDR, Zentralinstitut für Kybern.
        und Informationsprozesse,1086 Berlin, Kurstraße 33

Introduction: The continuous satisfaction of demands of the
national economy and international market requires a higher
quality of marketing. In case of supply missing (material,
cooperation) often the exactly observance of the production
and sales plans is not possible in all details. By means of
the programsystem CAPS, an interactive computer aided production and sales decision support system (DSS-CAPS) using
the algorithm of REH /Straubel-85-1/,/Straubel-86-1/, it is
possible to compensate or better to reduce deviations of the
production plan and its consequences on the fulfilment of the
overall targets of the enterprise.

Description: The fundamental aim was to create a simple and
robust model of production and sale. The computer based
system has to be easy by understood and applied by the user
and particularly by the (top) management. Past experience
hase shown us that even the best model/system will never be
used if it is not comprehensible and if the user (top
management) cannot grasp its function. The above mentioned
system is used for decision making for the direction of
production and delivery within the current year of planning.
The starting point for the current year is that the production and delivery plans are fixed for the considered period
of time. The relation can be illustratted thus:

$$Balance = Plan = Contracts \ (delivery).$$

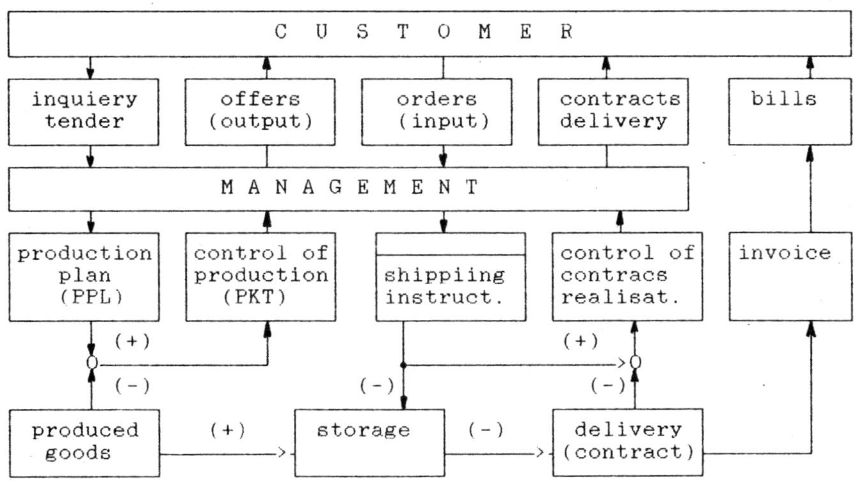

Pic. R1.1. Modelpart I- Data bank system of RABS1

The production plan as well as the stocks which result from
that are fixed regarding time and the stocks are largly
linked with contracts.
The thus chosen solutions for production and delivery of
goods present efficient alternatives from computer- aided and
determined compromise solutions (illustration R1.1 and
R1.2). The respective complexes and the existing connections
and their interdependences will not be described in this
article. As for the modelpart RABS1 we refer to /Schmidt-87-
1/. The applied basic model is elucidated in pictures R1.1
and R1.2.

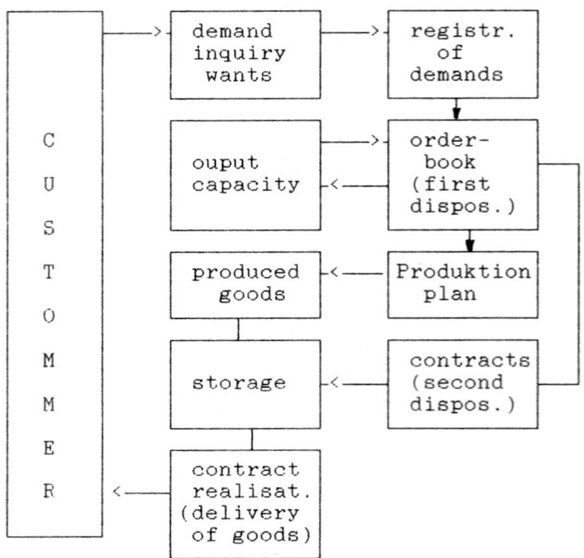

Pic. R.1.2. Modelpart II of DSS-CAPS

Picture R1.1 illustrates the flow of information within
operative working of customers´orders. The results of this
model form the bases (a priori information) for the decision
support system CAPS.

The   production managemant is faced with the task of making
optimum use of the existing production capacity.That means to
realise the plannend production targets , which are assigned
by top management, with a minimal expenditure of material;
working time and man-power. A necessery consequence of this
is , among others, the reduction standstills of machineries
and its most efficient employment especially with regard to
optimal production quantity.
The sales management is confronted with the task   of
satisfying the needs of the market in such a way that the
position in the market can be guaranteed in future and its
own share in the market can be increased lucratively and
profitably.
The application of the DSS-CAPS realises for the first time
the coherence or better interdependence of targets of pro-

duction with those of the sales management. The production
realisation according to the given production plan depends
mainly on the real production capacity at the respective
period of time . Unintentional production deficits can be
caused for example by supply which has gone missing, been
delayed or which is of poor quality both from the home market
and abroad and by unintentional demages.   This
influences on the production process can affect the temporal
conditions as well as shifts in the assortment
of the produced goods. As a result negative effects can come
into existence as to the realisation of contracts and the
total turnover- policy of the profit- seeking enterprise.
The fundamental requiement of the enterprise however reads :
fulfillment of the monthly and cumulative economic targets by
application of the mini-max-method.
By means of CAPS it is possible to compensate or better to
reduce deviations of the production plan and its consequences
on the fulfilment of the overall targets of the enterprise
within the current period of planning (current year of
planning). We proceed from the idea that the production
management is able to offer various alternatives of production for the respective types of products (either - or).
The choice of an efficient production alternative cannot and
must not decided by the production managemant. It has not
got either the necessary survey nor the insight into the
complexity of marketing processes. From this follows the
necessity for the production management as well as the sale
management to suggest variants of decisions to the top
management of the enterprise. The variants should be pareto-
optimal solutions of a set of compromise points which are
calculated by means of CAPS on the bases of the following
targets:

- backlog of commodityproduction cumulative and   ---> Min
  monthly (ad valorem)
- backlog of produced goods according to the   ---> Min
  production plan - backlog of sold goods (ad
  valorem)
- backlogs of fulfilment of contracts   ---> Min
- backlog of the sold commodity production   ---> Min
  cumulative until the end of the currently year
- difference between the date of realisation   ---> Min
  fixed in the contract and the actuel date of
  delivery
- difference between the final stock according   ---> Min
  to the production plan and the actual stock
  availabel
- deviations from the assortment according to   ---> Min
  the production plan which cannot bo sold or
  linked with contracts within the current
  period of planning

What alternative is given priority by the top management
depends on the subjective preference structure of the
decisions makers and thus it is the responsibility of the top
management himself. The decision maker has the possibility
for assigning to the customers different priorities. The
customers and following from that the contracts are

prioritised taking into consideration economical and
commercial aspects. The procedure CAPS offers the possibility
of involving state orders as restrictions too. But this
approach is left to top management s discretion.

A description in detail of the algorithm of REH you find in
/Straubel-83-1/, /Straubel-85-1/, /Straubel-86-1/ and
/Wittmüß-83-1/.

Experience/Conclusions: The DSS-CAPS realises the connection
between the targets of production and sales management with
the aim for realizing the highest possible profit.
Consequently CAPS is a multi-criterical DSS using the
algorithm of REH By means of CAPS the decision maker and
thus the top management has the possibility to decide for one
of the pareto-optimal solution of a set of compromise points
relative to produced and sold goods. In dependence from the
choosen alternative all target results are shown to the
decision maker (record).
The data bank system RABS1 is used in more than 50 enterprices. The DSS- CAPS will be introduce into this enterprices
in 1989. Among other things one result and success is the
improvement in the field of management in quality and
quantity.

References

/Schmidt-87-1/ Schmidt,R. RABS1- rechnergestützter Absatz 1.
               Stufe, Neue Technik im Büro 4/87
               VEB Verlag Technik

/Schmidt-87-2/ Schmidt,R. Data Bank Systems in the Management
               Neuse Technik im Büro 6/87
               VEB Verlag Technik
               lecture,no publeshed manuscript

/Straubel-85-1/ R. Straubel, A. Wittmüß, R. Rosenmüller in
                Proceedings of the international symposium
                held in Berlin (GDR) 1985, edited by Achim
                Sydow, Manfred Thoma and Robert Vichnevetsky
                Akademie Verlag Berlin 1985,Math. research 27

/Straubel-86-1/ R. Straubel, A. Wittmüß, Das Programmsystem
                REH zur rechnergestützten Entscheidungshilfe
                ZKI Information 2/1986
                Akademie der Wissenschaften der DDR

/Straubel-83-1/ R. Straubel, A. Wittmüß, R. Rosenmüller
                eine rechnergestützte Entscheidungshilfe bei
                makroökonomischen Planansatzrechnungen
                ZKI Information 4/1983 p. 106- 156

/Wittmüß-83-1/ A. Wittmüß ,
               Ein sequentielles Entscheidungsverfahren zur
               Bestimmung einer mehrkriterialen "optimalen"
               Steuerung
               ZKI Information 2/1983, p. 94- 113

Modelling and Optimization

Schmelovsky, Karl-Heinz [1]

This paper is an attempt to get physical modeling and mathematical optimization somewhat closer together, using the concept of enlarged state space. Here, every influence, assumed to be neither exactly known nor totaly random, is described by state variables. This leads automatically to markoff processes in that state space. Furthermore state models can always be formulated so that observables depend only on actual state and - possible - on white time discrete noise.

$$1 \qquad y_j = y(z_j \; ; \; N_j) \quad ; \quad z_j = z(t_j) \quad ; \quad \langle N_j \cdot N_k \rangle = \delta_{j,k}$$

so, optimal time discrete (closed loop) control can always be separated into two open loop tasks:

1) Find the best actual probability statement R_k from recursive processing observations by Bayes' estimate regarding control actions done hitherto (state estimation).

2) Compute a decision rule (strategy function) $u(R_k)$, thus that iterative application of this rule optimizes expected total aim.

Obviously, control is most effective if observation is immediately followed by decission using this information. In this case times t_k and t_j coincide, and R_k is equal to posteriory statement $R_{P,j}$ immediately after observation. But it should be mentioned, that some reordering of algorithm is necessary in order to realize this condition technically (s. b.). besides, in some cases, observation is delayed physically. Than in any case extrapolated statement must be used [1].

Of course equ. 1 does not mean a unique mapping of state on observation, regardless whether N_j disappeares or not. In all real cases dimension of observation is less than that of state, mostly it is scalar. Hence, state components can only be estimated from a sequence of observations, and here statistical disturbance prevents exact state knowledge.

Starting point is the primary model, defining state space and setting starting conditions. Possible refinements by autonomeous learning are regarded as state estimation whithin this frame.

In this concept changing models and strategies are local approximations to a (unaquirable) globel optimum. The width of this frame is only restricted by the complexity that can be handled. But this tends not to a "black box" model. A central feature of optimal strategies is the convexity if aim function in probability space [2,3]. That means, that neglection of real as well as introduction of fictious information lowers efficiecy of control. Hence the primary model should, ideally,

----

Institut f. Kosmosforschung d. AdW d. DDR, 1199 Berlin, Rudower Chaussee

describe all that and only that, what is known on the process in the very beginning of the task. Practically , things are a little more complicated.

In order to get a mathematical model at all, some idealization is un-avoidable. For mathematical rigour it seems best to idealize until the remaining problem can be brought down to numerical algorithm in full strength. For real exactness I can't agree. Certainly, it loocks better to neglect e.g.a statistical influence than to use a pi*tumb estimate. But zero is an even worse pi*tumb estimate for a positive definite value. Certainly, the error of gaussian approximation in nonlinear problems can't be estimated sufficietly. But how to estimate the error introduced by neglecting an essetial nonlinearity? In my opinion it's more exact to regard an essential influence in any approximation than to neglect it's totality, to accept an existing uncertainty than to pose a voluntary hypothesis.

Modern investigations [4,5] in statistcal optimization have revealed a fact, that was hitherto hidden under the flowers of "sufficiently simp-le"models. Aside from other problems of idealization: if one puts only the main features of real problem into mathematical model, optimisation gets practically unsolvabel. Asking for optimal decision based on the information obtainable from incomplete observation of a statistically disturbed process, starting from incomplete primary modell, ends up with really phantastic demands on processing power. So it seems reasonable to ask for technical optimum, the best strategy obtainable with existing process knowledge and available processing power. Due to the fact, that human processing power is also limited, it seems unprobabel that this conditioned optimization can be solved directly. But a kind of hill climb technique can be used. Let A be the most complex model for which a realizable optimal strategy $S\{A\}$ can be found, and B a more realistic , hence more complex, model of the same process. Than a realizable strate-gy $S'\{B\}$ is obviously a step toward the mentioned optimum if it can be shown that its better if B holds. Please, regard that optimality of $S'\{B\}$ is not demanded and cant be demanded, due to precondition. Obviously, in order to use this method, the model must not be restricted unduely in its essential features.

Core of the model (fig. 1) is the mathematical description of the very process in its classical state variables z^c.

This is more a problem of physics (in general meaning) then of system theory, as fare as physical description is possible.

For the remainder, it is true, it is more a heuristic description in causal language. It's the strength of state language, that it can use physical information but cope with more vague ones if necessary. The fact, that from the few interactions regarded explicitely the reaction of the process can t be predicted exactely, can only be described by statistical disturbances, the large differences in predictibility by their intensities. Only this qualitativ aspect can and must be regarded.

This is even more true for spectral characteristics.

Mathematically its no problem to describe them by filtered white noise, introducing state variable z^m, if adequate primary information exists. Regardles whether spectral characteristic is regarded, the deciding quantity is the spectral intensity, i.e. the integral of AKF. So besides dispersion a typical time is needed [6].

Mostly some essential parameters ar'nt known sufficiently. They can be modeled by state variables z^a (a for adaption).

For their motion the simplest diffusiv model (free or restricted by backdriving force) is more then sufficient. Again the problem of typical time arises.

Describing the technical tracts from sensing point to in port (ADC) and from out port (DAC) to physical actuator by state variables z^b and z^u separates the enlarged object of control exactly from controlling intelligence. Condition Equ 1 is trivialy fullfilled because the output of observation tract is a component of state z^o; $y_1 = z^o$.

Regard that observation is formaly noisefree, whithout any consequences on knowledge of z^c.

If all is regarded, that can be regarded in the given frame, the remainder is pure chance, the white process. So always a system of differential equations with white disturbances is obtained. The process is markovian in enlarged state space obtained by concatenation of all subspaces mentioned above.

$$\frac{dz}{dt} = \mathcal{F}(z, u_j) + B\,\xi \quad ; \quad \mathcal{F} = \left\{\mathcal{F}^v\right\} \quad ; \quad \xi = \left\{\xi^v\right\}$$

2

$$\left\{\xi^v(t)\,\xi^\mu(t')\right\} = \delta_{v,\mu} \cdot \delta(t-t')$$

From here one gets directely the Fokker-Planck-Equation

$$\frac{dp(z)}{dt} = -\sum \frac{\partial}{\partial z^v}\left\{\mathcal{F}^v(z, u_j)\,p(z)\right\} + \frac{1}{2}\sum_v \sum_\mu \varkappa^{v,\mu}\,\frac{\partial^2}{\partial z^v \partial z^\mu}\,p(z)$$

3

$$t_j < t < t_{j+1} \quad ; \quad \mathcal{R} = \left\{\varkappa^{v,\mu}\right\} = B \cdot B^T$$

Of course the maximal model must be reduced in every concret case. So e. g. if sensing noise ξ^m is white and sensing chanel has no essential inertia observation of H.z throught-than optimal- integrate/dump can by approximated by

$$y_j = \frac{1}{\Delta t}\int_{t_{j-1}}^{t_j}\left(H \cdot z + \beta_m \cdot \xi^m\right)dt = H \cdot z_j + N_j$$

4

$$\langle N_j \cdot N_k \rangle = N_o\,\delta_{j,k} \quad ; \quad N_o = \frac{\beta_m^2}{\Delta t}$$

(see 6), saving one dimension.

Under these precondition the posteriory probability is obtained by the recursion [7]

$$5a \qquad R_A(z_j) = \int_{E_Z} R(z_j/z_{j-1} \; ; \; u_{j-1}) \, R_P(z_{j-1}) \, d\overset{\circ}{z}_{j-1}$$

$$d\overset{\circ}{z}_{j-1} = dz^1_{j-1} \cdot dz^2_{j-1} \cdots\cdots dz^n_j \; ; \; A = Apriory$$

$$5b \qquad R_P(z_j) = \frac{1}{C} \, R(y_j/z_j) \, R_A(z_j)$$

$$5c \qquad R(y_j) = \int R(y_j/z_j) \, R_A(z_j) \, d\overset{\circ}{z}_j$$

The likelyhood $R(y_j/z_j)$ is obtained from equation 1 and statistics of noise N_J. The transition probability is the Greens-function of F.-P.-E. So 5a can also be regarded as solution of F.-P.-E. at t_j with posteriory probability as starting condition. For simplicity t_j and t_k coincide here, generally the problem can also be solved for a sequence of control actions between observations. Equ 5c is only a renormation in state estimation, but it's the probability to get a certain y_j in next step for extrapolation.

Real computers can only handle a finite set of variables. This set, used for description of probability density in the frame of a pregiven function, is the probability statement for decision. Normally the set $\hat{z}$ = expectation vector; K = dispersion matrix of a Gaussian distribution or even subsets are the best we can do.

$$6 \qquad R_{P,j} = \left\{ \hat{z}_{P,j} \; ; \; K_{P,j} \right\} \quad or \quad = \hat{z}_{P,j}$$

The main problem is to substitute 5a by a recurency in $\left\{ \hat{z} ; K \right\}$
First step is to multiply F.P.E. successively with all z^ν and all $z^\nu \cdot z^\mu$, and integrate over state space. Derivations are eliminated by partial integration, densities under integral are gaussian approximated. One gets

$$7a \qquad \frac{d\hat{z}^\nu}{dt} = \int F^\nu(z; u_j) \, R_G(z-\hat{z}; K) \, d\overset{\circ}{z}$$

$$7b \qquad \frac{dk^{\mu,\nu}}{dt} = \int z^\mu F^\nu(z; u_j) R_G(z-\hat{z}; K) d\overset{\circ}{z} - \hat{z}^\mu \int F^\nu(\cdot) R_G(z-\hat{z}; K) d\overset{\circ}{z}$$

$$+ konj \cdot Terms \; (\mu, \nu \; exchanged) + k^{\nu,\mu}$$

Further proceeding is demonstraded for the case, that $F^\nu(z; u_j)$ contains one nonlinear term $f(z^\mu)$. Integrating this term first on all components besides dz^μ

8
$$\int P(z^\eta)\, R_G(z^\eta - \hat{z}^\eta;\, k^{\eta,\eta})\, dz^\eta = \tilde{P}(x;k)/_{x=\hat{z}^\eta}^{\;k=k^{\eta,\eta}}$$

$$\tilde{P}(x,k) = \frac{1}{\sqrt{2\pi k}}\int P(\xi)\exp\left[\frac{(\xi-x)^2}{2k}\right]d\xi$$

In 7b we integrate respective term first on all components besides dz^η and dz^μ.

Probability shrinks to

9
$$R(z^\mu;z^\eta) = R(z^\mu/z^\eta)\,R(z^\eta)$$

So one gets

$$\int z^\mu\, P(z^\eta)\, R_G(z-\hat{z};K)\, d\hat{z} = \int\left[\int z^\mu R(z^\mu/z^\eta)\,dz^\mu\right]P(z^\eta)R(z^\eta)\,dz^\eta$$

10
$$= \int \langle z^\mu/z^\eta\rangle\, P(z^\eta)\, R(z^\eta)\, dz^\eta = \hat{z}^\mu \int P(z^\eta)\,R(z^\eta)\,dz^\eta +$$

$$+\, k^{\eta,\eta}\cdot\frac{1}{k^{\eta,\eta}}\int (z^\eta - \hat{z}^\eta)\, P(z^\eta)\, R_G(z^\eta - \hat{z}^\eta;\, k^{\eta,\eta})\, dz^\eta$$

First summand cancels with second one in 7b, second one can be expressed by

11
$$\frac{1}{k^{\eta,\eta}}\int (z^\eta - \hat{z}^\eta)\, P(z^\eta)\, R_G(\cdots)\, dz^\eta = \frac{d}{dx}\,\tilde{P}(x,k)/_{x=\hat{z}^\eta}^{\;k=k^{\eta,\eta}}$$

The convolution function depends only slight on $k^{\eta,\eta}$ mostly a rought estimation $\tilde{k}^{\eta,\eta}$ sufficies, e. g. stationary value from straight linearization. Developing at $\hat{z}_P$ gives the linearized form.

12a
$$\frac{d\hat{z}}{dt} = A\left(\hat{z}_{P,j}\right)\hat{z} + C\left(\hat{z}_{P,j}\right)u_j + D\left(\hat{z}_{P,j}\right)$$

12b
$$\frac{dK}{dt} = A\left(\hat{z}_{P,j}\right)\cdot K + K\cdot A^T\left(\hat{z}_{P,j}\right) + R$$

where the term stemming from nonlinear function are

12c
$$\alpha^{\nu,\eta} = \frac{\partial}{\partial x}\,\tilde{P}(x,k)/_{x=\hat{z}_{P,j}^\eta}^{\;k=\tilde{k}^{\eta,\eta}}\;;\quad d^\nu = \tilde{P}(\hat{z}_{P,j}^\eta;\,\tilde{k}^{\eta,\eta}) - \alpha^{\nu,\eta}\,\hat{z}_{P,j}^\eta$$

Rem: P belongs to $\mathcal{F}^\nu$

From this the Kalman-Bucy-Equisation is obtained by integration over Δt

13a
$$\hat{z}_{A,j+1} = \alpha\left(\hat{z}_{P,j};\Delta t\right)\cdot\hat{z}_{P,j} + C\left(\hat{z}_{P,j};\Delta t\right)u_j + D\left(\hat{z}_{P,j};\Delta t\right)$$

13b
$$K_{A,j+1} = \alpha\,(\cdot)\cdot K_{P,j}\,\alpha^T(\cdot) + R\left(\hat{z}_{P,j};\Delta t\right)$$

Where α is the homogenous and ℓ and $\mathcal{D}$ are the inhomogenous basic solutions of 12a for D = 0 respectively $u_j = 0$.

Matrix $\mathcal{R}$ is the basic inhomogenous solution of 12b. Mostly here the dependance on $\hat{z}_{p,j}$ can be neglected. Second K.B.E. is obtained straight from 5b.

$$
14a \qquad \hat{z}_{p,j} = \hat{z}_{A,j} + \frac{H K_{A,j}}{N_0 + H K_{A,j} \cdot H^T} \left(y_j - H \hat{z}_{A,j} \right) = \hat{z}_{A,j} + M_j \left(y_j - H \hat{z}_{A,j} \right)
$$

$$
14b \qquad K_{p,j} = K_{A,j} - \frac{K_{A,j} H^T H K_A}{N_0 + H K_{A,j} \cdot H^T}
$$

This apparently cumbersome way to K.B.E. is needed to use them as approximation for nonlinear problems and to keep Δt free, either to use it as optimization parameter [8], ore to cope with given but changing Δt [1].

One main problem is nonlinear interaction between parameters z^a and physical state z^c. Adaption is inherently a nonlinear problem. Only in special cases — e. g. if respective state kept constant — straigt linearization is possible. We restrict intermediately space to $z = \{z^c, z^a\}$, writing $z^c = x$; $z^a = \Theta$.

$$
15a \qquad \frac{dx}{dt} = A_1 \cdot (\Theta) \cdot x + C_1 (\Theta) u_j + B_1 \, \xi^1
$$

$$
15b \qquad \frac{d\Theta}{dt} = A_2 \cdot (\Theta) + B_2 \, \xi^2
$$

Matrices of K.B.E. are:

$$
16a \qquad \alpha = \begin{Bmatrix} \alpha_1 (\hat{\Theta}_{p,j}) & ; & \alpha_{1,2} (\hat{x}_{p,j} \; ; \; u_j) \\ \alpha_2 & ; & \theta \end{Bmatrix}
$$

$$
\begin{Bmatrix} \ell_1 (\hat{\Theta}_{p,j}) \\ \theta \end{Bmatrix} \qquad \mathcal{R} = \begin{Bmatrix} \mathcal{R}_1 & \theta \\ \theta & \mathcal{R}_2 \end{Bmatrix}
$$

where $\alpha_1 ; \alpha_2 ; \ell_1 ; \mathcal{R}_1 ; \mathcal{R}_2$ are defined similar to before and

$$
16b \qquad \alpha_{1,2}^T = \frac{\partial}{\partial \Theta} \left\{ \alpha_1 (\Theta) \cdot \hat{x}_{p,j} + \ell_1 (\Theta) u_j \right\}^T \Big/ {}_{\Theta = \hat{\Theta}_{p,j} \approx \tilde{\Theta}}
$$

Dependance of $\alpha_{1,2}$ on $\hat{\Theta}_{p,j}$ can mostly by neglected.

For strategy its only necessary to find respective next step. But in order to get it, its necessary to regard its influence on consecutive steps. This meens that not only the reaction of object on a given sequence of control steps but the sequence itself must be extrapolated.

Dynamical programming solves this problem by assuming, that consecutive decisions are optimal also. The problem is, that "optimal" means realy "as good as possible" with than existing knowledge and processing means.

Aim of optimization is here to minimal total loss, expressed by the sum (more exact the integral) of immediate losses

$$17 \cdot \quad \left\langle \sum_{jA}^{j_E-1} \phi\left(z_{j+1} \; ; \; u_j\right) \right\rangle$$

Immediate loss can mostly be expressed by control error loss and direct control costs (e.g.energy)

$$18 \quad \phi\left(z_{j+1} \; ; \; u_j\right) = f\left(x_{j+1} - s_{j+1}\right) + k\left(u_j\right) \; ; \quad s_j \text{ wanted trajectory.}$$

The basic "deterministic" approximation neglects statistical influence in this task, introducing the value of a state $G(z)$ i.e. the maximal total gain obtainable starting from this state in remaining steps. We use this terminius also, in spite of that we treat a minimum loss problem, i. e. real value has opposite sign. Value function is obtained by minimizing sum of immidiate loss and value of consecutive state recursively [9].

$$19 \quad G_{j-1}\left(z_{j-1}\right) = \underset{u_{j-1}}{\text{Min}} \left\{ \phi\left(z_j \; , \; u_{j-1}\right) + G_j\left(z_j\right) \right\}$$

Together with state transform obtained from core model. Strategy $u\left(z_{j-1}\right)$ is the argument of the minimum. Only this can be used, the sequence of further steps is an artefact of a. m. neglections and without real meaning.

The control algorithm obtained by inserting $\hat{z}_{P,j}$ for unknown state z_j is mostly the best that can be realized, apart from some immprovements m.b. For exact statistical optimization the expectation of imediade loss plus expected value of consecütive probability statement must be minimized [2,3]

$$20 \quad G_{j-1}\left(R_{P,j-1}\right) = \underset{u_{j-1}}{\text{Min}} \left\{ \left\langle \phi\left(z_j \; , \; u_{j-1}\right) \right\rangle + \int R\left(y_j\right) G_j\left[R_{P,j}\left(y_j\right)\right] d\overset{o}{y} \right\}$$

state transform is replaced by equ's 5a to 5c.
In Gaussian approximation one gets [5]

$$21 \quad \begin{aligned} G_{j-1}\left(\hat{z}_{A,j-1}; K_{P,j-1}\right) &= \underset{u_{j-1}}{\text{Min}} \left\{ \int\int \phi\left(z_j, u_{j-1}\right) R_G\left(z_j - \hat{z}_{A,j} \; ; K_{A,j}\right) d\hat{z}_j + \right. \\ &\quad \left. G_j\left(\hat{z}_{A,j} \; ; K_{P,j}\right) + Tr\left[\left(K_{A,j} - K_{P,j}\right) \cdot D\right] \right\} \end{aligned}$$

where D is the Jakobian of G at point $\hat{z}_{A,j}$; $K_{\mathcal{P},j}$

State transform is substituted by K.B.E. - 13a, 13b, 14b (but not 14a). For linear processes and linear observation dispersion gets independend on control and is a precalculable function of j. For both side asymptotical case (fare from end and begin) K_A and $K_{\mathcal{P}}$ get fixed matrices the later gets a dummy argument of G. Third term depends only slightely on control. So, for strategy, the only difference is the first term, beeing a precalculable function of $\hat{z}_{A,j}$. Due to smother character of it this is even more simple to solve than 19. For asymetric error loss functions strategy deviates from deterministic one by going "cautiously" toward easier side increasingly for increasing uncertainity. Going back some steps, one gets good starting strategies for problems whith high initial uncertainity and asymmetric risk.

Only if immediate loss is quadratic additionally, strategies 19 and 21 coincidide. Deterministic losses stemming from b_j and statistical ones are fully separable. Value-function gets quadratic also, and strategy gets linear

$$22 \qquad u_j = L_j \cdot \hat{z}_{\mathcal{P},j}$$

For calculation of matrix of value-function G and strategy vector L see e. g. [4] .

In general case equ. 21 is practically unsolvable. But full K.B.E. in realtime gets realisable in an increasing number of cases and strategy-vector $L(\Theta)$ can be calculated for every fixed set $\hat{\Theta}_{\mathcal{P},j}$. Doing this operativly with actual value of parameter estimate one gets a learning strategy. Praktically it seems sufficient to do one step backward in strategy-recursion for every step forward in estimate. The problem is, that most of the calculation can only by done after getting y_j . But using the one step back strategy $L(\hat{\Theta}_{\mathcal{P},j-1})$ this can be overcome. Now $\hat{x}_{A,j}$; $L(\hat{\Theta}_{\mathcal{P},j-1})$; $H \cdot x_{A,j}$; $K_{A,j}$ vector M_j and scalar $\alpha_j = L^T(\hat{\Theta}_{\mathcal{P},j-1}) \cdot M_j$ can be calculated before y_j is known.

$$23 \quad \begin{aligned} u_j &= L(\hat{\Theta}_{\mathcal{P},j-1}) \cdot \hat{x}_{\mathcal{P},j} = L^T(\hat{\Theta}_{\mathcal{P},j-1}) \cdot \hat{x}_{A,j} + L^T(\hat{\Theta}_{\mathcal{P},j-1})(\hat{x}_{\mathcal{P},j} - \hat{x}_{A,j}) = \\ &= L(\hat{\Theta}_{\mathcal{P},j-1}) \cdot \hat{x}_{A,j} + L^T(\hat{\Theta}_{\mathcal{P},j-1}) \cdot M_D (y_j - H\hat{z}_{A,j}) = \\ &= L(\hat{\Theta}_{\mathcal{P},j-1}) \cdot \hat{x}_{A,j} + \alpha_j (y_j - H\hat{z}_{A,j}) \end{aligned}$$

So in time between observation and decision only one scalar mult and two sub/adds are necessary. By similar development also for vectorial control/observation and nonlinear optimization in X this critical processing load can be reduced to small percentage of overall load. Due to the fact that inserting a more recent estimate can only increase efficienty, it can be combined freely with any strategy. Passiv learning is certainly not an optimal strategy for parametrical model, because this

contains experimental learning see (lit c.a.). But its a good example
for a suboptimal one, beeing certainly better than the optimal one for
"sufficiently simple"(fixed parameter)model.
To show the differences between deterministic and statistic strategies
lets go back to full state. Firstly both are statistic in that sense
that they must be coupled with statistical state estimate in order to
close the loop.State space in estimate is defined by the very problem.If
inertias in sensing are essentially z^m must be included,if disurbance
is essentially nonwhite z^d,e.t.c.. But in strategy of deterministic
approx. only those components can appear that influence x.Further, with
disturbance disappears z^d and parameters behave here like constants.
So strategy space reduces to $z = \{z^C ; z^u\}$ maximally. For statistic
strategies probability statement in full space is the argument. But,
regarding that observation is totally neglected in lower approximation,
it seems legitime to approximate observation chanel here always like
e.qu.4 excluding z^m, z^r. So even in linear quadratic case the equivalence of remaining e.qu. 13a to state transform is only formally. It can
contain z^d and even z^Q in the linearizable cases m.a. Besides it's a
tranformation from posteriory to consecutive apriori expectation. This
is sufficient for strategy but the sequence of further steps can only be
calculated by inserting $\hat{z}_{A,j}$ for $\hat{z}_{P,j}$ illegitimely. So the
additional condition "use only first step" is obtained automatically.
Statistic strategies can only by calculated in the vicinity of linear
problem. Deterministic ones can by used far more freely for problems
nonlinear in physical state, but can't cope with essential statistical
effects. So only the combination can give effective control strategies
in sophisticated applications.

The sequence of reasonable models for a given process is not continous,
and algorithmic complexity increases exponentially with complexity of
model. So almost always the available processing power is not used up by
the simpler model and by fare not sufficient for the strict optimization
of the next one. Oversimplification is to high a prize for mathematical
exactness, but wild heuristic is also no constructive way. Hill climb,
combaining different approximations in quite controled way, seem a
wayout.

Most important abbreviations
z = state-vector; $\hat{z}$ = expectation of state-vector;
K = dispersion-matrix
lower indizes: A = apriori, immediately before observation
               P = posteriori, immediately after observation
               j, k, l reference time, can be shifted for conveniance

Upper indizes: greek letters (or numbers) components; latin letters partial vectors; double upper indizes at matrices accordingly.

All vectors, besides H, are column-vectors.

$\{a^{\nu}\}$ = vector of components a^{ν} ;

$\{a^{\nu,\mu}\}$ = matrix of components $a^{\nu,\mu}$.

$\langle \cdots \rangle$ = expectation of ...

$\mathcal{R}_G(z-\hat{z}, K)$ = Gaussian compound distribution with parameters $\hat{z}, K$

Literature

/1/ Rollenhagen, F.: Ein Beitrag zur Synthese und Realisierung adaptiver Regelungsalgorithmen Dissertation A, Berlin 1984, Akademie der Wissenschaften der DDR

/2/ Åström, K.J.: Optimal control of Markoff Processes with incomplete state information. Journ. of Mathematical Analysis and Application 10 (1965) 174-205

/3/ Åström, K.J.: Optimal control of Markoff Processes with incomplete state information. Journ. of Mathematical Analysis and Application 26 (1969) 403-406

/4/ Bar-Shalom, Y.: An Actively Adaptive Control for Linear Systems with Random Parameters via the Dual. IEEE Transaction on Automatic Control, vol. AC-18, No. 2, April 1973

/5/ Schmelovsky, K.-H.: Control and experiment in Markov processes with incomplete state knowledge. Syst. Anal. Model. Simul. 2 (1985) 4, 283 - 307

/6/ Schmelovsky, K.-H.: Zustandsschätzung und adaptive Regelung mit mikroelektronischen Mitteln. Akademie-Verlag, Berlin 1983

/7/ Stratonowitsch, R.L.: Uslowie markovskije prozessi i ich primineme k teorie optimalnowo uprawlenie. Isd. MGU, Moskau 1966

/8/ Voß, H.: Analyseverfahren zur Abschätzung der Entwurfsparameter eines zeitdiskreten KALMAN-Filters. AdW 1987, Institut für Kosmosforschung, Bereich System- und Signalelektronik

/9/ Cooper, L.; Cooper, M.W.: Introduction to Dynamic Programming. Pergamon Press, New York 1981

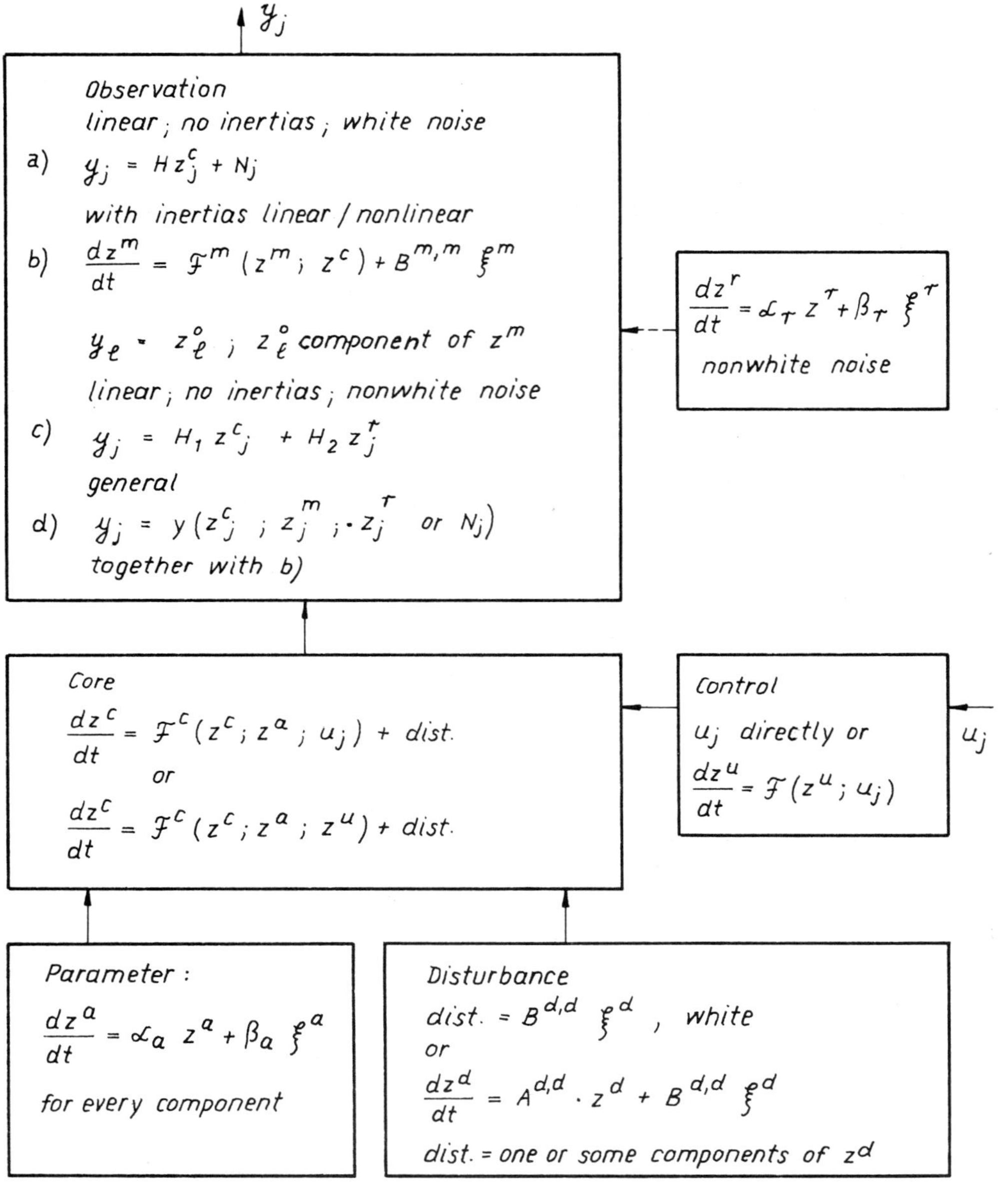

$\uparrow y_j$

Observation

linear ; no inertias ; white noise

a) $\quad y_j = H z_j^c + N_j$

with inertias linear / nonlinear

b) $\quad \dfrac{dz^m}{dt} = \mathcal{F}^m (z^m ; z^c) + B^{m,m} \, \xi^m$

$\quad y_\ell = z_\ell^o \; ; \; z_\ell^o$ component of z^m

linear ; no inertias ; nonwhite noise

c) $\quad y_j = H_1 z_j^c + H_2 z_j^r$

general

d) $\quad y_j = y(z_j^c \; ; \; z_j^m \; ; \cdot z_j^r \;\; \text{or} \; N_j)$

together with b)

$$\dfrac{dz^r}{dt} = \alpha_r \, z^r + \beta_r \, \xi^r$$

nonwhite noise

Core

$\dfrac{dz^c}{dt} = \mathcal{F}^c (z^c ; z^a ; u_j) + \text{dist.}$

$\qquad$ or

$\dfrac{dz^c}{dt} = \mathcal{F}^c (z^c ; z^a ; z^u) + \text{dist.}$

Control

u_j directly or

$\dfrac{dz^u}{dt} = \mathcal{F}(z^u ; u_j)$

$\qquad u_j$

Parameter :

$\dfrac{dz^a}{dt} = \alpha_a \, z^a + \beta_a \, \xi^a$

for every component

Disturbance

$\text{dist.} = B^{d,d} \, \xi^d \;$, white

or

$\dfrac{dz^d}{dt} = A^{d,d} \cdot z^d + B^{d,d} \, \xi^d$

dist. = one or some components of z^d

$$Z = \begin{Bmatrix} z^c \\ z^d \\ z^a \\ z^m \\ z^r \\ z^u \end{Bmatrix}$$

Self-Organizing Methods in Modelling and Clustering: GMDH Type Algorithms

Ivakhnenko A.G.

The complete bibliography, which provides a comprehensive list of articles and books on the theory and application of the Group Method of Data Handling (GMDH) algorithm from 1970 to 1982, was published in the book [12]. In this book three types of algorithms were described: the original GMDH algorithm (by Stanley J. Farlow), algorithm of adaptive learning networks related to GMDH (by Roger Barron, A.N. Mucciardi, F.G. Cook, Y.N. Croing and Andrew R. Barron) and modified GMDH algorithm (by Y. Sawaragi, Y.S. Ikeda and others). These three types of algorithms for modelling by sifting many models-candidates according to external criteria, as it appears, were applied to solution of many practical problems successfully. It was shown by papers published in the considered book [12] and in the many other papers [11].

To be short let us suppose that reader knows all these types of sifting algorithms, pointed out above. Let us review the further developments of the theory and applications of the sifting type GMDH algorithms since 1982 till now. We shall try not only to point out some new papers and books but to consider the new ideas in further developments and let us try to explain the reasons why GMDH algorithm are often so successful in practical applications.

The attempts to compare the existing theory of identification for the purposes of automatic control, represented in hundreds of papers and books (the paper by Piter Eikhoff, for example [1]) and the theory of model structure identification by the sifting algorithms of GMDH [3, 10] show the very essential difference in goals:

Identification for automatic control purposes is directed to find the physical model and more complex overfitted models. Model structure identification by GMDH algorithms is directed to find physical model too. But computer by sifting the set of model-candidates shows that we can reach this model only when we have available complete and accurate set of the factors (predictors) without noise. The special means of extrapolation of Geometrical Locus of Minima (GLM), or the use of canonical form of criterion is needed to find the physical model [5,9,13]. Having the short sample of noisy data without special means we obtain only the underfitted non-physical models which are the optimal by many criteria for the short time prediction of processes. Only sifting

Glushkov Institute of Cybernetics, Kiev, USSR, Academika Glushkova Prospect, 20.

methods of GMDH can find these biased and underfitted models and use them of prediction purposes. Theory of identification for automatic control purposes does not know these models. Theoretically underfitted models correspond to optimal communication systems, working in presence of noises (second Shannon's theorem [3]). Without Shannon's approach is impossible to understand why the models for structure identification and models for prediction of processes are to be different and why the rule "the more complex is the model – more accurate it is" is not correct.

The new development is to apply the computer sifting algorithms for planing of experiments [9]. The different plan-candidates can be compared by the criterion which estimates the resulting model. Shannon's approach is valid here too: the bigger are noises – the more simple plan is to be. The theory of experiment planing does not know this idea.

The new development is to apply the computer sifting algorithms for clustering and pattern recognition [13]. There is no principle difference between the modelling and clusterization. The difference is only in the degree of detailization of mathematical language. The language used for clusterization is a more fuzzy one. All these ideas and developments are possible only when the unimodality of the characteristic "criterion-complexity of model (or plan, or clusterization)" is provided. Complexity is to be ranked on information bases: first – the information of any one column of data sample is used, second – of any columns and so on. Such way ranking was used only in the last days developments of GMDH algorithms particularly in Objective System Analysis (OSA) algorithm [2,4] and Objective Computer Clusterization Algorithm (OCC) [6]. Such type of complexity ranking is close to the idea of implicite patterns (from difference equations theory) which are recommented for application. At every step of sifting we use the implicite pattern i.e. a system of equation but not one equation only as it is usually in modelling. The OSA and OCC algorithms lead us to the models in the form of system of equations system too, which is much accurate than single system or one equation, anyways.

The convergence of iterative processes in multilayered GMDH algorithms are carefully studied [1]. But it is not clear enough the general disposition of points to which processes are converging particularly upon the different external criteria. The last days problem is to make choise between the so-called accuracy and robust approaches. One can provide the accuracy type approach only: he tries to find model, plan or clusterization which are the most accurate for the given sample of data.
But computer, programmed by sifting algorithms can find the model which is the most noncontradictive one: "robust" means does not changed very much at any part of data sample. Noncontradictive and balance type criteria points by their minimums out the models, plans or clusteriza-

tions which are the most unbiassed or noncontra- diction. But these criteria of differential type are manivalued: i.e. they need some means of
regularization. The last idea is to use for sifting the external criterion equal to sum of several noncontra- diction criteria computed for
different divisions of data sample to two equal one to another parts
and   :

The robust approach in GMDH algorithms is to be recommented only if
it will be useful applied for solution of many practical problems, the
number of which is not large enough yet. It seems the robust approach
will have advantages for short samples of noisy data which we obtain
very often in the practical problems.

Literature

[1] Eikhoff P. Evaluation of Parameters an Structural Identification
    (Survey). Avtomatika (1987)6, 6-15

[2] Ivakhnenko A.G. Generalization of the Dialogue Language as a Means
    to Decrease the Degree of Man Participation in Solving Problems of
    System Analysis, Prediction and Control. Avtomatika (1983)5, 3-11

[3] Ivakhnenko A.G., Karpinsky A.M. Computer-Aided Self-Organization of
    Models in Terms of the General Communication Theory (Information
    Theory). Avtomatika (1982)4, 7-15

[4] Ivakhnenko A.G., Kostenko Yu.V. System Analysis and Long-term Pred-
    iction on the Basis of Model Self-organization (OSA algorithm).
    Avtomatika (1982)3, 11-19; (1983)3, 3-11

[5] Ivakhnenko A.G., Kritzky A.P. Restoration of a Signal (or Physical
    Model) by Experimentation of the Geometrical Locus of Criterion
    Minimum. Avtomatika (1986)3, 27-35

[6] Ivakhnenko A.G., Stepashko V.S. Noise Immunity of Modelling. Nauka
    Dumka, Kiev 1985

[7] Ivakhnenko A.G., Luh J., Semina L.P., Ivakhnenko G.A. Objective
    Computer Clusterization. Part 2. Avtomatika (1987)1, 1-16

[8] Ivakhnenko A.G., Yurachkovsky Yu.P. Complex System Modelling by
    Experimental Data. Radio i Swiaz, Moscow 1987

[9] Ivakhnenko A.G., Yurachkovsky Yu.P. On Sampling in a Set of Output
    Variables and GMDH Application for Passive and Active Experiment.
    Avtomatika (1988)1, 92-94

[10] Mueller J.A., Ivakhnenko A.G. Selbstorganisation von Vorhersage-
    modellen. VEB Verlag Technik, Berlin 1984

[11] Sawaragi Y., Soeda T., Tamira T. et al. Statistical Prediction of
    Air Pollution Levels Using Non-Physical Models. Avtomatika 15
    (1976), 441-451

[12] Farlow S.J. (ed.) Self-organizing Methods in Modelling GMDH Type
    Algorithms. University of Main, USA, 1983

[13] Yurachkovsky Yu.P. Analytical Division of Optimal Quadratic
    Discriminating Criteria. Avtomatika (1988)1, 3-11

Bilinearization of Nonlinear Systems

H. Schwarz, H. T. Dorissen and L. Guo[1]

1. Introduction

In many control problems of practical interest, the cause-effect relation of a plant has to be modelled by a class of nonlinear differential equations. Because of the difficulty in dealing with general nonlinear equations, it is a normal way to linearize it regarding the operating steady state, using Taylor series expansion, and to deal with the resulting linear system. However, the linearized model is sometimes inadequate, and there is a growing need for a better approximation of the substantial nonlinear system when robust behaviour of nonlinear systems is required over the complete operating range, what means, that the nonlinear dynamics of the plant must be handled. One method to deal with systems of which the physical cause-effect relations may be modelled by ordinary differential equations of this type

$$\dot{x}(t) = f(x(t)) + \sum_{i=1}^{m} g_i(x)\, u_i(t) \left.\right\}$$
$$y(t) = g(x) \qquad\qquad\qquad\qquad \tag{1.1}$$

where $x \in R_n$, $u \in R_m$ and $y \in R_p$

is "bilinearization", with the goal to obtain approximations of the second order of the nonlinear equations, while linearization leads to first order approximations (Yousuf, 1984; Beater, 1987). The method of bilinearization will yield considerably better results for all those physical and technical systems, where the most relevant nonlinearity consists in a multiplicative connection of an energy and/or mass flow with the control variable. An essential advantage of this procedure is, that the system theory of bilinear systems

$$\dot{x}(t) = Ax(t) + \sum_{i=1}^{m} N_i x(t) u_i(t) + Bu(t) \left.\right\}$$
$$y(t) = Cx(t) \qquad\qquad\qquad\qquad\qquad \tag{1.2}$$

where $x \in R_n$, $u \in R_m$ and $y \in R_p$ and the system matrices $\{A, B, C, N_i\}$ of appropriate order, is well established (Rugh, 1981; Schwarz, 1987). But on the other hand very few applications of this theoretic procedure for technical plants are known only.

---

[1] Department of Measurement and Control, Faculty of Mechanical Engineering, University of Duisburg, 4100 Duisburg 1, F.R.G., Lotharstr. 65. This research was supported by the VW-foundation, Hannover, and the Deutsche Forschungsgemeinschaft, Bonn, F.R.G. .

The contribution of this paper is to show how bilinear models of essentially nonlinear technical systems may be constructed by two methods. The first method starts from the nonlinear system equation and a bilinear model is computed from two linear models for two suitably choosen operating points. The second method uses system-realization theory for bilinear systems. By stimulating a nonlinear discrete time model - or real system - by a particular series of unit impulses a bilinear model is evaluated via Markov parameters. This procedure is suitable for simulation and systemidentification of real technical systems as well. Simulation results and results of systemidentification of hydraulic drives in our laboratory are presented, respectively.

2. Bilinearization of a Translational Hydraulic Drive

First we present the nonlinear mathematical model of the translational drive being sketched in Fig. 2.1 .

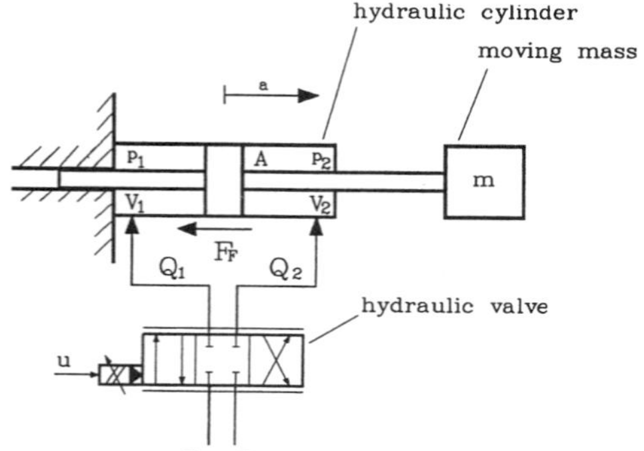

Fig. 2.1 Sketch of a hydraulic drive

The state variables are defined as:

x_1 = position of the piston
x_2 = velocity of the piston
x_3 = pressure in volume A
x_4 = pressure in volume B
x_5 = position of the valve spool
x_6 = velocity of the valve spool

The dynamics of this system can be modelled by these nonlinear state space equations

$$
\begin{bmatrix} \dot{x}_1 \\ \dot{x}_2 \\ \dot{x}_3 \\ \dot{x}_4 \\ \dot{x}_5 \\ \dot{x}_6 \end{bmatrix} = \begin{bmatrix} x_2 \\ -\dfrac{1}{m}[F_R(x_2) - A_A x_3 + A_B x_4 + F_{st}] \\ \dfrac{E_{\ddot{O}l}(x_3)}{V_A(x_1)}[Q_A - A_A x_2 - Q_{Li} - Q_{LA}] \\ \dfrac{E_{\ddot{O}l}(x_4)}{V_B(x_1)}[Q_B + A_B x_2 + Q_{Li} - Q_{LB}] \\ x_6 \\ -\omega_0^2 x_5 - 2 D \omega_0 x_6 - x_6 - \omega_0^2 F_R(x_6) \end{bmatrix} + \begin{bmatrix} 0 \\ 0 \\ 0 \\ 0 \\ 0 \\ \omega_0^2 \end{bmatrix} u(t) \quad (2.1)
$$

$$y(t) = [\ 1\ \ 0\ \ 0\ \ 0\ \ 0\ \ 0\ ]\, x(t)$$

where $F_R(x_u)$ is the friction force (Fig. 2.2):

$$F_R(x_2) = k_v \cdot x_2 + \mathrm{sgn}(x_2)[F_1 + F_{ex} \cdot \exp(-c|x_2|)] \quad (2.2)$$

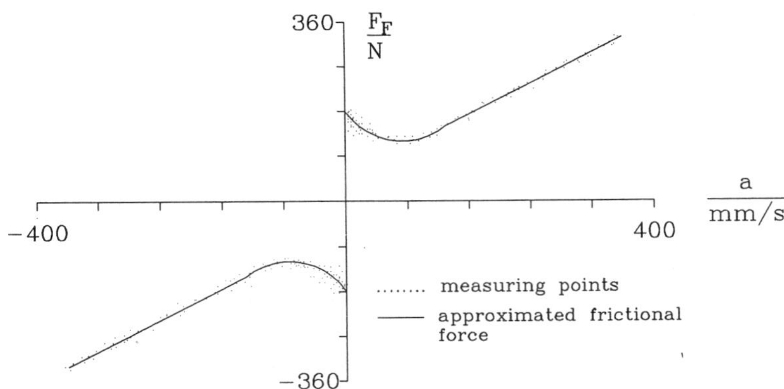

Fig. 2.2 Friction force characteristic

Beater (1987) proposed this bilinearization method: define two linear models for appropriate operating points of the nonlinear state model:

$$\dot{\underline{x}}_1(t) = \underline{A}_1\,\underline{x}_1(t) + \underline{b}_1\,\underline{u}_1(t) \quad \text{and} \quad \dot{\underline{x}}_2(t) = \underline{A}_2\,\underline{x}_2(t) + \underline{b}_2\,u_2(t) \qquad (2.3)$$
$$y(t) = \underline{c}^T\,\underline{x}_1(t) \qquad\qquad\qquad y(t) = \underline{c}^T\,\underline{x}_2(t) \quad ,$$

and compute a bilinear state space model via comparison of the coefficients of the matrices in (2.3) and those of a bilinear model (1.2):

$$\dot{\underline{x}}_1(t) = \underline{A}_1\underline{x}_1(t) + \underline{b}_1\underline{u}_1(t) = (\underline{A} + \underline{N}u_1(t))\,\underline{x}_1(t) + \underline{b}u_1(t)$$
$$\dot{\underline{x}}_2(t) = \underline{A}_2\underline{x}_2(t) + \underline{b}_2\underline{u}_2(t) = (\underline{A} + \underline{N}u_2(t))\,\underline{x}_2(t) + \underline{b}u_2(t) \qquad (2.4)$$

From the step responses in Fig. 2.3 one may realize how well the bilinear approximation model fits the dynamic behaviour of the essentially nonlinear physical system.

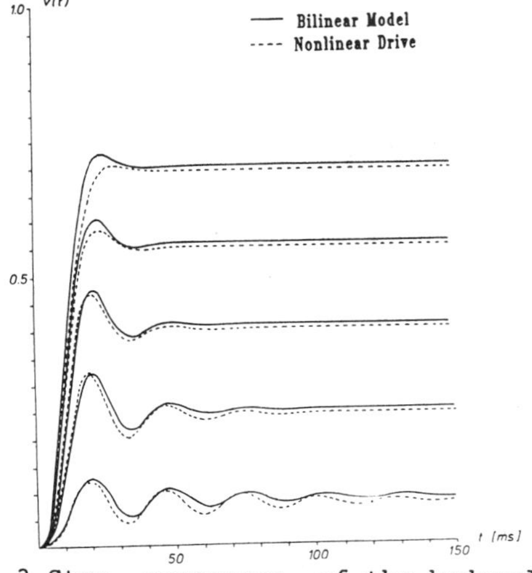

Fig. 2.3 Step responses of the hydraulic drive and its bilinear approximation

3. Bilinearization of a Rotational Hydraulic Drive

A recent control scheme for rotational hydraulic drive is "Drive control by variable displacement motors" as shown in Fig. 3.1 (Marner and Ulm, 1988)

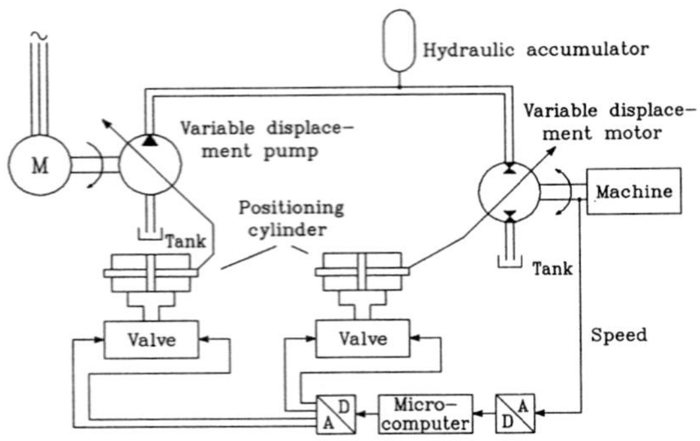

Fig. 3.1 Drive control scheme

The dynamics of the drive (valve, positionary cylinder, hydraulic motor) are governed by these state equations:

$$
\begin{aligned}
\dot{x}_1 &= x_2 \\
\dot{x}_2 &= -f_{RV}(x_2) - K_1 x_1 - K_2 x_2 - K_6 x_3 - K_7 x_4 + V_{st} u(t) \\
\dot{x}_3 &= x_4 \\
\dot{x}_4 &= K_3 x_5 - f_F(x_3) - f_{RS}(x_4) \\
\dot{x}_5 &= f_3(x_1, x_5) - K_5 x_4 \\
\dot{x}_6 &= -f_{RT}(x_6) + K_{14} p_A x_3
\end{aligned}
\tag{3.1}
$$

where the state variables $x_i(t)$ have these physical meanings:

x_1 = position of the pilot piston (part of the valve)
x_2 = velocity of the pilot piston
x_3 = position of the piston of the position cylinder
x_4 = velocity of the piston of the position cylinder
x_5 = difference pressure at the positioning piston
x_6 = rotational speed of the hydraulic motor

$$
\begin{aligned}
f_F &= K_9 x_3 + K_{10} \mathrm{sgn}(x_3) \tag{3.2}
\end{aligned}
$$
f_F = characteristics of the compressed springs
$$
\begin{aligned}
f_3 &= K_{13} \{\ \mathrm{sg}(x_1)\mathrm{sgn}(\Delta p_s - x_5)\ \ 0.5\ |\Delta p_s - x_5| \tag{3.3} \\
&\quad - \mathrm{sg}(-x_1)\mathrm{sgn}(\Delta p_s + x_5)\ \ 0.5\ |\Delta p_s + x_5|\ \}
\end{aligned}
$$
f_3 = pressure dynamic of the positioning unit
$$
f_{RV} = K_{18} x_2 + \mathrm{sgn}(x_2)\ (K_{19} + K_{20}\ \exp(-|x_2|/K_{21})) \tag{3.4}
$$
f_{RV} = characteristics of the friction of the pilot piston (Fig. 2.2)
$$
f_{RS} = K_{22} x_4 + \mathrm{sgn}(x_4)\ (K_{23} + K_{24}\ \exp(-|x_4|/K_{25})) \tag{3.5}
$$
f_{RS} = characteristics of the friction of the positioning piston
$$
f_{RT} = K_{27} x_6 + \mathrm{sgn}(x_6)\ (K_{28} + K_{29}\ \exp(-|x_6|/K_{30})) \tag{3.6}
$$
f_{RT} = characteristics of the friction of the hydraulic motor
p_A = pressure difference at the motor
Δp_s = pressure difference of the positioning unit
V_{st} = gain of the plant
all coefficients K_i, $i = 1, 2, \ldots, 32$ are defined to be constant.

A complete system consisting of a pressure station, rotational drive and a local engine was studied very carefully within a teststand in our hydraulic laboratory. Fig. 3.2 shows exemplarily the rather nonlinear behaviour of the system referring to the operating pressure (one input variable).

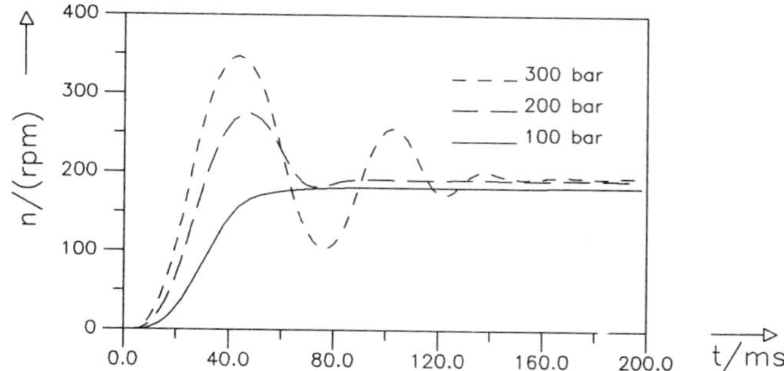

Fig. 3.2 Step responses of the speed variable for n=200 rpm and operating pressure 100, 200 and 300 bar

For the hydraulic motor (Mannesmann-Rexroth A4V/HS) all coefficients and nonlinear relations have been measured carefully. That means, simulations of the system by integration of (3.1) give results which fit measured curves rather precisely. This is demonstrated in Fig. 3.3 where measured step functions are plotted together with those of the nonlinear model (3.1) and a bilinear model (3.7).

$$\dot{x}(t) = Ax(t) + Bu(t) + \mathrm{sgn}(u_1)N_1 xu_1 + N_2 xu_2 + k\,\mathrm{sgn}(x_6) \qquad (3.7)$$
$$y(t) = c^T x(t)$$

$$A = \begin{bmatrix} 0 & 0.1\cdot10^1 & 0 & 0 & 0 & 0 \\ -0.4\cdot10^6 & -0.2\cdot10^4 & -0.2\cdot10^7 & -0.5\cdot10^4 & 0 & 0 \\ 0 & 0 & 0 & 0.1\cdot10^1 & 0 & 0 \\ 0 & 0 & -0.2\cdot10^5 & -0.4\cdot10^4 & 0.5\cdot10^4 & 0 \\ 0.7\cdot10^6 & 0 & 0 & 0 & -0.2\cdot10^5 & -0 \\ 0 & 0 & 0.5\cdot10^4 & 0 & 0 & -0.1\cdot10^0 \end{bmatrix}, \quad B = \begin{bmatrix} 0 & 0 \\ 0.2\cdot10^7 & 0 \\ 0 & 0 \\ 0 & 0 \\ 0 & 0 \\ 0 & 0 \end{bmatrix}$$

$$N_1 = \begin{bmatrix} 0 & 0 & 0 & 0 & 0 & 0 \\ 0 & 0 & 0 & 0 & 0 & 0 \\ 0 & 0 & 0 & 0 & 0 & 0 \\ 0 & 0 & 0 & 0 & 0 & 0 \\ -0.4\cdot10^6 & 0 & 0 & 0 & 0 & 0 \\ 0 & 0 & 0 & 0 & 0 & 0 \end{bmatrix}, \quad N_2 = \begin{bmatrix} 0 & 0 & 0 & 0 & 0 & 0 \\ 0 & 0 & 0 & 0 & 0 & 0 \\ 0 & 0 & 0 & 0 & 0 & 0 \\ 0 & 0 & 0 & 0 & 0 & 0 \\ 0 & 0 & 0 & 0 & 0 & 0 \\ 0 & 0 & 0.3\cdot10^2 & 0 & 0 & 0 \end{bmatrix}$$

$$k^T = [\; 0 \quad 0 \quad 0 \quad 0 \quad 0 \quad -0.3\cdot10^3 \;], \qquad c^T = [\; 0 \quad 0 \quad 0 \quad 0 \quad 0 \quad 1 \;]$$

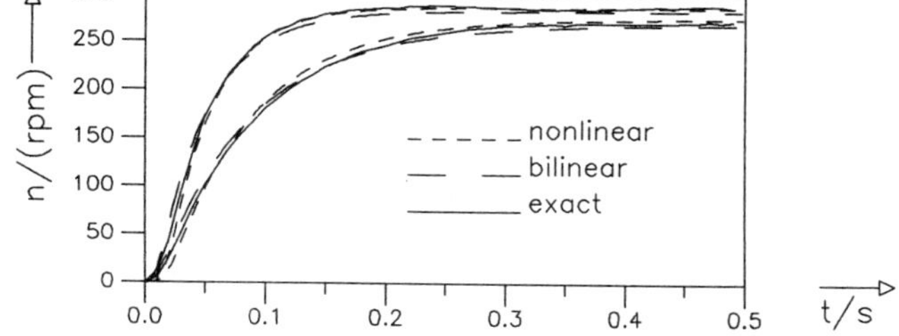

Fig.3.3   Step responses of the motor speed for the set point n=300 rpm and u_1=150 and 250 bar.

4. Bilinearization by Computer Experiments

If a nonlinear mathematical model of a nonlinear plant is not known, one
may try to find a system description by suitable experiments with the
system to find some input output relations which may describe the dynamic
behaviour of the system. In linear system theory this is done by frequency response method, for instance. Schwarz and Dorissen (1988) have published a method for experimental bilinearization of a nonlinear system.
The proposed method consists in stimulation of the system under consideration by particular discrete time impulse signals and evaluating the
Markov parameters of a discrete time bilinear system model (BLS). Using
systemrealization algorithms at first a discrete BLS may be computed from
which a continuous BLS can be found. The unknown plant has to be connected with a process-computer (Fig. 4.1) and then the system is excited by
a sequence of discrete unit impulses (Fig. 4.2).

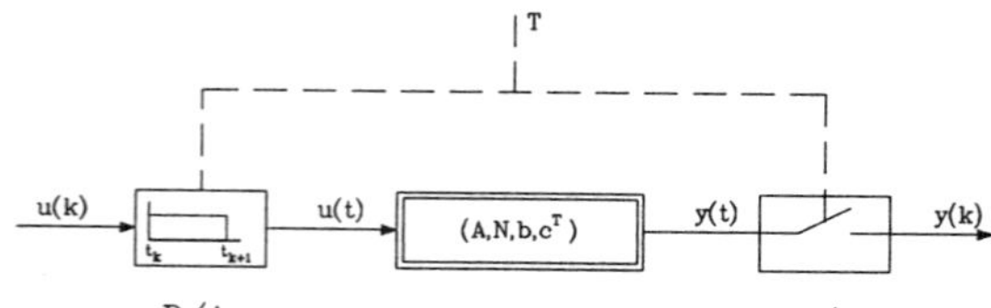

Fig. 4.1 Blockdiagram of the equivalent discrete time system

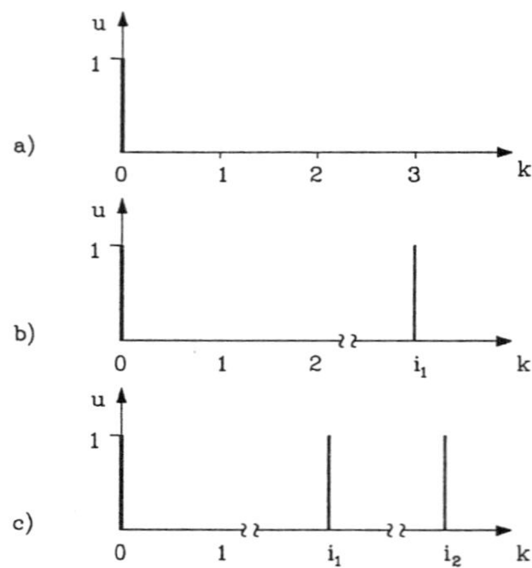

Fig. 4.2 Discrete time input signals to be used for systemidentification
for BLS

From the output signal series for each experiment (1st experiment one
impulse, 2nd experiment double impulse and so on) Markov parameters of a
BLS will be defined within the scheme (4.1). From these Markov parameters
a state space model of a BLS can be computed using system realization
methods (more details may be found in Schwarz and Dorissen, 1988) due to
lack of space in this paper).

$$
\begin{bmatrix} y(1) \\ y(2) \\ y(3) \\ y(4) \\ y(5) \\ y(6) \\ \vdots \end{bmatrix} =
\begin{bmatrix}
M_0 & 0 & 0 & 0 & 0 & 0 & 0 & 0 & \cdots \\
M_1 & M_0 & M_{0,0} & 0 & 0 & 0 & 0 & 0 & \cdots \\
M_2 & M_1 & M_{0,1} & M_0 & M_{0,0} & M_{1,0} & M_{0,0,0} & 0 & \cdots \\
M_3 & M_2 & M_{0,2} & M_1 & M_{0,1} & M_{1,1} & M_{0,0,1} & M_0 & \cdots \\
M_4 & M_3 & M_{0,3} & M_2 & M_{0,2} & M_{1,2} & M_{0,0,2} & M_1 & \cdots \\
M_5 & M_4 & M_{0,4} & M_3 & M_{0,3} & M_{1,3} & M_{0,0,3} & M_2 & \cdots \\
& & & & \vdots & & & &
\end{bmatrix} \cdot
\begin{bmatrix}
u(0) \\ u(1) \\ u(0)\,u(1) \\ u(2) \\ u(1)\,u(2) \\ u(0)\,u(2) \\ u(0)\,u(1)\,u(2) \\ u(3) \\ \vdots
\end{bmatrix}
\tag{4.3}
$$

By this procedure the translational hydraulic drive of sec. 2 was
treated. The system with the blockdiagram in Fig. 4.3 was <u>experimentaly</u>
bilinearized.

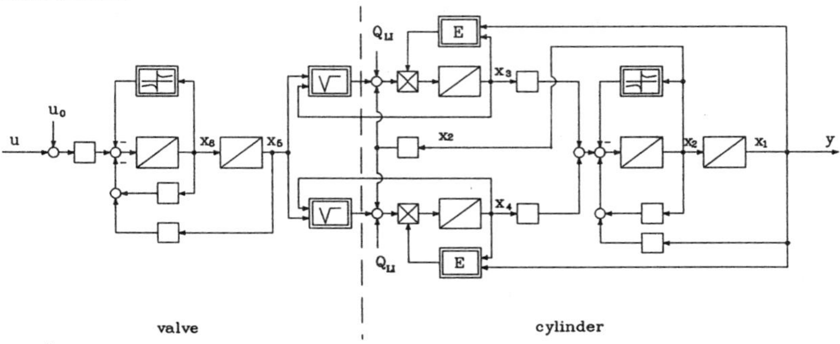

Fig. 4.3 Blockdiagram of the hydraulic drive

In Fig. 4.4 the step responses of the hydraulic drive in Fig. 4.3 and
those of a bilinear model which was found by experimental bilinearization
are plotted together. Once again it can be demonstrated that the dynamics
of a suitably defined bilinear model may fit those of a real rather
nonlinear technical system.

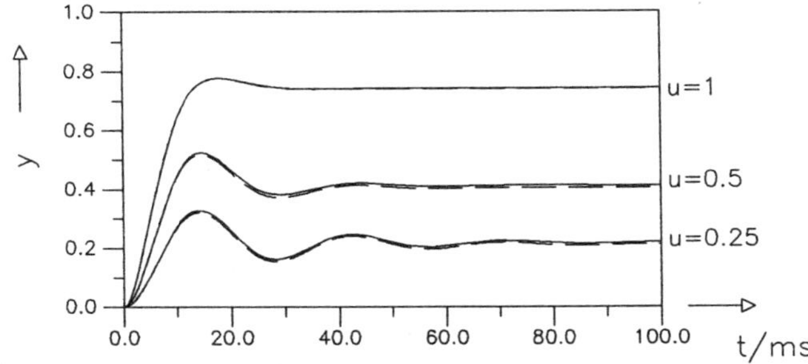

Fig. 4.4 Step responses of system in Fig. 4.3 and the via systemidentification algorithm computed system (dashed line)

Main advantages of the bilinearization of real systems dynamic equations
are:

- better approximation of the system dynamics for a wider range of
  signal amplitudes
- possibility of synthesis of robust controllers for nonlinear technical systems due to the fact, that BLS theory is meanwhile very well
  established.

6. Concluding remarks

The main idea of this paper is to suggest the method of bilinearization to improve the analysis of nonlinear control systems and the synthesis of controllers. The bilinearization can be done by experimental systemidentification either within a simulation scheme or at a real plant. For this purpose a processcomputer should be used to identify in a first step a discrete time BLS by means of a series of experiments to evaluate the Markov parameters, which create the base of a sytemrealization algorithm. If the real system is a bilinear system we find the (input signal dependent) equivalent discrete time BLS. But if the real system is actually a nonlinear system like the above mentioned hydraulic drive we get a bilinear approximation with an accuracy depending on how many Markov parameter chaines are identified by the experimental procedure. At this moment we could regard only the results for purely deterministic signals i.e. we did not deal with input and measurement noise. This was possible because in laboratory experiments noise problems could mostly be avoided. The next research objective is the inclusion of stochastic at least stochastic corrupted signals and to improve the computation algorithms.

References

Beater, P. (1987). Zur Regelung nichtlinearer Systeme mit Hilfe bilinearer Modelle. Phd-Thesis University Duisburg (FRG). Düsseldorf: VDI-Verlag.

Dorißen, H.-T. (1986). Zur mathematischen Realisierung bilinearer Systeme. Diploma-Thesis, MSRT, University Duisburg.

Guo, L. (1987). Bilineare Modellbildung und Messung an dem sekundärgeregelten hydraulischen Rotationsantrieb (A4VSO40HS). Research-Report, MSRT, University Duisburg.

Isidori, A. and Ruberti, A. (1973). Realization Theory of Bilinear Systems. in Mayne, D.Q. and Brockett, R. (editors). (1973). Geometric Methods in System Theory. Boston: Reidel.

Isidori, A. (1985). Nonlinear Control Systems: An Introduction. Berlin: Springer.

Marner, W. and Ulm, M. (1988). A Modern Concept for the Digital Control of Hydraulic Gears. 5th IAVD Congress on vehicle design components. 9-11 March 1989, Geneve (Switzerland).

Schwarz, H. (1987). Stability of Discrete-Time Equivalent Homogeneous Bilinear Systems. Control Theory and Advanced Technology, 3, 3, 263-269.

Schwarz, H. and Dorissen, H. T. (1988). Systemidentification of Bilinear Systems via Realization Theory and its Application. (Contributed paper under review for publication).

Yousuf, M. (1984). Regelung eines hydraulischen Antriebes mittels bilinearer Systemmodelle. Diploma-Thesis, MSRT, University Duisburg.

Analytical Analysis of a Stochastic Partial Differential Equation

Weijian Zhang

Department of Electrical Engineering
6731 Boelter Hall
University of California, Los Angeles
Los Angeles, CA 90024

ABSTRACT

This paper investigates the following stochastic partial differential equation

$$dX_t = -A\,X_t\,dt + G(X_t)\,dB_t \tag{a1}$$

where A is a self-adjoint, positive definite operator with eigensystem

$$\{\ \lambda_n;\ \xi_n;\ n = 0, 1, 2,\ \cdots\ \}$$

and $G(.)\colon H_{-\alpha} \to \sigma_2(H, H_{-\alpha})$ ($\alpha \geq 0$) is defined by

$$G(h)\,\xi_n = (h, \eta_n)_{-\alpha}\,\eta_n \qquad n = 0, 1, 2,\ \cdots \quad h \in H_{-\alpha}$$

with $\eta_n = \lambda_n^\alpha\,\xi_n$, $n = 0, 1, 2,\ \cdots$ and $\{\ H_\beta, -\infty < \beta < \infty\ \}$ is a Hilbert scale with generating operator A. And $\{\ B_t,\ t \geq 0\ \}$ is a cylindrical Brownian motion on $H = L^2(D)$ where D is a domain in R^d.

The existence and the uniqueness of the solution of (a1) is proved and its stability is discussed. By computing the kernels of the integral representation of the solution, the analytical expression of the solution is obtained as

$$X_t = e^{-\frac{t}{2}} \sum_{n=0}^{\infty} c_n\,e^{-\lambda_n t}\,\eta_n\,e^{w_n(t)} \tag{a2}$$

where $X_0 = \sum_{n=0}^{\infty} c_n\,\eta_n \in H_{-\alpha}$, and $\{\ w_n(t) = B_t(\xi_n), n = 0, 1, 2,\ \cdots\ \}$ is a sequence of independent one-dimensional Brownian motion. The second order statistics of the solution is fully studied by determining the covariance operator.

The approach is along the line of the Wiener-Ito direct sum decomposition of $L^2(E^* \to H_{-\alpha})$ by generalizing the notions of multiple Wiener integral and iterated stochastic integral.

1. INTRODUCTION

Let $H = L^2(D)$, D is a domain in R^d and $T = [0, \infty)$. Let $A: H \to H$ be a self-adjoint operator such that $A^{-1} \in \sigma_2(H)$ and suppose the eigensystem of A is given by

$$\{ \lambda_n: n = 0, 1, 2, \cdots \} \qquad \{ \xi_n(\sigma); n = 0, 1, 2, \cdots \}$$

such that $0 < \lambda_0 < \lambda_1 < \cdots$ and $\{ \xi_n; n = 0, 1, 2, \cdots \}$ forms an orthonormal basis (ONB) in H.

Define: $\eta_n = \lambda_n^\alpha \xi_n$, $n = 0, 1, 2, \ldots$, $\alpha > 0$. Then $\{ \eta_n; n = 0, 1, 2, \cdots \}$ is an ONB in $H_{-\alpha}$ which is a Hilbert scale space [2] with generating operator A and inner product $(\xi, \eta)_{-\alpha} = (A^{-\alpha}\xi, A^{-\alpha}\eta)_H$.

In this paper, we consider the following bilinear stochastic partial differential equation

$$dX_t(\sigma) = - AX_t \, dt + G(X_t) \, dB_t \qquad \sigma \in D, \quad t \in T \tag{1.1}$$

where $B_t = B(t; h)$ is a cylindrical Brownian motion on H [5] and $G: H_{-\alpha} \to \sigma_2(H, H_{-\alpha})$ is defined by

$$G(h)\xi = \sum_{n=0}^{\infty} (\xi, \xi_n)_H \, (h, \eta_n)_{-\alpha} \, \eta_n \tag{1.2}$$

for any $h \in H_{-\alpha}$ and $\xi \in H$, i.e.,

$$G(h)\xi_n = (h, \eta_n)_{-\alpha} \, \eta_n \qquad n = 0, 1, 2, \cdots \quad h \in H_{-\alpha}$$

And we also have for any $h \in H_{-\alpha}$,

$$\|G(h)\|^2_{\sigma_2(H, H_{-\alpha})} = \sum_{n=0}^{\infty} \|G(h)\xi_n\|^2_{-\alpha}$$

$$= \sum_{n=0}^{\infty} |(h, \eta_n)_{-\alpha}|^2 = \|h\|^2_{-\alpha} \tag{1.3}$$

Similar equations, but somewhat different equations of this type, have been investigated by many authors (A. V. Balakrishnan [1], D. A. Dawson [3], A. Shimizu [7]), however, the present approach is along the line of the decomposition of $L^2(E^* \to H_{-\alpha})$ which provides an explicit solution of (1.1). In this paper we adopt the white noise space (E^*, B, μ) as the basic probability space, i.e.,

$$\int_{E^*} e^{i\langle \eta, \omega \rangle} \, d\mu(\omega) = e^{-\frac{1}{2} \|\eta\|^2}$$

where $\|\eta\|^2 = \int_{D \times T} |\eta|^2 \, dx$, $E \subset L^2(D \times T) \subset E^*$ is a Gelfand triple and $\langle ., \rangle$ is the canonical bilinear form connecting E and E^*.

Stochastic partial differential equations have been discussed in connection with quantum field theory, theory of partial differential equations involving random terms and filtering theory in electrical engineering.

2. RESULTS

We start with the statements concerning the existence, uniqueness and continuity of the evolution-solution [5] of (1.1).

THEOREM 1: Eq (1.1) has a unique e-solution X_t in $H_{-\alpha}$ for any given $X_0 \in H_{-\alpha}$.

PROOF: By Theorem 3.3 of [4], it is sufficient to prove that G is a Lipschitz continuous mapping from $H_{-\alpha}$ to $\sigma_2(H, H_{-\alpha})$.

In fact, for any $h, k \in H_{-\alpha}$, one has

$$\|G(h) - G(k)\|^2_{\sigma_2(H, H_{-\alpha})}$$

$$= \sum_{n=0}^{\infty} \|G(h)\xi_n - G(k)\xi_n\|^2_{-\alpha}$$

$$= \sum_{n=0}^{\infty} |(h-k, \eta_n)_{-\alpha}|^2$$

$$= \|h - k\|^2_{-\alpha} \qquad\qquad QED.$$

THEOREM 2: The unique e-solution of (1.1) is continuous.

PROOF: By Theorem 3.4 of [6], it is sufficient to show that the semigroup $\{ T_t = e^{-At}; t \geq 0 \}$ and $G(.)$ are commutative.

In fact, for any $\xi \in H$, $h \in H_{-\alpha}$, one has

$$T_t\, G(h)\xi = T_t \left[\sum_{n=0}^{\infty} (\xi, \xi_n)_H \, (h, \eta_n)_{-\alpha} \, \eta_n \right]$$

$$= \sum_{n=0}^{\infty} e^{-\lambda_n t} \, (\xi, \xi_n)_H \, (h, \eta_n)_{-\alpha} \, \eta_n$$

$$= \sum_{n=0}^{\infty} (\xi, \xi_n)_H \, (h, T_t\eta_n)_{-\alpha} \, \eta_n$$

$$= \sum_{n=0}^{\infty} (\xi, \xi_n)_H \, (T_t h, \eta_n)_{-\alpha} \, \eta_n$$

$$= G(T_t h)\, \xi \qquad\qquad QED.$$

THEOREM 3: Let $\Phi_n(t; x_1, \cdots x_n; \sigma)$ be the kernel of the integral representation of the solution $X_t(\sigma)$ of (1.1). Then $\{ \Phi_n; n = 0, 1, 2, \cdots \}$ satisfies the following functional equation

$$\Phi_{n+1}(t; x_1, \cdots x_{n+1}; \sigma) = \frac{1}{n+1} \chi_{[0, t]}(t_{n+1})$$

$$\times \sum_{k=0}^{\infty} (\Phi_n(t_{n+1}; x_1, \cdots x_n; .), \eta_k)_{-\alpha} \, \xi_k(y_{n+1}) \, e^{-\lambda_k(t-t_{n+1})} \, \eta_k(\sigma)$$

$$0 \leq t_1 \leq \cdots \leq t_n \leq t_{n+1}, \qquad n = 0, 1, 2, \cdots; \quad x_j = (y_j, t_j) \in D \times T$$

$$\Phi_0(t; \sigma) = (T_t X_0)(\sigma) \qquad\qquad (2.1)$$

To prove this theorem we need the following

LEMMA (Y. Miyahara [5])

Let $S \in L^2(T \times E^* \to \sigma_2(H, H_{-\alpha}))$ and assume that $S(t; \omega)$ satisfies the following conditions:

1. $S(t; \omega)$ is B_t-adapted, where $B_t = \sigma\{ <\omega, \eta>; supp(\eta) \subset [0, t] \times D \}$;

2. $S(t; .) \in L^2(E^* \to \sigma_2(H, H_{-\alpha}))$ and its kernel is given by $S_n(t; x_1, \cdots x_n; \sigma) \in L^2((D \times T)^n \times D \to H_{-\alpha})$.

Then the $\tilde{H}_{n+1}(H_{-\alpha})$-component of $\int_0^t S(s)\, dB_s$ is equal to $\int_0^t S_n(s; x_1, \cdots x_n; y)\, dB_s(y)$ and its kernel is given by

$$\frac{1}{n+1} \chi_{[0, t]}(t_{n+1}) S_n(t_{n+1}; x_1, \cdots x_n; y_{n+1}) \qquad t_1 \leq \cdots \leq t_{n+1} \qquad (2.2)$$

PROOF OF THEOREM 3: For any $\xi \in H$,

$$[\, G(X_t)\,]_n \xi = [\, G(X_t)\xi\,]_n$$

$$= \sum_{k=0}^{\infty} (\xi, \xi_k)_H \,(\, I(\Phi_n),\, \eta_k)_{-\alpha}\, \eta_k$$

$$= I[\, \int_D \{\, \sum_{k=0}^{\infty} (\Phi_n(t;\, x_1,\, \cdots\, x_n;\, .),\, \eta_k)_{-\alpha}\, \xi_k(\sigma')\, \eta_k(\sigma)\, \}\, \xi(\sigma')\, d\sigma'\,] \tag{2.3}$$

Then it follows that

$$[\, T_{t-s}\, G(X_s)\,]_n \xi = [\, T_{t-s}\, G(X_s)\xi\,]_n$$

$$= I[\, \int_D \{\, \sum_{k=0}^{\infty} (\, \Phi_n(s;\, x_1,\, \cdots\, x_n;\, .),\, \eta_k)_{-\alpha}\, \xi_k(\sigma')\, e^{-\lambda_k(t-s)}\, \eta_k(\sigma)\, \}\, \xi(\sigma')\, d\sigma'\,] \tag{2.4}$$

Therefore the kernel of $T_{t-s}\, G(X_s)$ is given by

$$\sum_{k=0}^{\infty} (\, \Phi_n(s;\, x_1,\, \cdots\, x_n;\, .),\, \eta_k)_{-\alpha}\, \xi_k(y)\, e^{-\lambda_k(t-s)}\, \eta_k(\sigma) \tag{2.5}$$

$$n = 0,\, 1,\, 2,\, \cdots$$

Then, by the above Lemma and the relation

$$X_t = T_t\, X_0 + \int_0^t T_{t-s}\, G(X_s)\, dB_s \tag{2.6}$$

we obtain the functional equation (2.1). $\qquad\qquad$ QED

In the next theorem, the explicit solution of (1.1) is provided.

THEOREM 4: 1. For given $X_0 = \sum_{n=0}^{\infty} c_n \eta_n \in H_{-\alpha}$, the unique solution of the functional equation (2.1) is given by

$$\Phi_0(t;\, \sigma) = T_t\, X_0 = \sum_{n=0}^{\infty} c_n e^{-\lambda_n t}\, \eta_n(\sigma);$$

$$\Phi_n(t;\, x_1,\, \cdots\, x_n;\, \sigma) = \frac{1}{n!} \chi_{[0,\, t]}(t_n) \sum_{k=0}^{\infty} c_k \prod_{j=1}^{n} \xi_k(y_j)\, \eta_k(\sigma)\, e^{-\lambda_k t}$$

$$t_1 \le\ \cdots\ \le t_n;\quad n = 1,\, 2,\, \cdots \tag{2.7}$$

2. The unique e-solution of (1.1) is given by

$$X_t = e^{-\frac{t}{2}} \sum_{n=0}^{\infty} c_n\, e^{-\lambda_n t}\, \eta_n\, e^{w_n(t)} \tag{2.8}$$

where $w_n(t) = B_t(\xi_n)$, $n = 0,\, 1,\, 2,\, \cdots$ is a sequence of independent one-dimensional Brownian motions.

Furthermore, we have

$$\|X_t\|_{L^2(E^* \to H_{-\alpha})} = e^{\frac{t}{2}}\, \|T_t\, X_0\|_{-\alpha} \tag{2.9}$$

PROOF: 1. Using the expression $\Phi_0(t;\, \sigma) = \sum_{n=0}^{\infty} c_n\, e^{-\lambda_n t}\, \eta_n(\sigma)$, we can solve for $\Phi_n(t;\, x_1,\, \cdots\, x_n;\, \sigma)$, $n = 1,\, 2,\, \cdots$ inductively from the functional equation (2.1). The uniqueness is due to the uniqueness of $\Phi_0(t;\, \sigma)$ for a given $X_0 \in H_{-\alpha}$.

2. Claim:

$$\|\sum_{n=0}^{\infty} I(\, \Phi_n(t;\, x_1,\, \cdots\, x_n;\, \sigma)\,)\|_{L^2(E^* \to H_{-\alpha})}^2 = e^t\, \|T_t\, X_0\|_{-\alpha}^2 < \infty \tag{2.10}$$

Suppose the above claim is true, then the uniqueness of the integral representation gives

$$X_t = \sum_{n=0}^{\infty} I(\, \Phi_n(t;\, x_1,\, \cdots\, x_n;\, \sigma)\,)$$

$$= T_t\, X_0 + \sum_{n=1}^{\infty} n!\, \hat{I}(\, \Phi_n(t;\, x_1,\, \cdots\, x_n;\, \sigma)\,) \tag{2.11}$$

while we have

$$n!\, \hat{I}(\, \Phi_n(t;\, x_1,\, \cdots\, x_n;\, \sigma)\,)$$

$$= \hat{I}(\, \chi_{[0,\, t]}(t_n) \sum_{k=0}^{\infty} c_k \prod_{j=1}^{n} \xi_k(y_j)\, \eta_k(\sigma)\, e^{-\lambda_k t}\,)$$

$$= \sum_{k=0}^{\infty} c_k\, e^{-\lambda_k t}\, \eta_k(\sigma) \int_0^t \int_0^{t_1} \cdots \int_0^{t_{n-1}} \prod_{j=1}^{n} \xi_k(y_j)\, dB_{t_n}(y_n) \cdots dB_{t_1}(y_1)$$

$$= \sum_{k=0}^{\infty} c_k\, e^{-\lambda_k t}\, \eta_k(\sigma) \int_0^t \int_0^{t_1} \cdots \int_0^{t_{n-1}} dB_{t_n}(\xi_k) \cdots dB_{t_1}(\xi_k)$$

$$= \sum_{k=0}^{\infty} c_k\, e^{-\lambda_k t}\, \eta_k(\sigma) \tag{2.12}$$

Then (2.11) and (2.12) give

$$X_t = \sum_{k=0}^{\infty} c_k\, e^{-\lambda_k t}\, \eta_k\, [\, \sum_{n=0}^{\infty} \frac{(\frac{t}{2})^{\frac{n}{2}}}{n!}\, H_n(\, \frac{B_t(\xi_k)}{\sqrt{2t}}\,)\,]$$

$$= \sum_{k=0}^{\infty} c_k\, e^{-\lambda_k t}\, \eta_k\, e^{-\frac{t}{2} + B_t(\xi_k)}$$

$$= e^{-\frac{t}{2}} \sum_{n=0}^{\infty} c_n\, e^{-\lambda_n t + w_n(t)} \tag{2.13}$$

In (2.13) we have used the generating function relation of Hermite polynomial [3]:

$$\sum_{n=0}^{\infty} \frac{t^n}{n!}\, H_n(x) = e^{-t^2 + 2tx} \tag{2.14}$$

Finally, we need to prove the claim (2.10). In fact,

$$\|\, \sum_{n=0}^{\infty} I(\, \Phi_n(t;\, x_1,\, \cdots\, x_n;\, .)\,)\,\|^2_{L^2(E^* \to H_{-\alpha})}$$

$$= \|T_t\, X_0\|^2_{-\alpha} + \sum_{n=1}^{\infty} \|I(\, \Phi_n(t;\, x_1,\, \cdots\, x_n;\, .)\,)\|^2_{L^2(E^* \to H_{-\alpha})}$$

$$= \|T_t\, X_0\|^2_{-\alpha} + \sum_{n=1}^{\infty} (n!)^2\, \|\hat{I}(\, \Phi_n(t;\, x_1,\, \cdots\, x_n;\, .)\,)\|^2_{L^2(E^* \to H_{-\alpha})}$$

$$= \|T_t\, X_0\|^2_{-\alpha} + \sum_{n=1}^{\infty} (n!)^2 \int \cdots \int_{t_1 \leq \cdots \leq t_n} \|\Phi_n(t;\, x_1,\, \cdots\, x_n;\, .)\|^2_{-\alpha}\, dx_1...dx_n \tag{2.15}$$

We also have

$$(n!)^2 \int \cdots \int_{t_1 \leq \cdots \leq t_n} \|\Phi_n(t;\, x_1,\, \cdots\, x_n;\, .)\|^2_{-\alpha}\, dx_1 \cdots dx_n$$

$$= \int \cdots \int_{t_1 \leq \cdots \leq t_n \leq t} \sum_{k=0}^{\infty} c_k^2 \prod_{j=1}^{n} \xi_k^2(y_j)\, e^{-2\lambda_k t}\, dx_1 \cdots dx_n$$

$$= \sum_{k=0}^{\infty} c_k^2\, e^{-2\lambda_k t} \int_0^t \int_0^{t_1} \cdots \int_0^{t_{n-1}} dt_n \cdots dt_1$$

$$= \frac{t^n}{n!}\, \|T_t\, X_0\|^2_{-\alpha} \tag{2.16}$$

Therefore, by (2.15) and (2.16) we obtain

$$\left\| \sum_{n=0}^{\infty} I(\Phi_n(t; x_1, \cdots x_n; .)) \right\|^2_{L^2(E^* \to H_{-\alpha})}$$

$$= \|T_t X_0\|^2_{-\alpha} \sum_{n=0}^{\infty} \frac{t^n}{n!}$$

$$= e^t \|T_t X_0\|^2_{-\alpha} < \infty \tag{2.17}$$

(2.9) is obvious based upon (2.11) and (2.17). $\qquad$ QED

In the corollary which follows, the stability of (1.1) is discussed. The stability we consider is in the following sense:

DEFINITION 1: The eq (1.1) is said to be stable if for every $X_0 \in H_{-\alpha}$, the corresponding e-solution X_t satisfies:

$$\int_0^{\infty} \|X_t\|^2_{L^2(E^* \to H_{-\alpha})} \, dt < \infty$$

Since, by part 2 of Theorem 4,

$$\|X_t\|^2_{L^2(E^* \to H_{-\alpha})} = e^t \|T_t X_0\|^2_{-\alpha}$$

$$= \sum_{n=0}^{\infty} c_n^2 \, e^{-2(\lambda_n - \frac{1}{2})t} \tag{2.18}$$

we can obviously conclude

COROLLARY: Eq (1.1) is stable if and only if

$$\lambda_0 = \min_n \lambda_n > \frac{1}{2}.$$

Next, we will study the second order statistics of the solution X_t. Let $(X_t)_n$ be the $\tilde{H}_n(H_{-\alpha})$- component of X_t, i.e.

$$(X_t)_n = I(\Phi_n(t; x_1, \cdots x_n; \sigma))$$

let us define

DEFINITION 2: The self-adjoint, nonnegative operator S_n determined by

$$(S_n \xi, \eta)_{-\alpha} = \int_{E^*} (\xi, (X_t)_n)_{-\alpha} ((X_t)_n, \eta)_{-\alpha} \, d\mu(\omega) \tag{2.19}$$

for any $\xi, \eta \in H_{-\alpha}$, is called the covariance operator of X_t of degree n.

THEOREM 5: For (1.1), the covariance operator S_n of X_t of degree n is given by

$$(S_n \xi)(\sigma) = \frac{t^n}{n!} (v(t; \sigma, .), \xi)_{-\alpha}$$

$$\xi \in H_{-\alpha}, \quad n = 1, 2, \cdots \tag{2.20}$$

where $v(t; \sigma, \sigma') = \sum_{n=0}^{\infty} c_n^2 \, e^{-2\lambda_n t} \, \eta_n(\sigma) \, \eta_n(\sigma')$.

And consequently,

$$\int_{E^*} (\xi, X_t)_{-\alpha} (X_t, \eta)_{-\alpha} \, d\mu(\omega)$$

$$= (e^t - 1) (S\xi, \eta)_{-\alpha} + (T_t X_0, \xi)_{-\alpha} (T_t X_0, \eta)_{-\alpha} \tag{2.21}$$

where $(S \xi)(\sigma) = (v(t; \sigma, .), \xi)_{-\alpha}$.

PROOF: For any $\xi, \eta \in H_{-\alpha}$, we have

$$\mathbf{E}\,(\xi, (X_t)_n)_{-\alpha}\,(\,(X_t)_n, \eta)_{-\alpha}$$

$$\mathbf{E}\,(A^{-\alpha}\xi, I(\,A^{-\alpha}\,\Phi_n\,))_H\,(\,I(\,A^{-\alpha}\,\Phi_n\,),\,A^{-\alpha}\eta)_H$$

$$= (n!)^2\,\mathbf{E}\,(A^{-\alpha}\xi, \hat{I}(\,A^{-\alpha}\Phi_n\,))_H\,(\,\hat{I}(\,A^{-\alpha}\Phi_n\,),\,A^{-\alpha}\eta)_H$$

$$= (n!)^2\int_D (A^{-\alpha}\xi)(\sigma)\int_D\int_{0\le t_1\le\ldots\le t_n}\!\!\cdots\int (A^{-\alpha}\Phi_n(t;\,x_1,\,\cdots\,x_n;\,.))(\sigma)$$

$$(A^{-\alpha}\Phi_n(t;\,x_1,\,\cdots\,x_n;\,.))(\sigma')\,dx_1...dx_n\,(A^{-\alpha}\eta)(\sigma')\,d\sigma'\,d\sigma$$

$$= (A^{-\alpha}\xi,\,A^{-\alpha}S_n\eta)_H$$

$$= (\xi,\,S_n\eta)_{-\alpha} \tag{2.22}$$

where

$$(S_n\eta)(\sigma)$$

$$= (n!)^2\int_D\int_{0\le t_1\le\ldots\le t_n}\!\!\cdots\int \Phi_n(t;\,x_1,\,\cdots\,x_n:\,\sigma)\,(A^{-\alpha}\Phi_n(t;\,x_1,\,\cdots\,x_n;\,.))(\sigma')\,dx_1...dx_n$$

$$(A^{-\alpha}\eta)(\sigma')\,d\sigma'$$

$$= (A^{-\alpha}\phi_n(t;\,\sigma,\,.),\,A^{-\alpha}\eta)_H$$

$$= (\phi_n(t;\,\sigma,\,.),\,\eta)_{-\alpha} \tag{2.23}$$

where ϕ_n denotes

$$\phi_n(t;\,\sigma,\,\sigma')$$

$$= (n!)^2\int_{0\le t_1\le\ldots\le t_n}\!\!\cdots\int \Phi_n(t;\,x_1,\,\cdots\,x_n;\,\sigma)\,\Phi_n(t;\,x_1,\,\cdots\,x_n;\,\sigma')\,dx_1\,\cdots\,dx_n$$

$$= \int_{0\le t_1\le\ldots\le t_n}\!\!\cdots\int\,[\,\sum_{k=0}^{\infty} c_k\prod_{j=1}^{n}\xi_k(y_j)\,e^{-\lambda_k t}\,\eta_k(\sigma)\,]\,[\,\sum_{l=0}^{\infty} c_l\prod_{i=1}^{n}\xi_l(y_i)\,e^{-\lambda_l t}\,\eta_l(\sigma')\,]\,dx_1\,\cdots\,dx_n$$

$$= \int_0^t\!\int_0^{t_1}\cdots\int_0^{t_{n-1}}\sum_{k=0}^{\infty} c_k^2\,e^{-2\lambda_k t}\,\eta_k(\sigma)\,\eta_k(\sigma')\,dt_n\,\cdots\,dt_1$$

$$= v(t;\,\sigma,\,\sigma')\,\frac{t^n}{n!} \tag{2.24}$$

Combining (2.23) and (2.24), we obtain (2.20). $\hspace{2cm}$ QED.

ACKNOWLEDGEMENT: Special acknowledgement goes to Dr. A. V. Balakrishnan for his encouragement and support. This work was supported by NASA under the grant NSG 4015.

References

[1] A. V. Balakrishnan, " Stochastic bilinear partial differential equations ", *Proc. of the 2nd U.S.-Italy Seminar on Variable Structure Systems,* May, 1974; *Lecture Notes in Systems Theory 111,* Springer-Verlag, 1975, ed. A. Ruberti and R. R. Mohler.

[2] A. T. Bharucha-Reid, *Random Integral Equations,* Academic Press, New York, 1972

[3] D. A. Dawson, " Stochastic evolution equation ", *Mathematical Biosiences ,* Vol 15, 1972, pp 287-316.

[4] Y. Miyahara, " Stability of linear stochastic differential equations in Hilbert space ", *Information, Decision and Control in Dynamic Socio-Economics ,* ed. H. Myoken, Bunshindo / Kinokuniyo, Tokyo, 1978, pp 237-252

[5] Y. Miyahara, " Infinite dimensional Langevin equation and Fokker-Planck equation ", *Nagoya Math. J.* Vol. 81, 1981, pp 177-223

[6] Y. Miyahara, " Stochastic differential equations in Hilbert space ", *OIKONOMIKA ,* Vol. 14, No. 1, June, 1977, pp 37-47

[7] A. Shimizu, " Construction of a solution of linear stochastic evolution equations on a Hilbert space ", *Proc. of the Interna. Symp. on Stoch. Differen. eq ,* Kyoto, 1976, pp 385-395

Is Floating-Point Arithmetic Still Adequate?

Gerd Bohlender[1]

Summary: For complicated numerical problems, the error analysis has to be performed by the computer. Several methods for automated error analysis are known. Floating–point arithmetic has to be augmented and programming languages for scientific computation have to be provided (PASCAL–SC and FORTRAN–SC) for that purpose.

1. Introduction

Floating–point arithmetic has been successfully implemented on a vast range of computers, from supercomputers down to microcomputers. It is used intensively in most scientific and engineering applications. So, what are the problems with floating–point arithmetic?

First, execution speeds are still growing dramatically. Therefore larger and larger problems can be treated by the computer. But as the complexity of the problems increases, an error analysis of the computed results can no longer be performed by the user alone. Instead, the computer should provide tools which assist the application programmer in error analysis. Second, performance of computer systems should not only be measured in terms of execution speeds, but as well in terms of the power of its operations. Powerful operations allow the user to program on a much higher level. Third, the interface between the arithmetic unit and main memory or the host processor frequently is a bottleneck: operations can be executed much faster in the arithmetic unit than operands can be transferred over the interface. Therefore operands should be kept inside the arithmetic unit as long as possible.

These observations lead to the conclusion that floating–point arithmetic has to be augmented by more powerful operations. These operations can be higher numerical operations, such as interval operations (which are already supported by the IEEE standard for floating–point arithmetic), complex operations, matrix and vector operations. These operations should compute results with maximum accuracy in order to facilitate an error analysis.

2. Verification of results

Numerical algorithms in the set of real numbers usually fall into one of three classes: *finite exact methods* (e.g. Gaussian elimination), *iterative methods* (e.g. Newton's method), or *approximative methods* (e.g. rational approximation of a standard function, discretization of a differential equation). On a computer, a termination criterion has to be introduced for an iterative method which converts it into an approximative method. Ordinary floating–point computation delivers only approximations in each of these cases; no information about the precision of the result is given by the computer. The floating–point result does not even prove the existence or uniqueness of the true result.

[1]Institut für Angewandte Mathematik, Universität Karlsruhe, Kaiserstr. 12, D–7500 Karlsruhe, West Germany

Several methods have been developped to compute error bounds for the result:

☐ *manual forward or backward analysis* is very complicated in large problems and may easily afford more work than the solution of the proper problem.

☐ *symbolic manipulation* can be employed (at compile time) in order to reduce the complexity of formulas or to make them numerically more stable; afterwards (at execution time) the result can be computed using e.g. *rational / longreal evaluation.*

☐ *"numerical experimentation"*: the problem is run with slightly different data or in single and double precision; according to the different results the user guesses the number of correct digits.

☐ *perturbation/permutation method* [7]: the same algorithm is run several (typically three) times; each operation in the algorithm is perturbed by a random rounding and possibly the sequence of operations is modified; the number of correct digits of the result can be estimated by statistical methods.

☐ *interval methods* [1]: simply replacing floating–point numbers with intervals in an algorithm frequently leads to unsatisfactory wide intervals for the results (naive interval computation, dependent intervals); therefore special interval methods have to be employed. Iterative methods in interval mathematics usually work as follows: starting with a wide interval I_0 that contains the true solution x,

a converging sequence of intervals

$$I_k \supset I_{k+1} \supset \ldots \supset \{x\}$$

is computed which all contain the true solution. In a computer, the intervals are inflated by rounding errors. Therefore the sequence remains constant after a finite number of steps.

☐ *inclusion methods* [5] start with an arbitrary floating–point approximation x of the true result y; afterwards a verification step is appended which uses interval arithmetic: a sequence of intervals I_k is computed (starting with $I_0 = [x,x]$) which has the property

$$\text{if } I_{k+1} \subset \text{interior } (I_k) \text{ then } y \in I_{k+1}$$

3. PASCAL–SC extensions for scientific computation

The different methods described above can be supported by an appropriate programming language – or ideally by the hardware of the computer. In PASCAL–SC [4, 6] the standard language PASCAL has been extended with a large number of extensions for the computation of error bounds and for the verification of results:

Directed roundings are already included in the IEEE standard on floating–point arithmetic [2], but in traditional programming languages they cannot be accessed efficiently. In PASCAL–SC the operations +< and +> compute a lower and upper bound for the sum, resp. This allows an easy implementation of interval operations.

Dot products can be computed with least significant bit accuracy – even if cancellation occurs – using a new data type dotprecision. Expressions for this data type are marked by the symbol # and a rounding mode. Let e.g. r be a floating–point number and x,y,u,v be vectors of floating–point numbers; then

$$\#< (x^*y + u^*v - r)$$

computes the scalar product s := x*y + u*v − r with infinite precision and rounds it to the next floating–point number which is less than or equal to s. These dotprecision expressions are implemented for floating–point numbers, vectors, and matrices. If implemented in hardware, the exact evaluation of dotprecision expressions turns out to be **faster** than ordinary floating–point arithmetic.

Arithmetic libraries with least significant bit accuracy are included in the language for all customary numerical data types (real and complex floating–point, real and complex intervals, vectors and matrices for all of these). These higher arithmetic operations reduce the influence of rounding errors in an algorithm and at the same time make an error analysis much simpler. If implemented in hardware, such higher arithmetic operations can reduce the amount of operand transfers significantly; e.g. in a matrix multiplication only $3*n^2$ floating–point numbers have to be transferred in contrast with $2*n^3+n^2$ using ordinary floating–point operations and accumulation in a register.

Numerical libraries which compute verified results are available for a large class of standard numerical problems (linear and nonlinear systems, eigenvalues and eigenvectors, evaluation and zeroes of polynomials, evaluation of rational expressions, differential equations, etc. see [5, 6]). Let us for example consider the solution of a linear system:

```
procedure SolveLinearSystem            (var A: RealMatrix; var b: RealVector;
                                        var x: IntervalVector; var okay: boolean);
  var
    R: RealMatrix;                      {an approximate inverse of A}
    B: IntervalMatrix;                  {an interval inclusion of I–R*A}
    I: RealMatrix;                      {the identity matrix}
    y: IntervalVector;                  {x and y are iterates of the interval method}
    z: IntervalVector;                  {an interval inclusion of R*b}
    Eps: IntervalVector;               {a small interval vector around 0}
    i,dim,k: integer;                   {some counters}
  begin
    dim := ubound (b) − lbound (b);
    for i := 1 to dim do
      Eps[i] := intval (−1E−99, 1E−99);                    {depending on floating–point system}
    R := {approximate inverse of A, using e.g. Gaussian elimination};
    I := mrid (dim);
    z := ## (R*b);                      {compute R*b exactly and round it to interval}
    B := ## (I − R*A);                  {compute residue exactly, round it to interval}
    x := z; k := 1;                     {start iteration}
    repeat
      y := x + Eps;                     {inflate interval x slightly}
      x := z + B*y;                     {interval iteration}
    until (x < y) or (k > 10);          {x is subset of interior of y, or max is reached}
    okay := k <= 10;                    {if max reached, then result could not be verified}
  end;
```

Longreal and rational arithmetic can be implemented easily in PASCAL–SC using operator declarations, the module concept, and dynamic arrays. A longinteger module and a simple module for rational arithmetic are described in [4].

Randomly rounded operations which are needed for the perturbation / permutation method can be defined in PASCAL–SC by overloading the predefined floating–point operations with user–defined operators as follows:

```
module RandomRounding;
  function boolrand: boolean;
    begin boolrand := {...} end;
  operator + (a,b: real) plus: real;
    begin      if boolrand then plus := a +< b
               else plus := a +> b
    end;
{similarly −, *, / , comparisons, etc.}
end. {of module RandomRounding}
```

4. Conclusion

Ordinary floating–point arithmetic delivers only approximations of the solution. But it can be augmented with additional operations that simplify the verification of the computed results. This was demonstrated with PASCAL–SC above; the programming language FORTRAN–SC [3] possesses similar properties. Higher arithmetic operations and verification methods should be supported by hardware.

References

[1]   Alefeld, G., Herzberger, J.: Introduction to Interval Computations. Academic Press, 1983.

[2]   ANSI/IEEE Std. 754–1985, Binary Floating–Point Arithmetic, New York, Aug. 1985.

[3]   Bleher, J. H., Kulisch, U., Metzger, M., Rump, S. M., Ullrich, Ch., Walter, W.: FORTRAN–SC: A Study Of A FORTRAN Extension For Engineering / Scientific Computation With Access To ACRITH. Computing 39, Nov. 1987 (pp. 93–110).

[4]   Bohlender, G., Ullrich, Ch., Wolff v. Gudenberg, J., Rall, L. B.: PASCAL–SC: A Computer Language for Scientific Computation. Academic Press, Orlando, 1987.

[5]   Kaucher, E., Kulisch, U., Ullrich, Ch. (eds.): Computerarithmetic: Scientific Computation and Programming Languages. Teubner Verlag, Stuttgart, 1987.

[6]   Kulisch, U. (ed.): PASCAL–SC: A PASCAL Extension for Scientific Computation. Information Manual and Floppy Disks. Teubner Verlag, Stuttgart, 1987 (version for Atari ST) / Wiley–Teubner Series in Computer Science, 1987 (version for IBM PC).

[7]   Vignes, J., Alt, R.: An Efficient Stochastic Method for Round–Off Error Analysis. In: Miranker, Toupin (eds.): Accurate Scientific Computation. Springer Verlag, Lecture Notes CS 235, 183–205, 1986

Using Systems of Incomplete, Often Inconsistent, Models

Robert H. Adams[1]

SUMMARY

Is it possible to obtain major technological benefits from the limitations imposed by Goedel's Incompleteness Theorem? Can we configure our physical laws as filters stimulated by an immensely complex pseudo-random function in phase-space? Developing methods to partially define this function from experimental data as seen through the various filters will produce new and extremely useful models of our physical universe.

DISCUSSION

CLOCKS -- A COMPLETE NUMBER THEORY

A clock consists of an oscillator, a zero detector, and perhaps a counter. The oscillator is defined by a time-dependent differential equation. Designate a specific clock as our standard of time. Let the state of its oscillator be sampled using a system of synchronized test clocks. No matter how many test clocks are used, we cannot show that time is either continuous or discrete [1].

If a clock is being used as the standard of time, then time is being used to define time. Consequently, a complete number theory is formed. Not being able to prove that time is either continuous or discrete is a practical manifestation of Goedel's Incompleteness Theorem [3] [4].

SPECIAL THEORY OF RELATIVITY - A MODEL

The Physical Sciences are built on a rich fabric of models of reality created from our many languages (including mathematics). Goedel demonstrated that there is no consistent system which is complete enough to be self-defining or self-proving [3] [4]. Consequently, our searcn for those physical laws which define our universe will be perpetual with each major discovery resulting in a significantly improved model.

----

[1]   11123 Stagg St., Sun Valley, Ca. 91352, U.S.A.

Attempts to determine the absolute velocity of the earth in the "ether" by several investigators near the turn of the century resulted in the Special Theory of Relativity[2]. All of these methods involved a special interferometer and an estimate of the speed of light.

The experiments designed to measure the speed of light superimposed information on the light in order to measure its travel time over a specified distance. In 1849, Fizeau obtained a value of 3.15e8 meters/second using a shutter as a modulator. Foucault improved the technique with a rotating mirror modulator. Between 1878 and 1935, Michelson, Pease, and Pearson obtained a value of 2.997e8 using an improved rotating mirror system. These experiments did not measure the speed of light. They measured the transmission velocity of information superimposed upon light.

The most famous interferometer designed to measure the absolute velocity of the earth was developed by Michelson. This device was constructed in the form of a cross with one light path parallel and the other light path perpendicular to the direction of the earth's velocity. The transit times for the light along each of the paths are given by the following equations:

$$\text{Parallel Path:} \qquad T_1 = L/(C-V) + L/(C+V) \qquad \text{1)}$$

$$\text{Perpendicular Path:} \qquad T_2 = 2*L/\text{SQRT}(C^2-V^2) \qquad \text{2)}$$

Because of the non-linear relation between the two paths, a non-zero velocity, V, would have caused a change in the fringe patterns when the device was rotated 90°. No changes were observed.

The Special Theory of Relativity cannot be considered a "theorem." To qualify as a theorem, one must prove it to be true through the rigorous application of previously proven theorems and universally accepted axioms. "C", in equations 1) and 2) is the transmission speed for information superimposed upon light. Michelson, in his experiment to evaluate the absolute velocity of the earth, used an unmodulated light source. Consequently, there was insufficient rigor in the sequence of experiments to prove or even derive the Special Theory of Relativity as a theorem.

You say, "So what?" The Special Theory of Relativity has been experimentally validated by many investigators. The attempts at measuring the absolute velocity created serious doubts in Newtonian Mechanics and consequently, provided a fertile ground for the birth of Relativity. Newtonian mechanics was the best model of the behavior of

----
[2] The absolute value of the speed of light might suggest the presence of a complete number theory.

our physical universe during that period. Relativity must be considered another model. We must not assume that it will not be superseded by still another model.

THE CLOCK -- A BLUEPRINT FOR DISCOVERY

A model of a clock consistent with Goedel's Theorem was developed [1]. Using the fact that every consistent number theorem must have an undecidable proposition [3] [4], the clock was modeled as a two-pole resonant filter stimulated by random source. A random generator was found to be unacceptable as a source due to the fact that time as defined by the output of the filter was still being used to define the sequence of random numbers. A random function in phase-space with no time dependency produced an excellent clock.

We detect an orbiting "UFO" (Unidentified Flying Object). Furthermore, according to our standard clock, it appears at precisely the same time each day (a probability distribution of zero width) at each of the points of observation. Due to the stability of the orbit, we can conclude that there is either some form of system and/or intelligent life attempting to maintain the orbit. This object is equivalent to our standard clock. Consequently, we can learn very little about its operation or capabilities. Even comparing its stable trajectory with our knowledge of decaying orbits will give us, at most, one point in its defining function.

Now, let us suppose that the sightings form a finite random probability distribution with respect to the expected time of observation. Again, there is no evidence of orbital decay. These sightings along with our understanding of orbital mechanics can now give us many points in its defining functions.

A PROPOSAL

The engineering and scientific community has greatly benefited from the real number system [2]. We cannot prove either experimentally or theoretically that it exists. Despite this, we need it, accept its existence, and use it much of our technological activity.

We must accept Goedel s incompleteness theorem and treat our laws of physics as greatly simplified models of our physical universe. Furthermore, I propose that the physical universe is pseudo-random, deterministic, and sufficiently complex to always appear truly random (without any rule). We cannot prove or disprove these propositions. However we can greatly benefit from their acceptance.

To be consistent with Goedel's Incompleteness Theorem, we must configure our physical laws as filters through which we view a highly complex, pseudo-random, physical universe. Then we can extract some small portion of the pseudo-random function definition from the data collected at the "output" of these filters. A small portion of the definition of the pseudo-random function will create more questions than it would answer. Such questions would drive the production of new models, some of which will produce significant opportunities for technological advancement.

Finally, when one designs a instrumentation system, one must be very careful to select a robustly non-singular system of defining linear equations. Similarly, the availability of a collection of widely differing models relative to the same phenomenon can greatly assist the process of defining the underlying pseudo-random function. I believe that the greatest variety of models on any given phenomenon can be found in the international scientific organizations such as IM-ACS.

REFERENCE

[1] Adams, R.H. "Goedel's Incompleteness Theorem - A Practical Application." Proceedings-IMACS 12'th World Congress, (1988).

[2] Feferman, S. The Number System-Foundations of Algebra And Analysis. Addison-Wesley, 1964.

[3] Goedel, K. On Formally Undecidable Propositions. New York; Basic Books, 1962

[4] Hofstadter, D.R. Goedel, Escher, Bach. Vintage Books, 1980.

Top-Down Modelling of Complex Systems
by means of Word Bond Graphs

W. Borutzky[*]

Abstract

Development of dynamic models for complex systems often does not exploit the hierarchical structure of the system design. Rather, equations describing physical effects of interest are set up, manipulated, and combined at an early stage of the modeling process. In contrary, by applying a structured bond graph approach to an electrohydraulic servo valve as an example of a complex multi energy domain system it is shown how word bond graphs, known since bond graphs were invented by Paynter but seldom used, account for the hierarchical structure of a system design in a natural way and allow for systematic top-down development of a consistent simulation model on a *graphical* basis. Formulation of dynamic system equations is the last step in this modeling procedure and can be easily carried out by inspection of the *final* bond graph. By this way a state space model was prepared for the servo valve, coded as a FORTRAN subroutine of a general ODEs solver, and the system simulated on a workstation.

1   Introduction

For multi energy domain systems bond graphs [3] have proven to be a powerful modeling tool. As an excellent example of such a system a two stage electrohydraulic servo valve is modeled by starting from a first, coarse decomposition into subsystems and a representation of power flows between them. Modeling power flows in the first instance accounts for the physical system structure. However, by deciding on cause and effect at each multiport of the resolved graph results in a causally augmented bond graph, which in addition to the physical structure reflects the computational system structure.

By inspection of this *final* bond graph dynamic system equations can be easily formulated in a systematic manner as an ordered set of assignment statements. The latter may be input in a general purpose continuous system simulation program like ACSL by almost no additional effort, or simply coded e.g. as a FORTRAN subroutine of a general purpose package for solving ordinary differential equations (ODEs) like e.g. LSODE [2]. Although some years ago CAMP [1], a preprocessor for automatic equations generation from *flat* bond graph descriptions was developed, a methodology for easy and systematic system equations formulation in the format of assignment statements is of growing importance, since more and more designers will have workstations of ever increasing computing power at their own disposal but perhaps not sophisticated software packages for bond graph modeling, respectively simulation.

2   The system

During the last decades the dynamic behavior of servo valves has been investigated by many workers, whereby the operation of the servo valve often was treated linear and modeled by a low order transfer function neglecting several inherent nonlinear effects. This is sufficient if the global dynamic response of an entire feedback control system is of interest, if the servomechanism is lightly loaded and the amplitude of the forcing signal is small. In such cases the frequency response characteristic of the load is usually much inferior to those of the valve. However, in applications, in which the frequency response of the load is comparable to those of the valve high frequency lags of the latter become important. Thus, the valve has to be analyzed in more detail accounting for inherent nonlinearities like, for instance flow reaction forces on the valve spool. To this end some good work has been carried out in the past [4], [5]. It is interesting to note however, that the modeling process in most cases is flat and strongly equations oriented. That is, the hierarchical structure of a servo valve design is not exploited. If the bond graph approach is not used equations are set up directly, manipulated, and combined in order to model physical effects. Obviously, such an approach is not very systematic and not very flexible. Refinements or substitution of a component model cannot be done without repeating a lot of modeling steps.

---

[*]The author is with Gesellschaft für Mathematik und Datenverarbeitung (GMD) P. O. Box 1240, 5205 St. Augustin, FRG

3 Top-Down Modeling

Complex systems normally are designed hierarchically. In a top-down approach a system to be designed first of all is decomposed into subsystems, which in turn are further partitioned into functional building blocks. Designing each component and its interconnection to others on the same level results in an implementation of the entire system. Now, the basic idea is to follow this way in developing a dynamic model for system simulation by top-down stepwise refinement of word bond graphs according to the hierarchical structure of the system design. Starting from a section view, omitted in this paper due to the lack of space, first of all, the two stage servo valve to be modeled, may be decomposed into the following four functional building blocks:

- an electromechanical transducer (torque-motor),

- a mechanical-hydraulic amplifier,

- a hydraulic power stage (four-way valve),

- and a mechanical feedback.

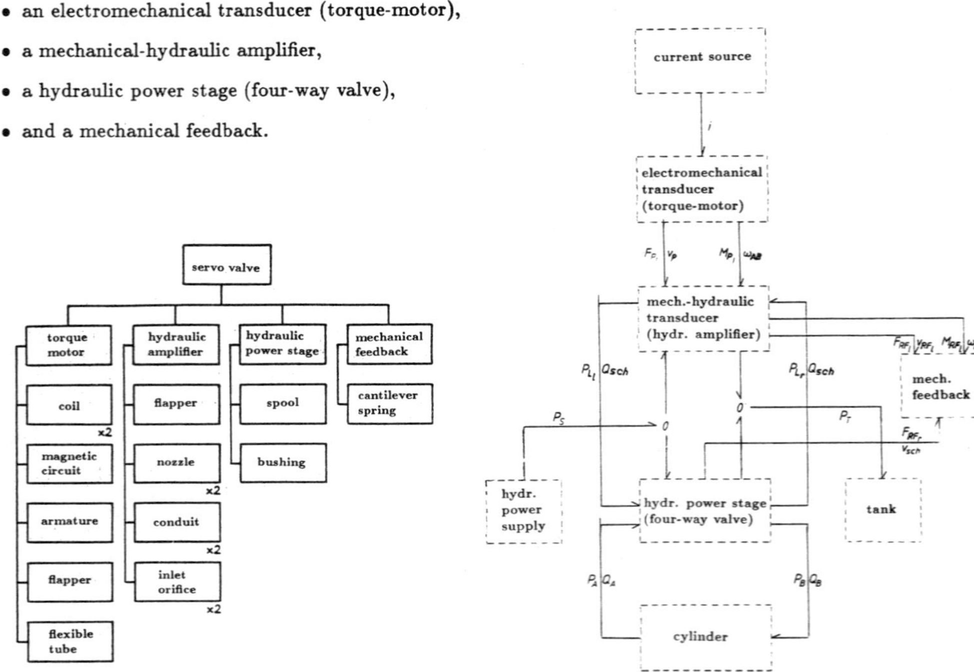

Fig. 1: Structure tree

Fig. 2: Initial word bond graph

Each building block can be further subdivided as shown by the structure tree of Fig. 1. For instance, the torque-motor consists of two coils, a magnetic circuit (containing all parts of the torque-motor that conduct the magnetic flux), the armature as a mechanical (rigid) body, the upper part of the flapper and a flexible tube, on which the armature is mounted. The above structure tree serves as a guide for systematic, stepwise development of a bond graph model. In order to derive a dynamic model for the valve we have to describe the interconnections of components at each level of refinement. This can be done by using word bond graphs to model power flows between components. The power exchange between top level components is shown by Fig. 2, in which bonds may be associated with real physical connections like hydraulic conduits, mechanical links, (flapper - feedback cantilever spring - spool) or an electrical wire (current source - torque motor).

Once the overall physical system structure is fixed, we can concentrate on the top level components themselves by setting up a word bond graph for each one, representing power flows between their subcomponents listed in the structure tree. As an example, Fig. 3 shows a word bond graph for the flapper-nozzle preamplifier. A major advantage of the top-down modeling approach is that bond graph models for the subcomponents e.g. the conduits may be modified or replaced as needed without effecting the physical structure of the upper level component.

Finally all lowest level components, e.g. the conduits of the hydraulic amplifier have to be resolved into bond graphs containing only primitive elements. It must be emphasized, that this step is not governed by writing continuity equations or force balances as in other modeling approaches. Rather, by assigning bond graph elements to physical effects that have been decided to be of importance for global system dynamics, a graph is set up in a formalized manner following a standard procedure. Furthermore, if the *type* of the constitutive law is known for each element, signal flow information may be added by simple physical reasoning on cause and effect at each multiport. (E.g. normally a force acting on a rigid body is considered to be the cause for motion.) Neglecting effects of minor importance led to causal conflicts in the bond graphs for the torque-motor and the hydraulic power stage. In order to achieve at a final mathematical model with explicit differential equations, these conflicts were resolved by simplifying the primary bond graphs. As a result, the final global causal bond graph for the servo valve is depicted in Fig. 4 showing a strong similarity to the initial word bond graph in Fig. 2. As can be seen from the subgraph of the torque-motor, the four air gaps between pole pieces and armature are modeled by a single two port capacitor $C_{L_{ges}}$ with a resulting magnetic force F_A on the armature. Also, neglecting oil compressibility in the hydraulic power stage leads to algebraic dependencies between the four volume flows of the resistor bridge. The conflict was removed by substituting the bridge by a single equivalent resistor R_{ers}. Since the piston of the high precision metering cylinder has a small inertia and friction is very low the load pressure $p_A - p_B$ is approximately zero. Hence, the cylinder may be omitted and ports A and B shortened. In fact, the bond graph of the hydraulic power stage simply accounts for the total shunt volume flow Q_0 from hydraulic power supply to the tank, controlled by the displacement y of the spool, and the force balance at the spool, including the sum of flow reaction forces $F_{Ström}$.

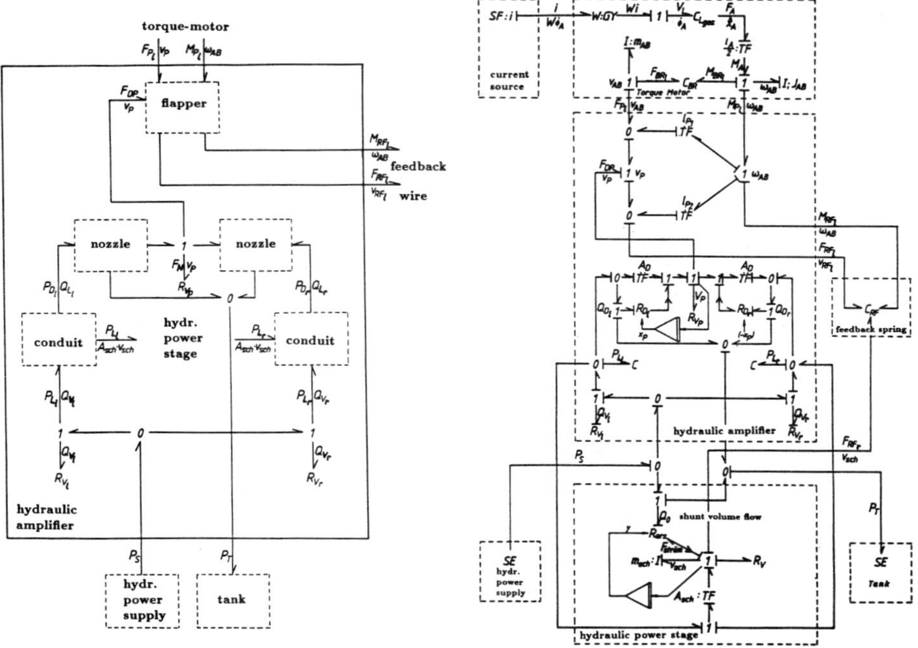

Fig. 3: Word bond graph of the hydr. amplifier   Fig. 4: Final system bond graph

4 Formulation of equations

Except one causal violation in the torque-motor model the final bond graph contains no causal conflicts and no algebraic loops. The one-junction exhibiting the causal conflict represents the undefined magnetic flux in the armature, which can be easily eliminated from equations. The flux is undetermined, since all magnetic losses have been neglected. All storage elements have integral causality. Hence, we can conclude before formulating any equation, that the dynamic behavior of the servo valve is represented by a (nonlinear) state space model of 13 explicit differential equations. In order to derive an equivalent ordered set of assignment statements by inspection of the bond graph, first we choose for state variables the output variables of the storage elements and the kinematic displacement of flapper and spool. Now, for setting up system equations the following straightforward steps of a more general procedure have to be carried out:

1. Specify the time history for all (independent) sources.

2. Since their are no algebraic dependencies between dissipators their output variable may be written as a function of input and state variables. This has to be done for the resistors of each component and can be carried out in any sequence.

3. Expressing the time derivative of the output variable of each storage element by input and state variables results in a set of simultaneous explicit differential equations, in which the outputs of dissipators are auxiliary variables.

The resulting assignment statements were directly coded as a FORTRAN subroutine of an ODEs solver. The nonlinear time domain model was analyzed on a workstation by varying the frequency of the electrical input signal. For comparison the frequency response of the servo valve was measured by using a two channel analyzer (Solatron - Schlumberger, type 1170). The Bode plot of Fig. 5 below shows that there is a good agreement between performance measurements of three servo valves of the same type and simulation results.

5 Conclusion

The paper briefly presents a top-down modeling approach based on word bond graphs. The latter conveniently supports systematic and flexible development of dynamic models for simulation of complex inherent nonlinear multi energy domain systems. By modeling an electrohydraulic servo valve it is shown that the method is even applicable to a single complex device. The system to be modeled need not to be e.g. an entire hydraulic plant. Advantages are the possibility of exploiting the hierarchical structure of a system design and the flexibility in modifying or substituting subcomponent models without affecting word bond graphs of calling components. Furthermore, the approach concentrates on a *graphical* representation of power flows. As in flat bond graph modeling equations formulation is postponed to the end of the modeling process. It is pointed out that even for a composed system assignment statements may be easily set up manually in a systematic manner by inspection of the final global bond graph.

6 Acknowledgement

The author is indebted to his former research adviser Prof. Dr. A. Richter, Technical University of Braunschweig, for encouraging bond graph modeling of hydraulic systems. Discussions with Dr. Müller, Feinmechanische Werke Mainz, on the servo valve are also gratefully acknowledged.

References

[1] *Granda, J.:* Computer-Aided Modeling Program (CAMP): A Bond Graph Preprocessor, Doctoral Thesis,
Univ. of California Davis, Department of Mechanical Engineering, Dec. 1982.

[2] *Hindmarsh, A.C.:* ODEPACK, a systemized collection of ODE-solvers,
Lawrence Livermoore National Laboratories, Aug. 1982.

[3] *Karnopp, D.; Rosenberg, R.:* System Dynamics: A Unified Approach,
Wiley & Sons, New York, 1975.

[4] *Nikiforuk, P.W. et. al:* Detailed Analysis of a Two-Stage Four-Way Electrohydraulic Flow-Control Valve
Journal Mechanical Engineering Science, Vol. 11, No. 2, 1969.

[5] *Rabie, G.; Lebrun, M.:* Modélisation par les Graphs à Liens et Simulation d' une Servovalve Electrohydraulic à Deux-Etages
R.A.I.R.O. Automatique, Vol. 15.2, pp. 97-129, 1981.

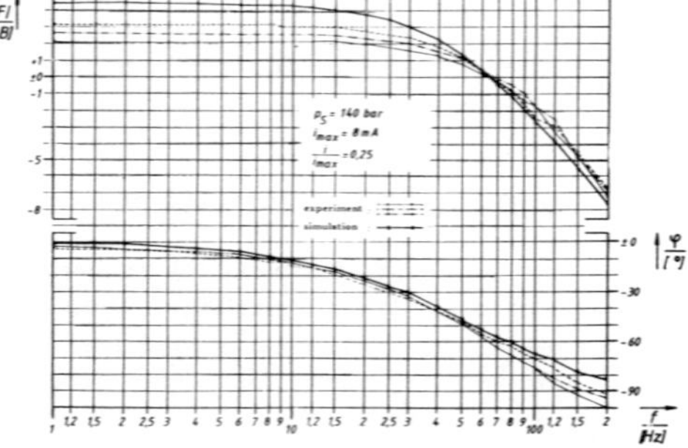

Fig. 5: Frequency reponse

An Approach of the Grey System Modelling and Simulation for Complex Systems

Yi Yunwen, Zhang Lu

Shenyang·Institute of Automation Academia Sinica
Shenyang 110003, P.R.China

ABSTRACT

There has been a long histroy of the System Dynamics developed by J.Forrester and its analysis of complex system behaviors. It is necessary to know the detailed system structure and a large amount statistical data to build the flow diagram when using this method. However, our quantitative analysis show that the integrated model can be seen as a combination of many abstracting grey system modeles which may be decomposed from complex system. In this way, the integrated model can be simulated on a computer. The approach of combining quantitative analysis with simulation has the following features,
    (1) A continuous differential model is established using grey generating numbers.
    (2) The integrated model can be run on a computer to simulate the system behavior using DYNA-MO language.
    In this paper, the approach of modelling the available abstracting grey dynamic systems and of simulating the integrated systems is described. The system analysis of experimental example show that grey system modelling is very useful in System Dynamics model.

1. INTRODUCTION

Among many methods for analysing social-economic system dynamic behavior the System Dynamics developed by J.Forrester and Grey System modelling theory advanced by J.Deng are significant. Analysing the dynamic variation of the social-economic system behavior is their common objective and the difference between them is that the System Dynamics is a computer simulation method while the Grey System modelling is an analytic method, both of them have some advantages and disadvantages.

In this paper, in order to describe the social-economic system behavior more presicely, a new approach which combines the two methods is proposed.

2. THE FEATURES OF SYSTEM DYNAMICS METHOD

The principle of System Dynamics shown that the quality of a complex system dynamic behavior depends upon internal structure and internal interactions of the system. The feedback loops correspond the linking of these relations when the system dynamic model is established . However, the causal relations within the system based on the analysis by different experts sometimes are quite defferent because of different views of experts. Even though the System Dynamics principle provides strict rules for building causal relations, the experts who analyses the same system structure usually obtain different results. Furthermore, there is no any good method for the consistency check of the real model and structure model so far.

3. THE DEVELOPMENT OF GREY SYSTEM THEORY

The concepts of Grey System are presented by professior J.Deng in 1982.The Grey System theory shows that the system for which the internal structure is known is deterministic system and the system for which the internal structure is unknown is non-deterministic system. So the systems between them are grey systems. A lot of non-engineering systems, for example, social, agriculture, ecology and environment systems are belong to study field of Grey System. These systems which lack physical models are called grey ones. That means the parameters of the system model are incomplete, or internal and external causal relations are indefinite. For instance, we have investigated a problem concerning the region grain product forecasting. For such a system, the relations between the input and the output are very difficult to be described precisely .The effect of soil fertile capability, the stochastic changing of the weathers, the farming labor expended on grain product etc. are all indefinite. the changing area of farming land are also imprecised. As such problem, the grain production value are considered as the sum of multi-factor data and also a result of multi-factor interaction .Therefor, it is not necessary to give the detailed expression which precisely describes the mathematical function in very difficulty.And we summarized the multi-elements of the system into a abstracted single input and a real output system. For instance, a grain production system model that is described by natural and social factors can be considered as an energy system, in which the input variable is the summary of every factor representing the abstract cause and the output variable is result, that is a observable time series data, thus a complex system is simplified. The method of establishing differential equation on the base of descrete data is proposed by reference [1].

4. A COMBINATION METHOD OF GREY SYSTEM MODELLING AND SIMULATION

A complex system may be divided into some connected sub-systems which are represented with abstract model blocks. The parameters of the model blocks are defined from their relations between inputs and outputs, then each model block is connected to a integrated net model and expressed using the low order differential equation. Obviously, the equation order of the output variable are usually more than one. Thereby, solving the equation of integrated net model is difficult and the System Dynamics simulation provided an excellent method to solve such a problem, DYNAMO language is an effective tool.

(1) One order forecasting model.

The output of an abstract system may be considered as the interactive result of the summarized multi-input factors. Suppose the original output data can be arranged as time series as

$$\{x(ti)\}=\{x(t1),\ x(t2),\ \ldots,\ x(tn)\}$$

The new single accumulated series are obtained by

$$\{\ x(ti)\ \}=\{\ x(t1),\ x(t2),\ \ldots,\ x(tn)\ \}$$

Each number of this series can be given by the formula

$$x^{(1)}(ti)\ \ =\sum_{j=1}^{i}\ x(tj)$$

Two interesting features of the method can be found. The first is that the fluctuation of the new data is diminished obviously after the processing as a result of the accumulation. Furthermore, the generated new series can be conceived as an exponential function. The Grey System model is a simple differential equation,

$$\frac{dx^{(1)}(t)}{dt} + x^{(1)}(t)=u \tag{1}$$

where the a and u are unknown parameters. To estimate the a and u, we define a new vector

$$\hat{\varphi} = [a, u] \tag{·2}$$

As suggested by [1], the $\hat{\varphi}$ can be estimated by

$$\hat{\varphi} =[\ B^T\ B^{-1}\ ]\ B^T\ Yn \tag{3}$$

where

$$B = \begin{bmatrix} -0.5\,[x^{(1)}(t_2)+x^{(1)}(t_1)] & 1 \\ -0.5\,[x^{(1)}(t_3)+x^{(1)}(t_2)] & 1 \\ \cdot & \cdot \\ \cdot & \cdot \\ -0.5\,[x^{(1)}(t_n)+x^{(1)}(T_{n-1})] & 1 \end{bmatrix}$$

which is a (n-1)×2 matrix and

$$Yn = [\ x(t2)\quad x(t3)\quad \ldots \quad x(tn)\ ]$$

is a vector.

The model can be used to combine with TABLE function in System Dynamics as a external variable. From equation (1) we obtain forecasting series value

$$\{\ \hat{x}(ti)\ \} = \{\ \hat{x}(t1),\ \hat{x}(t2),\ \ldots,\ \hat{x}(tn),\ \ldots,\ \hat{x}(t\ n+p)\ \}$$

To obtain the original forcasting series $\{\ x(ti)\ |\ i>n\}$, we conduct the following calculation

$$\hat{x}(t_i) = x^{(1)}(t_i) - x^{(1)}(t_{i-1}) \tag{4}$$

We can easily substitute the above forecasting values into TABLE function in DYNAMO.

(2) The transfering relation model of auxiliary variables.

Suppose the variable x1 is related with x2, x3, ..., xj variables and x1 is defined as a auxiliary. Let model being expressed as

$$x_1(ti) = \sum_{k=2}^{j} b_k\ x_k(ti) + b_1 \tag{5}$$

where the b1, b2, ..., bj are unknown parameters. The method of calculating the value of b1, b2, ..., bj can be found in [1]. Auxiliary variable equational expression in DYNAMO is shown by

$$A \qquad X1.K = B1 + B2 \ast X2.K + \ldots + Bj \ast Xj.K \tag{6}$$

(3) The level's dynamic model

The levle's dynamic model may be used to estimate the relations between the LEVEL variables and RATE variables.

Let variable x1 is related with x2, x3, ..., xj and x1 is defined as a level variable, thus

$$\frac{dx_1^{(1)}(t)}{dt} + a\ x_1^{(1)}(t) = \sum_{i=2}^{j} b_i\ x_1^{(1)}(t) \tag{7}$$

is a simple differential equation about LEVEL, in which unknown parameters a and bi are estimated by reference [1].

Let equation (7) being multiplied by dt and let dt=DT, and arrange the equation (7) again. we have

$$d\ x_1^{(1)}(t) = DT\ [-a\ x_1^{(1)}(t) + \sum_{i=2}^{j} b_i\ x_i^{(1)}(t)\ ] \tag{8}$$

where DT is the simulation time increment.

Supposing

$$d\ x_1^{(1)}(t) = x_1^{(1)}(t) - x_1^{(1)}(t-DT)$$

thus

$$x_1^{(1)}(t) = x_1^{(1)}(t-DT)+DT\,[-a\,x_1^{(1)}(t) + \sum_{i=2}^{j} b_i\ x_i^{(1)}(t)] \qquad (9)$$

Variable $x_1^{(1)}(t)$ is an accumulated value of $x_1^{(1)}(t)$ and its fluctuation is not obvious within one descrete time. So we approximate $x_1^{(1)}(t)$ to $x_1^{(1)}(t-DT)$.

Hence

$$x_1^{(1)}(t) = x_1^{(1)}(t-DT) + DT\ [-a\,x_1^{(1)}(t-DT) + \sum_{i=2}^{j} b_i\,x_i^{(1)}(t)\ ] \qquad (10)$$

Let $\qquad a\,x_1^{(1)}(t-DT)=R1(t)$

$$b_i\,x_1^{(1)}(t) = Ri(t) \qquad i=2,\ 3,\ \ldots,\ j$$

The equation (10) may be expressed by

$$x1^{(1)}(t) = x1^{(1)}(t-DT) + DT\ [\ -R1(t) +R2(t) + \ldots +Rj(t)\ ] \qquad (11)$$

We write equation (11) in DYNAMO as

```
L        X1⁽¹⁾.K = X1⁽¹⁾.J + DT [-R1.JK + R2.JK + ... +Rj.JK ]
R        R1.KL = A × X1⁽¹⁾.K
         .              .
         .              .
R        Rj.KL = Bj × X1⁽¹⁾.k
```

The effectiveness of the above three grey system models are illustrated by tested example for a long development planning of the land resouses in Liaoning province.

5. CONCLUTION

We have been atempted to use the analytical methods of the Grey System dividing a complex system and the typical dynamic modelling methods are proposed. Then we connect some dynamic models into a integrated net model and simulate the model on computer in DYNAMO. The tested results shown that analysis for a complex system behavior becomes more easily and precisely

Annotation: This reseach was under the special grant of Academia Sinica.

6. REFERENCE

[1]   Deng Ju-Long, Grey Systems, social and economic. National Defence Industrial Publishing, China, (1985).

[2]   Weng Qi-Fan, Huang Hai-Zhou, "Comparative reseach of the System Dynamics and the Grey System Theory", Symposium on chinese System Dynamics Asociation, (1986).

[3]   Yi-Yun-Wen, Li Hong-Xing, " The grey dynamic model for grain production and long-term forecasting", The 2nd Japan-Sino Sappro-Shenyang International Conference on Computer Application, (1986).

[4]   Zhang Lu, Yi Yun-Wen, "The application of Grey System modelling to structure analysis in System Dynamics", Being presented in the meeting of chinese Future Science Association,(1987).

Generic Modelling in Sonches

V.Wenzel , E.Matthaus, M.Flechsig[1]

ABSTRACT: The new relation between programming and modelling is discussed. Some important developments on the field of modelling and simulation support systems and their connection with AI are pointed out. Modelling of generic objects and processes is an efficient tool for complexity reduction of models. Generic modelling in SONCHES is realized by subcompartmentation.

KEY WORDS: Modelling language; simulation support system; object-orientation; branch-orientation; reusability; generic names, types, objects, subcompartments.

INTRODUCTION

Concerning the desirable appearance of a programme and, accordingly, a programming language, change of paradigm has been observed in last decade /1/. The first step towards a programming methodology – the structured /2/ – was followed by a second – the object-oriented programming /3/ preserving the benefits of the first. The structuring of programmes now is guided by the structure of problems itself, i.e. of the concrete or abstract objects together with their properties and relationships.

Advantages are numerous as it was pointed out e.g. in /1/. After the object-oriented approach has been spread, the question about relations between 'programming' and 'modelling' arises. In the following we want to elucidate the corresponding development for an important category of simulation tools and the mutual influences of both.

Depicture of generic objects and properties as a key notion for arranging reusability will be of particular interest.

SOME RECENT DEVELOPMENTS CONCERNING SIMULATION TOOLS

The idea of object-orientation was introduced when SIMULA offered its 'classes' for simulation purposes /4/. And mapping of objects plus behaviour is commonplace in past and contemporary simulation languages. From this point of view there is little difference between 'programming' and 'modelling' now. But from others the gap has become even larger. Indeed, the following hierarchy of languages corresponds to a procedure of successive abstraction with 'intelligent' translators between levels:

```
         natural, living language
                                       scientist
           language of a science/branch
                                       modeller
           simulation language
                                       translator
           higher programming language
                                       compiler
           assembler language
                                       assembler
           machine code
```

[1] *Central Institute of Cybernetics and Information Processes, Acad. of Sciences of GDR, 1086 Berlin, Kurstr.33*

Programming on the level of interacting objects economizes one step of this abstraction procedure - do to the ability of corresponding compiler, nowadays characterized as artificial intelligence (AI). In such way, a language of any level might move higher on the scale by incorporation of AI (theoretically, up to the ideal of a single compiler that understands English or German).

The level of simulation tools is related to this development at least as strong as the programming ones. Meanwhile it has been performed an analogous movement towards the real phenomena by creating branch-specific simulation systems /6,7/. The core of this development is the following idea:

The loss of arbitrariness by restriction to a special field of knowledge provides the freedom for further going support of users on the field of modeling and simulation via enrichment of support systems by AI.

In particular, as the list of properties of objects, of process performances, of structural patterns is restricted and of final content, a procedural implementation of the laws and rules of the supported scientific branch (ecology) could be realized.

On the other hand, this guarantees a more natural relation between users and simulation tools. To determine the appropriate tools the question is no longer: in which mathematical form the model can be represented ; but rather: what kind of problem on the base of which knowledge is to be solved. Problem solving, however, is quite more than preliminarily building a model in analogy to a programme. It implies also

- its debugging (verification) and validation;
- manifold changing of parameters,processes and structure;
- proving of hypotheses and conjectures;
- to enrich and, finally,reduce it successively within a procedure of incremental/decremental modelling;
- acquisition and maintenance of data and its procedural linkage with models;
- performance of sophisticated simulation experiments with high degree of automation; etc.

Consequently, modern (branch-oriented) simulation tools comprehend a conversational-maieutic modelling language together with a rich simulation environment and a result processing and (graphically) visualizing component within one consistent support system/7,8/.

REUSABILITY : GENERIC OBJECTS

The possibility to study more complex systems in more detail expresses most evidently the power of branch-oriented simulation systems. However, this is really true only if the corresponding models remain still readable, understandable and manageable. Features of modelling language that can help to reduce complexity concerning the model surface as well as all performances around the model - up to simulation itself - are of extraordinary importance.

The rationale of these features is to establish a high degree of reusability and different ways to organize it: An environment should be created, that allows definition of reusable objects and to bind them multiple and flexible into overall information flow necessary for solving the tasks.

This will be demonstrated by an example, which is general enough to be met on all language levels mentioned above, because it is characteristic of languages at all, namely, the generic names. A generic name (as e.g. the word 'beetle'- comprising all kinds of beetles) belongs to the level of natural language. On each subsequent level we meet another point of view and another notion to distinguish/comprise something generic, e.g.

stag-beetle (comprising all ontogenetic forms of it);

compartment of larvae (comprising larvae abundances of several areals);

generic type (comprising objects with certain abstract properties);

subscripted data of special type or Macro to be parameterized;

hardware-microprogramme.

In every case the generic name or type concerns a subset of objects with partly coinciding properties, varying with respect to one ore more others. Restricting now to simulation level we notice : prerequisite for efficient generic types is to distinguish object and its behaviour.

In SONCHES, objects are represented as compartments and (mathematically) realized as automata. They communicate with each other by inputs/outputs – may be through environmental factors, to be modelled as objects with less complicated behaviour.

A set of properties or attributs A_i $(i \in I)$, each varying within a range of states S_i, is assigned to objects, where S may be both discrete or continuous. $S = \Pi\, S_i$ $(i \in I)$ is the space of global states of the object.

The behaviour of an object is described by the overall rule R for state transition within S, where $R = (R_i)$, $(i \in I)$.

If two objects C,C' have a common hyperspace S^* of states :

$$S^* \subset S \cap S' \quad ; \quad S^* = \Pi\, S_j \ (j \in J \subset I)$$

and there exists K :

$$K \subset J \subset I \quad \text{with} \quad (R_k) = (R'_k), \ k \in K ,$$

then we say the index subset K of I defines a generic object $\mathbf{C}$, (to which belong at least C and C'). The elements of $\mathbf{C}$ are of equal behaviour with concern to the properties counted by help of K, but they are varying by those subscripted through $I \setminus K$.

REALIZATION OF GENERIC MODELLING IN **SONCHES**

A practical realization of generic modelling in simulation support systems must take into account the dichothomy of law and exception, of fix and flexible, of automated and optional. Essentially, this is realized and should be further realized in each component of the support system, in particular, for description of macro-simulation-experiments and the processing and presentation of corresponding results.

The most important way of generic modelling in **SONCHES** is that of subcompartmentation /7/. Objects with the same (ecological) function and properties (index set K) may be distinguished by age ($|I \setminus K|=1$). In this case subcompartments are the age classes of a population . Using a special sign (@) as a suffix of names of objects, processes like

     ZLARVAE@ : EGG@, TEMP[1]   * * MOD **AD**

describe the dynamical behaviour of a whole ensemble of objects. Creation and maintenance of age class structure is done by the system in connection with process type **AD** (ontogenetic advance) without further interventions of the user.

The distinction of structural equal objects according to space is of similar importance and methodologically analogous. Using another sign (&) the process

     ZLARVAE& : LARVAE&, TEMP&, RAIN&   * * MOD **MF**

will describe migration processes between all declared areals, where different living conditions are developing on the areals, influencing the migration rates computed by process type **MF** (migration/fluxes) , which is responsible for disribution in space.

Diversity, once established by process types **AD** and **MF** respectively, subsequently concerns all other process types too, e.g. abundance dependent mortality of larvae

     >LARVAE@& : LARVAE@&,...   * * MOD **AM**

where the superposition of both is used ($|I \setminus K| = 2$).

Determination of objects in space and time are the most important properties for playing the role of elements from $I \setminus K$ to produce generic objects. Therefore they are fixed and ready for use in **SONCHES**. Another special sign (%) serves for determination of optional generic objects of free interpretation by the user. Diversity according to this property is established or changed by processes of type **RC** (reconfiguration). Special structure can be introduced by the user through the mathematical realization of process **RC** depending on interpretation.

Literature:
/1/ *S.Cook: Languages and Object-Oriented Programming , Software*
   Engeneer. Journal , 1 (86) 2, S 73-80
/2/ *O.-J. Dahl, E.W. Dijkstra, C.A.R. Hoare: Structured Programming ,*
   Acad. Press 1972
/3/ *A. Goldberg, D. Robson : Smalltalk80 — The Language and its*
   Implementation , Addison-Wesley 1983
/4/ *G.M. Birtwistle et al.: 'Simula Begin' , Auerbach 1973*
/5/ *L.Steels : AI and Programming Languages , Inform. Process.86*
   H.-J.Kugler(ed.),Elsevier Science Publ. B.V.(North Holl.),IFIP 1986
/6/ *A. Knijnenburg, E. Matthaus, V. Wenzel: Concept and Usage of the*
   Interactive Simulation System SONCHES ,Ecolog. Model.26
   Elsevier North Holl. Publ. Amsterdam 1984
/7/ *V.Wenzel : Problemlosungen bei Entwurf und Realisierung eines*
   fachgebietsorientierten Simulationssystems (SONCHES)
   Dissertation A, Berlin, 1987
/8/ *M.Flechsig, E.Matthaus, V.Wenzel : Simulation Environment in*
   SONCHES, this Proceedings

[1]*Read: left depends on right and right and ...*
  math. realization of processes is left open here; see /7/

Statement and Tendencies of Models
for Complicated Technical Systems

Günter Hertel[1]

1. Statement

The paper deals with a new strategy of consideration of complicated systems. The main aim of the new strategy consists in overcoming the traditional manner of separate considerations of system efficiency and
system reliability. This step is very important because the increase in
system capacity and the application of redundant elements finally have
the same aim and results, resp. Please note we deal with systems which
can be characterized as complicated, performance-oriented or multi-
level systems. They can take different states, each of them having different system performance. If a failure occurs within the system then
the efficiency of the system is more or less reduced but it is not zero.
The great importance of those systems urgently demands the development
of mathematical models, which enables us to calculate the system efficiency by taking into account the reliability and maintenability features of the system elements. Please note the main feature of the evaluation of a complicated system can not be a reliability one but an efficiency feature. For this reason different mathematical models have been
developed based on the idea of integrating performance as well as reliability features in one model.

2. Mathematical models for complicated, multi-level systems

The mathematical models for multi-level systems with different efficiency in each of the levels have been developed in the following way
(see fig. 1 /1/):
1) Known reliability and maintenance models have been extended to performance considerations
2) Performance - Reliability - Model
3) Queueing/Reliability models
4) Classical queueing model with modified input and service
5) Classical queueing model with modified channel occupation time (MOT)

Advantages and disadvantages of the models will be described in the
table. There are many kinds of failures having different influences on
the inherent features of the system. Important is the existence

---

[1] Dr. sc. techn. Günter Hertel, Friedrich-List - University of Transport and Communications Dresden / GDR; Department Technical Traffic
Cybernetic

1-channel delay model. More theory is to be found in / 2 /.

2.4. Classical queueing model with modified input and service

The main idea is to consider the failure process as an additional arrival one with the highest breaking down and repair priority. In order to simplify the model the arrival and service processes are considered as "mixed" processes by "normal" demands and failures. The mixture can be achieved by "mixed" d.f. / 8 /. Such modified arrival interval and service times are characterized by their coefficients of variations. Comprehensive examples are shown in / 8 /.

2.5. Classical queueing model with modified occupation time (MOT)

The main idea is to consider the downtime due to a channel failure as a part of channel occupation time. GAVER was the first to develop the idea of a queueing model. Many authors applicated his idea to different kinds of failures /1/ and to limited repair capacity /2/. We calculate, beside the d.f. of MOT, the expected value $E(MOT)$ and the coefficient of variation $V(MOT)$, esp. Generally speaking, $V(MOT)$ is unequal 1 even if service, life and downtime are exponentially distributed. That's why we have found some good approximations of $GI/GI/c \& 1/O \& m \leq OO$ queueing models (review in /5/) which are useful for the analysis of systems with unreliable channels. Fig. 4 shows important relations between the quality feature EL_q and the traffic value ϱ_B normalized to the number c of channels and the maintenability of the channels repaired by different number of repairmen. The limitation of·the feature ϱ_B/c is simultaneously equal EL_D/c.

3. Tendencies

There are different models but no single model which unifies all the advantages. It seems that two models have a succesful future: PRM and MOT. The PRM has great advantage in application for any kind of system structure with monotonicity behaviour and independence of the system elements. The last limitation can occur due to cold-standby redundancy or limited repair capacity. Our latest researches seem to show that units of depending elements can be replaced by equivalent independant elements. In such a way also storage elements are to be taken into account. Nevertheless, the step to calculate the system performance features due to the system load has not been achieved yet by PRM.
The method MOT is quit universal. The d.f. of the first two moments or MOT are calculable for several kinds of failures and under the consideration of limited repair capacity. In any queueing model the service time will be replaced by MOT. In such a way all interesting system features (ISP and load-depending) are calculable. The single condition is only a method for this certain queueing model with nonexponential occupation time /5/. The simultaneous consideration of reliability and performance has well advanced but the important warehousing models are

anv limitations of the distribution fuction (d.f.) of the primary features (interarrival. service, life repair time). Dependencies between the elements exist if there are any cold-stanby redundancies or a limited repair capacity, for instance. Inherent performance features of the system are independent of the input (arrival demands). The customers demand a certain inherent system performance (ISP) and realize the system efficiency (like realized throughput) which can not be higher than the corresponding inherent system feature without the consideration of the input / 4 /.

2.1. Reliability model extended to performance consideration

The main idea is to create a multi-level model in which the different states are caused by the reliability, structure and capacity of the elements. That's why the system does not fail if an element fails but the system efficiency is less than the maximum. A trivial example is shown in fig. 2. If each of the system elements has the same performance l_D, and P_i (i=0,...,5) is the probability that i elements have failed then the d.f. of ISP $P(L_D{<}l_{D,x})$ is calculable (where L_D is the random variable of ISP and $l_{D,x}$ is the ISP of the demanded level x):
$P(L_D < 100\%) = 1 - P_0 - P_1$ or $P(L_D{<}75\%) = 1 - P_0 - P_1 - P_2$ etc.
A comprehensive example is shown in /3/.

2.2. Performance – Reliability – Model (PRM)

The application of Markov-multi-level-models to large-scale systems is too complicated, generally, because the number of different states increases quickly. That's why a model has been developed allowing the calculation of ISP for complicated structure like network or hierarchical one. A flowing model is connected by a multi-level monotonous reliability model (BOOLE). Structural relations between the elements are to be discribed as edges or nodes in a network. The different performances of the system elements are interpreted as flowing capacities. After the definition of the system failure the calculation of its probability is possible, on principle, but algorithms are not trivial (see bibliography /6/). The developed Reliability Branching Algorithm /9/ is relatively simple as well as quick. With a PASCAL-programme the d.f. of the ISP and secondary features (like EL_D) are calculable.

2.3 Queueing/Reliability models

The main idea is the connection of both a reliability model and a queueing one in a single model. The monography /7/ gives a comprehensive review of results and methodes of the construction of 1-channel delay models with unreliable channel. In the rule, the models are two-dimensional Markov-graphs with closed solutions of interesting queueing features depending on the input and service of the demands and on the reliability and maintenability of the channel. We have extended the method to delay-loss-systems and multi-channel-systems. Fig. 3 shows a

waiting for their integration! Of course, all the probability models
are only worth for a planning, no operating task. For this problem many new wishes are arising.

Table: Characteristics of models respecting common considerations of performance and reliability

	model				
	2.1.	2.2	2.3.	2.4.	2.5.
demanded system structure	simple	simple...difficult	simple	simple	anything
kinds of failure are taken into account	yes	no	yes	no	yes
algorithm	simple	difficult	simple...difficult	easy	simple...difficult
limitation to certain d.f. of primary features	expon., partly	no	exponen., completly	no	no
dependencies of elements are calculable	yes	no	yes	no	yes
input-depending system performance features are calculable	only approx.	only approx.	yes	yes	yes

4. References

/1/ Bär,M., K. Fischer and G. Hertel: Leistungsfähigkeit-Qualität-Zuverlässigkeit. transpress-Verlag, Berlin 1988

/2/ Bär,M., K. Fischer and G. Hertel: A Method of Calculation for Loss-Systems susceptible to Failures with any Number of Repair Channels and its Application to Delay Systems. Mathem. Research, Akademie-Verlag, Berlin 1985, Vol. 27, pp. 352-356

/3/ Fischer, K. and G. Hertel: Influence of Reliability, Inspection and Maintenance on Quality and Efficiency. EOQC QUALITY 30(1986)2,10-14

/4/ Grün, M. and G. Hertel: Qualität, Zuverlässigkeit und Leistung diskreter Systeme. I: Begriffe und Kenngrößen. II: Modelle und Methoden zur Ermittlung der Kenngrößen. Standardisierung und Qualität, Berlin 33(1987)7, 194-197 und 33(1987)8,228-231

/5/ Hertel, G.: Verteilungen von Warteschlangenlänge und Wartezeit in GI/GI/s/oo - Bedienungsmodellen mit störanfälligen Kanälen. Intern. Kongreß über Anwendungen der Mathematik in den Ingenieurwissensch., Weimar/GDR 1987.Berichte Heft 4, pp. 54-57

/6/ Hwang, G.L., T.A. Tillmann and S.H. Lee: System-Reliability Evaluation Techniques for complex/large Systems - a Review. IEEE Trans. on Rel. R-30(1981) Dec.

/7/ Kistner, Kl.-P.: Betriebsstörungen und Warteschlangen. Westdeutscher Verlag, Opladen 1974.

/8/ Kluge, P.-D. and W. Runge: Zufallsabhängige Fertigungsprozesse. Verlag Die Wirtschaft, Berlin 1984.

/9/ Rüger, W.: Leistungs- und Zuverlässigkeitsanalyse von Transportnetzen. Hochschule für Verkehrswesen "Friedrich List" Dresden/GDR. Dissertation 1985.

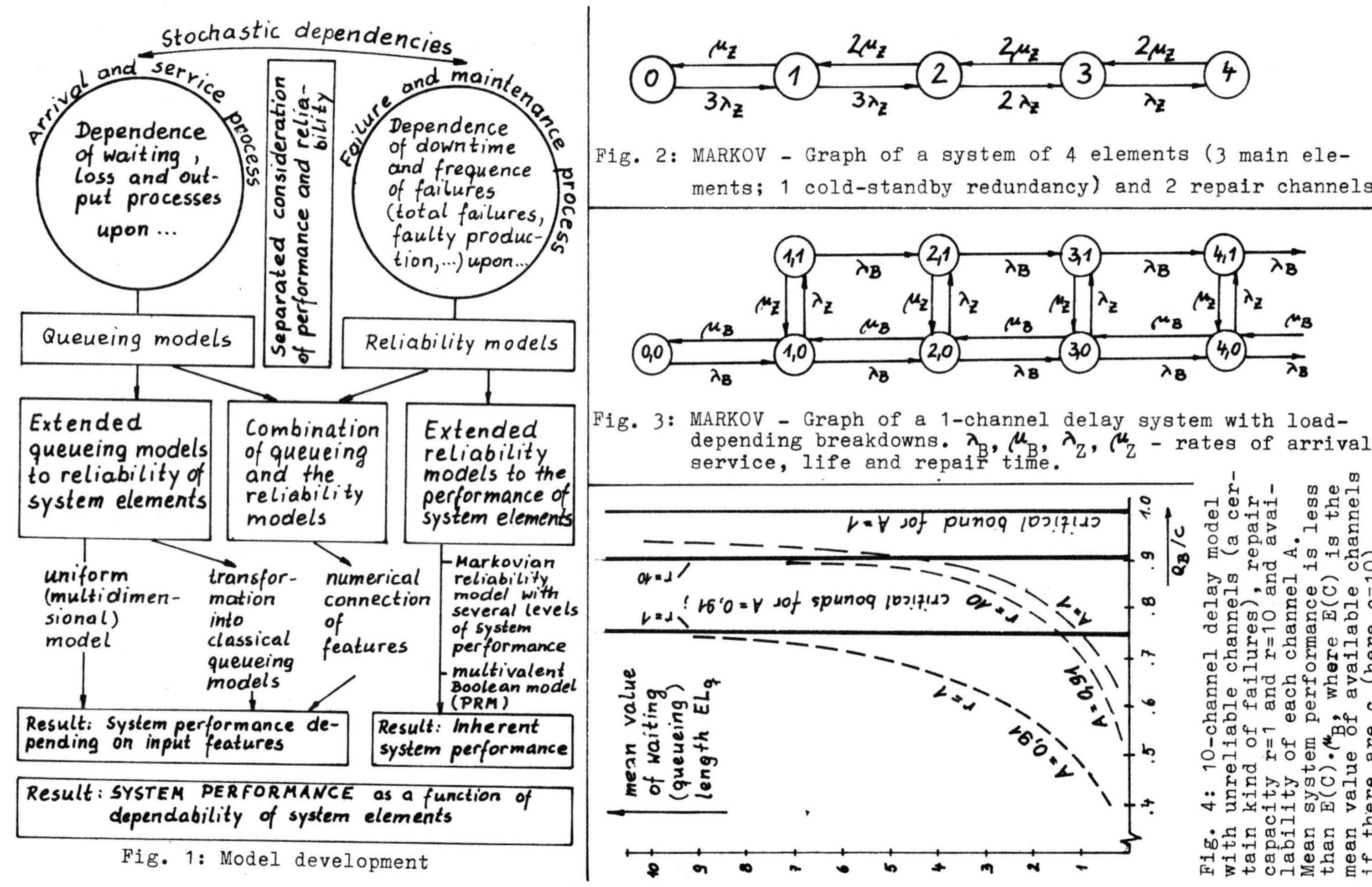

Fig. 1: Model development

Fig. 2: MARKOV – Graph of a system of 4 elements (3 main elements; 1 cold-standby redundancy) and 2 repair channels

Fig. 3: MARKOV – Graph of a 1-channel delay system with load-depending breakdowns. λ_B, μ_B, λ_Z, μ_Z – rates of arrival, service, life and repair time.

Fig. 4: 10-channel delay model with unreliable channels (a certain kind of failures), repair capacity r=1 and r=10 and availability of each channel A. Mean system performance is less than $E(C) \cdot \mu_B$, where $E(C)$ is the mean value of available channels if there are c (here c=10).

A Mathematical Model for Description of Random and Indefinite Factors from Unified Positions

Lebedev A.A. [1]

For needs of the operations research and of the analysis of complex systems we have a success (by weaking some axioms of the probability theory) to construct a theory which allows to describe random and indefinite factors, and also their combinations, from inified positions.
In the basis of the theory there are the axioms of the sublinear expectation which is an analogue of the probabilistic expectation.

Definition 1.   Let be given a measurable space $(\Omega, \mathcal{F})$, which can be interpretted as a space of elementary events (not necessary random). Let $\overline{E}_+(\Omega, \mathcal{F})$ be the set of non-negative measurable functions $\varphi: \Omega \to [0, \infty]$. Then a sublinear expectation (SE) is a functional on $\overline{E}_+(\Omega, \mathcal{F})$ satisfying the following properties:

1) $N(\chi_\Omega) = 1$ where χ_Ω is one identically;
2) $N(\alpha\varphi) = \alpha N(\varphi)$ if $\alpha \geq 0$;
3) $N(\varphi_1 + \varphi_2) \leq N(\varphi_1) + N(\varphi_2)$;
4) $N(\varphi_1) \leq N(\varphi_2)$ if $\varphi_1(\omega) \leq \varphi_2(\omega) \ \forall \omega \in \Omega$;
5) $N(\varphi_\kappa) \uparrow N(\varphi)$ if $\varphi_\kappa \uparrow \varphi$;
6) there exists a σ-finite measure $\overset{*}{\mu}$ for which $\overset{*}{\mu}(B) = 0$ implies $N(\chi_B) = 0 \ \forall B \in \mathcal{F}$ where χ_B is the characteristic function of the set B.

In problems of the operations research SE can be used to establish the preference relation between strategies quite so as the probabilistic expectation is used in the Bayesian theory of statistical decisions. Consider examples of SE.

Let be given a probability space $(\Omega, \mathcal{F}, P)$. The expectation

$$N(\varphi) = \int \varphi \, dP$$

is a sublinear expectation. The expectation is used usually if is random and its distribution is known.

Let be given a measurable space with a σ-finite measure $(\Omega, \mathcal{F}, \mu)$. The vrai maximum

$$N(\varphi) = \underset{\mu}{vrai\,max}\ \varphi(\omega) \triangleq sup\left\{ \alpha : \mu\left\{\omega : \varphi(\omega) > \alpha\right\} > 0\right\}$$

gives another example of SE. The vrai maximum is convenient to establish the preference relation between strategies when ω is an indefinite factor with a known set of possible values.

Let be given a dominated statistical structure $(\Omega, \mathcal{F}, \{P_\theta : \theta \in \Theta\})$. The functional

$$N(\varphi) = sup\left\{ \int \varphi \, dP_\theta : \theta \in \Theta\right\}$$

1) Lebedev A.A. Moscow Aviation Institute, USSR Moscow Volokolamskoye Highway, 4.

gives the third example of SE which is used when ω is a random factor with an indefinite destribution.

We can adduce an example of description of an indefinite factor with a random set of values and other examples with more complicated combinations of randomness and indefiniteness. In conclusion we adduce one exotical example.

Let be given a measurable space with a σ-finite measure $(\Omega, \mathcal{F}, \mu)$ and a measurable non-negative function $\lambda(\omega)$ for which $\mathrm{vraimax}\{\lambda(\omega)\} = 1$. The functional N defined on $\overline{\mathcal{E}}_+(\Omega, \mathcal{F})$ by

$$N(\varphi) = \mathrm{vrai\,max}_{\mu}\{\lambda(\omega)\cdot\varphi(\omega)\}$$

represents SE. If $\lambda(\omega)$ is a characteristic function of some set this functional coincides with the vrai maximum. In the general case $\lambda(\omega)$ can be interpretted as the function of belonging to some fuzzy set of possible values of ω .

The following theorem can be obtained as a cordlary from the Hahn-Banach theorem and the Daniell theorem.

Theorem 2. For every sublinear expectation N there exists a finite measure $\overset{o}{\mu}$ for which $\overset{o}{\mu}(B) = 0 \Longleftrightarrow N(\chi_B) = 0 \; \forall B \in \mathcal{F}$ and there exists a set A consisting of positive measures, dominated by $\overset{o}{\mu}$, and satisfying the property

$$N(\varphi) = \sup\left\{ \int \varphi\, d\mu : \mu \in A \right\} \quad \forall \varphi \in \overline{\mathcal{E}}_+(\Omega, \mathcal{F}). \qquad (1)$$

If some measure $\overset{*}{\mu}$ dominating SE is separable then there exists a countable set A satisfying the conditions of the theorem.

The measure $\overset{o}{\mu}$ will be called a supporting measure of SE and the set of measures A will be called a decomposition of SE (notice that measures of A are not necessarily probabilities i.e. such that $\mu(\Omega) = 1$).

The decomposition of SE is not unique but there exists the maximal decomposition which is a convex bounded closed subset of the cone $\mathcal{M}_+$ of positive measures from the Banach space $\mathcal{M}(\Omega, \mathcal{F}, N)$ of signed bourded measures dominated by the supporting measure $\overset{o}{\mu}$ of the sublinear expectation N (the norm in $\mathcal{M}$ is the total variation). The problem of existence of the minimal closed convex decomposition (MCCD) is solved successfully for two particular forms of SE: for the compactly decomposable SE and for the equally decomposable SE.

Definition 3. A sublinear expectation is called compactly decomposable if there exists a decomposition A whose measures are uniformly absolutely continuous with respect to the supporting measure $\overset{o}{\mu}$, i.e. $\forall \varepsilon > 0 \; \exists \delta > 0 \; \forall D \in \mathcal{F} \; \overset{o}{\mu}(D) < \delta$ implies $\sup\{\mu(D) : \mu \in A\} < \varepsilon$.

Definition 4. A sublinear expectation is called equally decomposable if there exists a decomposition whose all measures satisfy the property $\mu(\Omega) = 1$.

Theorem 5. For a compactly decomposable SE there exists MCCD which is bicompact in the weak topology $\sigma(\mathcal{M}, L_\infty)$.

Theorem 6. For an equally decomposable SE there exists MCCD which coincides with the closure of the convex hull for any of decompositions of SE into probabilities (i.e. into measures for which $\mu(\Omega) = 1$).

By leaning on Theorem 6 we can prove Theorem 7.

Theorem 7. SE is equally decomposable if and only if

$$N(\varphi + \chi_\Omega) = N(\varphi) + 1 \quad \forall \varphi \in \overline{\mathcal{E}}_+(\Omega, \mathcal{F}).$$

For construction of SE the concepts of the direct product and of the image are useful. Let be given two triplets $(\Omega_1, \mathcal{F}_1, N_1)$ and $(\Omega_2, \mathcal{F}_2, N_2)$ and let A_1 and A_2 be some decomposition for N_1 and N_2 respectively. Put

$$N_1 \times N_2(\varphi) = \sup\left\{\int_{\Omega_1 \times \Omega_2} \varphi \, d\mu : \mu = \mu_1 \times \mu_2, \ \mu_1 \in A_1, \mu_2 \in A_2\right\} \quad (2)$$

$\forall \varphi \in \overline{\mathcal{E}}_+(\Omega_1 \times \Omega_2, \mathcal{F}_1 \otimes \mathcal{F}_2)$. It turns out that SE defined by (2) does not depend on the choice of A_1 and A_2 for the fixed SE N_1 and N_2.

Definition 8. SE defined by (2) is called the direct product of the sublinear expectations N_1 and N_2.

Definition 9. Let $(X, \mathcal{G})$ be a measurable space and $h(\omega)$ is a measurable mapping $h: \Omega \to X$. Then the image of a sublinear expectation N on $(X, \mathcal{G})$ is the functional N_h defined on $\overline{\mathcal{E}}_+(X, \mathcal{G})$ by

$$N_h(g) = N(g(h(\omega))). \quad (3)$$

The image of a decomposition of N will be the set of the images of the measures from this decomposition. It turns out that N_h is a sublinear expectation and the image of a decomposition for N is a decomposition for N_h.

In conclusion we define the conditional SE. In the definition of the product of two elements of $\overline{\mathcal{E}}_+(\Omega, \mathcal{F})$ we agree that $0 \cdot \infty = 0$.

Definition 10. Let $\mathcal{B}$ be a sub-$\mathcal{G}$-algebra of $\mathcal{F}$, $f \in \overline{\mathcal{E}}_+(\Omega, \mathcal{F})$,

$$G(f) = \left\{g : g \in \overline{\mathcal{E}}_+(\Omega, \mathcal{B}), N(g\varphi) \geq N(f\varphi) \ \forall \varphi \in \overline{\mathcal{E}}_+(\Omega, \mathcal{B})\right\}. \quad (4)$$

Then the conditional sublinear expectation $N(f | \mathcal{B})$ of f with respect to $\mathcal{B}$ is the class of equivalent elements of $\overline{\mathcal{E}}_+(\Omega, \mathcal{B})$ defined by

$$N(f | \mathcal{B}) = \operatorname*{ess\,inf}_{\overset{\circ}{\mu}} G(f) \quad (5)$$

where $\overset{\circ}{\mu}$ is any of supporting measures for N.

In particular cases when N is a probabilistic expectation or a vrai maximum Definition 10 is reduced to the definition of the conditional probabilistic expectation or of the conditional vrai maximum respectively.

The Algorithms of Extremal Parameter Grouping

Gintautas DZEMYDA [1]

The structural methods for the empirical data processing are used widely in systems analysis. The method of extremal parameter grouping [1,2] belongs to this class of methods. It is devoted to the partition of the parameters $x_1,\ldots,x_n$ into a fixed number p of the unintersecting groups $A_1,\ldots,A_p$. The correlation matrix $R = \left\{ r_{x_i x_j}, \ i,j=\overline{1,n} \right\}$ characterizes the connections among the parameters ($r_{x_i x_j}$ is the correlation coefficient of parameters x_i and x_j). The covariance matrix may be used instead of the matrix R. However, the parameters with a greater dispersion will have greater significance in the analysis.

The problem of extremal parameter grouping includes maximization of the functional

$$I_1 = \sum_{L=1}^{p} \ \sum_{x_i \in A_L} r^2_{x_i F_L},$$

where F_L is the factor corresponding to the group A_L; $r_{x_i F_L}$ is the correlation coefficient of the parameter x_i and the factor F_L.

It is proved in [2] that

$$1. \quad F_L = \frac{\displaystyle\sum_{x_i \in A_L} \alpha_i^L x_i}{\sqrt{\lambda_L}},$$

$$r_{x_s F_L} = \frac{\displaystyle\sum_{x_i \in A_L} \alpha_i^L r_{x_i x_s}}{\sqrt{\lambda_L}},$$

where λ_L is the greatest eigenvalue of the matrix $R_L = \left\{ r_{x_i x_j}, \ x_i, x_j \in A_L \right\}$; α_i^L are the components of the normalized eigenvector of matrix R_L corresponding to λ_L.

$$2. \quad r_{x_s F_L} = \sqrt{\lambda_L} \ \alpha_s^L \text{ as } x_s \in A_L.$$

$$3. \quad I_1 = \sum_{L=1}^{p} \lambda_L.$$

---

[1] Institute of Mathematics and Cybernetics Lithuanian SSR Academy of Sciences, Vilnius, USSR

4. Let some partition of the parameters into the groups $A_1, \ldots, A_p$ be given and let the factors $F_1, \ldots, F_p$ be fixed. Let us analyse some parameter $x_s \in A_k$. By using the values of $r_{x_s F_L}$, $L = \overline{1,p}$, it is impossible to determine the group, where the value of the functional I_1 increases mostly after transfering the parameter x_s.

Note 1. Let some partition of the parameters into the groups $A_1, \ldots, A_p$ be given and let the factors $F_1, \ldots, F_p$ be fixed. Let us analyse some parameter $x_s \in A_k$. The value of the functional I_1 will increase after transfering the parameter x_s from the group A_k to the group A_L if we succeed in finding such a factor F_L ($L \neq k$) that

$$r^2_{x_s F_k} < r^2_{x_s F_L}$$

(see [1]).

Definition 1. By the local maximum of the functional I_1 we shall call its value coresponding to such a partition, where the correlation coefficient of any parameter with the factor, corresponding to the group including this parameter, is greater or equal to that of this parameter with the other factors.

A number of algorithms of maximization of the functional I_1 is proposed, which are based on Note 1 (see [1-3]). They are grounded on calculating the correlation coefficients of parameters with the factors and transfering the parameters from one group to another one depending on the values of these coefficients. Let us denote the algorithms of such a type by Al.

Denote:

1) λ_k^{-s} is the maximal eigenvalue of the matrix $R_k^{-s} = \{ r_{x_i x_j}, x_i, x_j \in A_k^{-s} \}$, where the group A_k^{-s} is obtained from the group A_k by eliminating the parameter x_s (note that $\lambda_k^{-s} = 0$ as $A_k^{-s} = \emptyset$);

2) λ_L^{+s} is the maximal eigenvalue of the matrix $R_L^{+s} = \{ r_{x_i x_j}, x_i, x_j \in A_k^{+s} \}$, where the group A_L^{+s} is obtained from the group A_L by adding the parameter x_s.

Note 2. $\lambda_k - \lambda_k^{-s} \geqslant 0$, $\lambda_L^{+s} - \lambda_L \geqslant 0$, $k = \overline{1,p}$, $L = \overline{1,p}$, $s = \overline{1,n}$.

Note 3. Let some partition of the parameters into the groups $A_1, \ldots, A_p$ be given. Let us analyse some parameter $x_s \in A_k$. The value of the functional I_1 will increase after transfering the parameter x_s from the group A_k to the group A_L if we succeed in finding such a value

λ_L^{+s} ($L \neq k$) that

$$\lambda_k - \lambda_k^{-s} < \lambda_L^{+s} - \lambda_L .$$

Note 4. Let some partition of the parameters into the groups A_1, ...,A_p be given. The value of the functional I_1 corresponding to the given partition is the local maximum in the case when for any parameter x_s (let $x_s \in A_k$)

$$\lambda_k - \lambda_k^{-s} \geqslant \lambda_L^{+s} - \lambda_L, \quad L = \overline{1,p},$$

is valid.

Notes 2-4 make the basis for the other type of algorithms of maximization of the functional I_1. These algorithms do not calculate the factors F_L, $L = \overline{1,p}$, and the correlation coefficients of individual parameters and the factors. In this way it is possible to find for any parameter x_s (let $x_s \in A_k$) the group, where the value of the functional I_1 increases mostly when transfering this parameter from the group A_k to this group. Such an algorithm is proposed in [2]. Let us denote it by A2.

The algorithm A2 requires considerably more expenditures of computation time in comparison with the algorithms of A1 type. It is the cause of the fact that A2 calculates the maximal eigenvalues of symmetrical matrices more often. However, A2 will be more effective than A1 in the sense of the average (greater values of the functional I_1 will be obtained), because in A1 the group for the transfering of the parameter x_s is determined by the values of the correlation coefficients of this parameter and the factors $F_1,...,F_p$. It is proved in [2] that in this manner it is impossible to determine the group, where the value of the functional I_1 increases mostly after transfering this parameter. Such a group is determined in A2.

In [2] some algorithms taking into account the results of Notes 1-4 taken together are proposed. They present a possibility to save the computation time in comparison with A2. But their efficiency is less than that of A2.

The initial partition of the parameters into p groups is necessary for the operation of the algorithms of maximization of the functional I_1. In [2] two algorithms of the initial partition are proposed and investigated. The number p of groups can be fixed preliminary or selected automatically.

The computer programs realizing algorithms of extremal parameter grouping are written in FORTRAN and can be used on the computer of the EC type. The properties of the algorithms are investigated in [2] on test and real data.

It was set in [4] that the algorithms of extremal parameter grouping may be used in cluster analysis. In this case the elements of the matrix R were chosen as the values of some potential function depending on the distance between a pair of points (objects) to be clustered. In [4] the results of comparison of the clustering algorithms, based on extremal parameter grouping, with the well known algorithms of the cluster analysis are presented.

REFERENCES

[1] Braverman, E.M. and I.B.Muchnik: The structural methods for the empiric data processing. Nauka, Moscow 1983.

[2] Dzemyda, G.: On the extremal parameter grouping. Teorija Optimaljnych Reshenij, 12, Vilnius (1987), 28-42. (in Russian)

[3] Dzemyda, G.: LP-search with extremal problem structure analysis. Stochastic Control: Proceedings of the 2nd IFAC Symposium, IFAC Proceedings Series, 1987, Number 2, Pergamon Press, New York, 1987, 499-502.

[4] Dzemyda, G. and J.Valevičienė: The extremal parameter grouping in cluster analysis. Teorija Optimaljnych Reshenij, 13, Vilnius (1988). (to appear in Russian)

A New Approach for Structural Modelling

Tatjana Lange[1]

Abstract. The main problem in structure modelling for large-scale systems in practice is to find out a suitable structure criteria. A criterion for indicating uncertainty conditions has been proposed and investigated in simulations. In the case of uncertainties special structure criteria are required. Two new structure criteria are proposed and investigated and compared with other criteria taken from literature, particularly GMDH-criteria.

Keywords. structure modelling, structure criteria, parameter estimation, choice of model structure, indication of uncertainty.

Three main types of modelling are known to exist:
1. **Analytical modelling**
2. **Simulation**
3. **Modelling of the so-called surface response**

This paper deals with the third modelling type. Models belonging to this type do not possess the precision of the 1st and 2nd type of modelling. They can be reliable only in a definite range of the independent variables with influence the system, and they can be built on the basis of experimental data whose obtaining may be limited due to the uniquness of the original system and the unpossibility to experiment with it (ecology, economy and others). These models can be employed for the preliminary investigations of complex systems and for the prediction of system behaviour. They can also be applied both independently and in combination with simulation modeels for optimization and control.

The author restricts his consideration to polinominal statistical and regressional models, but this limitation is not of principle charakter.

1) Central Institute of Cybernetics and Information Processes
   Kurstrasse 33, Berlin 1086, GDR

The task of construction of surface response models has 2 parts:

1. parameter estimation
2. choice of model structure

The availibility of a limited number of experimental datas often leads to considerable uncertainty of the task of complex system modelling because the realization of the statistical basic assumptions for regressional analysis in this case is questionable. Under these conditions especially the second part of the task, the choice of structure, is not investigated enough. Especially the choice of structure criteria is difficult.

The main practical problems are the following:

1. Obviously, given the fixed data volume, the conclusions of mathematical statistics based on the limit transition and probability convergence cannot be applied in every particular case. The question arises as to how actually can one decide whether the statistical methods of parameter estimation and RSS (residium sum of square) for structure choice can be applied in the given case characterized by a experimental data set.

2. When it is necessary to give up purely statistical methods in particular applying the theory of non-correct problems solution [1] for the regulation of statistical criteria the question arises: what kind of criteria for regulation (kind of stabilizing functional and the way of regulation parameter choice) of the estimated model parameters optimally (in the sence of prediction and so on) corresponds to the regulation for structure choice criteria.

3. In case the statistical methods are rejected and the surface response model is built in conditions of uncertainty, in particular with gnostic criteria [2] numerical problems appear: how is one to keep the advantage of the gnostic criteria being rubust when the precision of solution is sacrificed to the necessity to solve non-linear (i.e. multimodal) problems of optimization.

In search for a solution to the mentioned problems the author has
carried out a number of investigations.

1. The following criteria were suggested and computer aided
   tested:
   a) the criterion for indication of uncertainty (CIU) to deter-
      mine the degree of uncertainty of a modelling problem with
      the fixed range of data [3]

   b) two heuristic criteria for the choice of structure - the
      disturbance-criterion [3] and the criterion of non-linear
      sum [4][5]

2. Also was carried out a simulation investigation of the
   applicability of different criteria partly suggested by the
   author and partly taken from literature (particularly GMDH
   structural criteria [6][7]) in the conditions of various
   disturbances (Gaussian, multiplicative-additive, of the kind
   "exceptions" and others) [8].

3. A simulation analysis was performed of a combination of
   parameter criteria with structure criteria, in which concepts
   of gnostic parameter identification were applied to the
   structure criteria.

These simulation investigations showed the following:

1. In the preliminary analysis of systems CIU allows to select
   the method of identification with a high degree of
   reliability, but that requires time-consuming calculations and
   some experimental experience.

2. The disturbance criterion permits to find out the physical
   model if it is pressentamong the tested models even in the
   case of multiplicative-additive disturbance (up to 20 % of the
   actual signal).

3. KNS allows to select the model with the best properties for
   prediction in conditions of non-standard disturbance.

4. The parameter regulation along with structure regulation gives
   positive results under standard and multiplicative-additive
   disturbances but this is not so in the case of a disturbance
   of "exception"-kind.

5. Some positive results when gnostic criteria are applied in the
   case of disturbance of "exception"-kind require further
   confirmation.

References:

[1] Tichonow, A.; Arsenin, W.: Metody reshenija nekorrektnyx
    sadac. Moskwa, "Nauka", 1986 (in Russian)
[2] Kovanic, P.: Gnostical theory of small samples of real
    data problems of Control and Information Theory.
    Vol.13(5), pp. 303 - 319. Budapest, Publishing house of
    the hungarian academy
[3] Lange, T.; Gnatowski, K.; Umbreit, S.: Zur Wahl geeigneter
    Kriterien fuer die Struktursuche bei Modellierungsaufga-
    ben. Wiss. Z. TH Ilmenau 33(1987)2, S. 67 - 78 (in German)
[4] Lange, T.: Aggregation nichtlinearer Systeme.
    16. KdT/WGMA-Tagung "Grundlagen der Modellierungs- und
    Simulationstechnik", Rostock, 06. - 10.12.1987 (in German)
[5] Lange, T.: Methoden und Kriterien fuer die Modellbildung
    mit Struktursuche unter Unsicherheitsbedingungen.
    Wiss. Z. TH Ilmenau , Dec. 1987 submited (in German)
[6] Ivachnenko, A.G.: Dolgosrotchnoje prognostirovanije i
    upravlenie sloshnymi sistemami.
    Kiev, "Technika", 1975 (in Russian)
[7] Tamura, H.; Kondo, T.: Large-statial pattern identifica-
    tion of air pollution by a combinet of surcereceptor
    matrix and revised GMDH. Prod. IFAC-Symp. on Enzikemental
    Systems Planning, Design and Control 3 78, 1977
[8] Lange, T.: Multi-criteria desision approach to structural
    modelling with uncertainty conditions.
    Syst.Anal.Model.Simul., Berlin, Jan. 1988 submited

Problems of Qualitative Change of Parameters
in Different Hierarchical Levels

Apelt, E.; Apelt, D.[1]

Technical, economic and social connections can be modelled on different levels of structure. The interaction of aggregates within a production unit or the relation between systems of material processing within the economy are different examples of technological systems. For this, in general, models will be elaborated, in which all system-elements as well as all parameters, describing the system-elements, correspond a priori to the given aggregation level. In this case models of different aggregation level primarily are based on different databases. Two procedures are usual:

1. Extent and aggregation level of the model is fixed (inflexible models)

2. The model will be internally generated from database. The extent of the model can be changed by addition or omission of homogenous elements (quantitatively variable models)

An advantage in modelling complex connections in hierarchically built systems can be achieved, if it is possible to generate models from a homogenous database not only by addition or omission of single elements but also to choise the hierarchical level. The problem in this type of procedure is, that a change of hierarchical level is connected with aggregation or disaggregation of elements. This can lead to a change of properties of parameters.

The generation of models on different hierarchical levels on a common (undivided) database is a general problem of representation of hierarchical systems, it is not restricted on technical-economic network models.

At first we want to present an expert system, able to model a technological network on different structural levels and to optimize technical, economic and social aspects with regards one or more objectives.

----

1) Apelt, E.; AdW der DDR, Institut für chemische Technologie
Berlin, 1199, Rudower Chaussee
Apelt, D.; Humboldt-Universität Berlin, Bereich Asienwissenschaften
Berlin, 1020, Unter den Linden

Most detailed elements are processes,   for instance the lignite
briquetting. Elements of a higher level are coupled process systems,
for instance the aggregation of all liquid product processing in carbo-
chemistry. The highly aggregated elements in this model system are
material processing systems like carbo-chemistry or petrol processing
or power economy. The complete system includes all processes embracing
fossile organic carbon carriers.

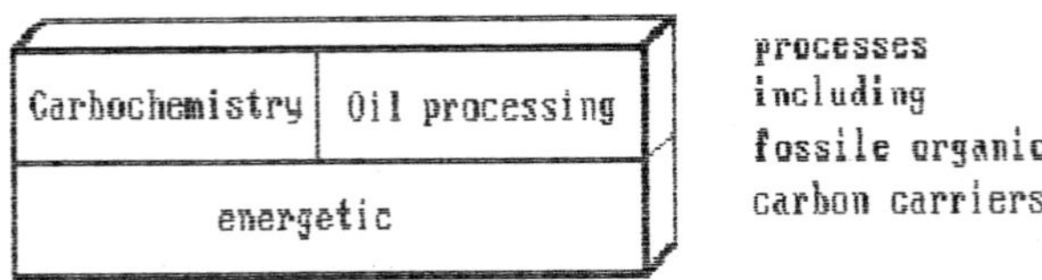

Fig.1: Complete model

Fig. 2 shows the structure of the expert system.
The modul "flexible model generation" creates the model from the single
elements and aggregates these elements up to the level choosen. This

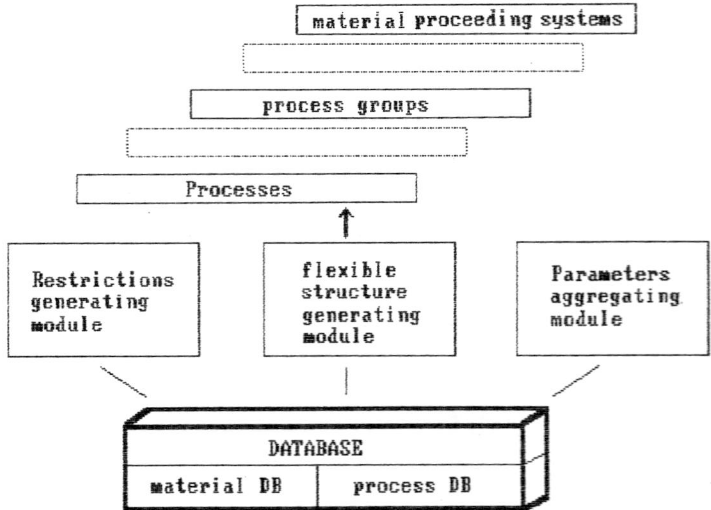

Fig. 2: Structure of the expert system

can be carried out by computer routines or manual and interactiv. Another modul generates the corresponding restrictions of balance and a
third one comprizes the parameters. To the structure of this expert
system and the software elaborated report on this conference A.
Barnikow/ U. Behrendt.

<u>On the aggregation of elements</u>

Aggregation is thought to be the integration of single elements to one
element on a higher level in the given system.

This can be put into practice internally by the computer. But this is
connected with theoretical problems. If the integration of elements is
done by according to biunique formation-rules, the properties of the
original elements are fixed rigidly in the newly occuring element. For
technological models these are

  - the working point of single elements (for instance the specific
    consumption index of procedures)

  - the proportionality of single elements (for instance the relation
    of material- or energyflows).

The internal flexibility, however, which has been existing between the
original elements, is lost during aggregation.

In optimizing calculation being in models aggregated in this way it is
possible to use all the effects of optimizing between the original
elements. A compromise is to combine single elements into a partial
model, to optimize it before aggregating according to given criteria.
All working points and proportions obtained have to be taken into the
higher levelled element. This is done internally by the computer. But,
it is not sure, whether these partial optima are the best values under
the aspects of the total system. The optimum of the total system does
not correspond, in some cases, to the optima of partial systems. Therefore it is interesting to get to know which errors can appear, when
working points and proportions of single elements are fixed during
aggregation.

In the system investigated these internal reserves are

    5 - 10 % of the capacity of processing and

    until 5 % of the income.

On the level of coupled chemical process systems these effects are more
significant.

The connection between the internal reserves resulting from the
flexibility of single elements and from the error of aggregation has to
be investigated for each hierarchical level. The error in technological

systems becomes smaller the more elaborated the technology is.

<u>On generating restrictions of balance</u>
The problem is investigated under the aspect of a needoriented network
model. Such a mode of consideration is charakteristic for centralised
planned economy. We start from that point, that working power, resources, basic fonds and volume of production are given. The system is
included in restrictions of balance.
Three levels of balance can be distinguished:
   - balance of incomes (portion of system of total production)
   - balance of resources (portion of existing resources)
   - balance of capacity.
In generating models has to be secured, that the restrictions
correspond to the limits of capacity balance, i.e. the planned model
extent. They are generated by a special modul.

<u>On the reversibility of aggregation</u>
It has been prooved to be useful to generate each model by aggregation
from the simplest elements. Disaggregation, i.e. splitting of elements
of a higher level, is not carried out. This would mean presupposition
of reversibility of aggregation. It is only possible, if aggregation
comes about by biuniquel formation-rules. As was mentioned above, this
is connected with the total loss of the flexibility of partial systems.
If models are used the aggregation of which comes about internally by
the computer, aggregation can be protocolled and repeated if wished. It
is, in certain sense, reversible.
In models the aggregation level of which is given a priori by formation rules and based on devided databases, this possibility does not
exist.

<u>On integration of parameters</u>
The abstraction levels of elements has to correspond to the parameters.
Therefore the internal aggregation of elements can be connected with
integration of parameters.
This can be done additively
                 multiplicatively
                 functionally.
In our model system the parameters are integrated, especially additive-
ly. Some economic sizes are modelled functionally, for instance the net

income. This size exist only on the highest aggregation level, and it replaces there the profit, which plays an important role on the plant level. It was not nesserary to integrate the parameters in strict dependance of the aggregation level. Therefore the parameters with fixed abstraction level are put into the database and can be called off for each aggregation level by computer-routines. In contrast to generating of the model structure a formal simplification was carried out. Thereby appears the problem, that economic indexes, which are required on the highest aggregation level already have to be define for the basic elements. This was difficult, especially for the net income.

Indeed there are possibilities for applicating hierarchical models in Archaeology. Here it is useful to investigate one or more object groups, i.e. pottery, tools, statues, reliefs etc., by hierarchical classifi-cation models, for instance Clusteranalysis, based on devided or undevided databases.
There are, in general, the same significant problems like aggregation of elements, reversibility of aggregation, integration of parameters, especially the redundancy of parameters on higher aggregated levels as in the economic network model presented in this paper. Indeed, the solution by computer is much easier, than of the network model, because neither optimization by various criteria is done nor single system elements of such a great complexity exist.

Altogether there are various possibilities to aggregate elements and to integrate parameters. The solution by the computer is mostly simple. But it is important to explain exactly the mode of procedure in the user's description. This should not restricted on a description of computer facilities. Even in hierarchically structured models of great complexity the significant consequences of the mode of procedures choosen must interpret.

Studying the Interactions Among Model Reduction Algorithms via CAD Technique

L. Fortuna, A. Gallo, G. Nunnari [1]

1. Introduction

Considerable attention has been devoted to the problem of deriving reduced order models for complex systems [7] . In this study we comment some aspects of model reduction problems based on considerations of empirical investigations carried out by using a useful tool for such aims : UMLLSR [2]. This is a Unified Interactive Modular Library of subprograms devoted to time-invariant linear system reduction. The development of a subprogram package is a very useful tool in view of the particular importance of Computer-Aided-Design in modular unified synthesis approach to control system design, but it also offers powerful support to investigate some theoretical aspects of the various approximation algorithms. UMLLSR consists of a set of modules implementing the various approximation techniques via a common data-base that allows the user to evaluate the performances of the different reduced models obtained. But the data-base also allows the exchanging of information among the various modules with the following aims:

1) to set further optimization in the synthesis of reduced models;

2) to study possible correlations among the obtained results using different approximation procedures in order to evaluate the efficiency of one algorithm respect to another, and possibly to predict the performance of a particular algorithm based on previous results;

3) to investigate numerical aspects of the algorithms in order to analyze their numerical stability (a) and to improve convergence of some iterative schemes (b).

Theoretical questions have been widely studied, numerical aspects less, but a universal method for model order reduction does not exist! Pratice and simulation play a basic role in model approximation. In this paper our attention is devoted to investigating the model reduction procedures based on the minimization of a quadratic error function [5]; looking at the balanced realization approach [6] we establish some useful conjectures from a practical point of view in model approximation.

2. Motivations and Preliminaries

Model approximation schemes based on optimization represented a natural way to get approximated models and standard minimization algorithms have been widely presented to solve the problem numerically [5] ; also balanced approaches [6] have been proposed obtaining a wide success in model reduction. Indeed very interesting properties as

[1] Istituto di Elettrotecnica ed Elettronica, Università di Catania
viale Andrea Doria 6, 95125 Catania, Italy

regards the parameter sensitivity have been recently [3] advanced to further appreciate the open-loop balanced realization schemes. Then why not couple the approaches, making optimal reduced models with the suitability of the balanced schemes?

The minimization of a quadratic error function J as an approximation procedure consists in determining a reduced rth order model S_r (A_r, B_r, C_r) from an original model S (A, B, C) (in minimal form) in such manner as to minimize for a class of particular inputs (impulses) a quadratic index depending on the discard between the model and original model outputs. An interesting algorithm has been proposed in [5] where the finding of the model S_r involves the iterative solution of matrix linear equations of $(n+r)$ order. The problem solved in [5] is optimal in the L^2 sense.

Let S ($\tilde{A}$, $\tilde{B}$, $\tilde{C}$) be an equivalent representation of S (A, B, C), if the observability Gramian and the controllability Gramian of S ($\tilde{A}$, $\tilde{B}$, $\tilde{C}$) are equal to a diagonal matrix Σ (σ_1^2, σ_2^2, ... σ_n^2), we can say that S ($\tilde{A}$, $\tilde{B}$, $\tilde{C}$) is a balanced realization of S (A, B, C) [6] if the internally balanced representation is partitioned as follows:

$$\tilde{A} = \begin{bmatrix} \tilde{A}_{11} & \tilde{A}_{12} \\ \tilde{A}_{21} & \tilde{A}_{22} \end{bmatrix} , \quad \tilde{B} = \begin{bmatrix} \tilde{B}_1 \\ \tilde{B}_2 \end{bmatrix} , \quad \tilde{C} = [\tilde{C}_1 \quad \tilde{C}_2] .$$

The reduced order model S ($\tilde{A}_{11}$, $\tilde{B}_1$, $\tilde{C}_1$) is a good representation of S ($\tilde{A}$, $\tilde{B}$, $\tilde{C}$) (in the sense of internal dominance), as regards its most controllable and observable part, if:

$$\left[\sum_{i=1}^{r} \sigma_i^4 \right]^{1/2} >> \left[\sum_{i=r+1}^{n} \sigma_i^4 \right]^{1/2} .$$

The following considerations derive from the computational experience of the authors using the UMLLSR to examine various model reduction problems.

4. Main Results

Conjecture 1

From various CAD experiences a good agreement between the original model and the reduced approximated model via balancing is reached when:

$$\rho = \frac{\left[\sum_{i=1}^{r} \sigma_i^4 \right]^{1/2}}{\left[\sum_{i=r+1}^{n} \sigma_i^4 \right]^{1/2}} > 50 \tag{1}$$

generically valid in the hypothesis of residuals of the same size.

This conjecture is true in the hypothesis of systems with eigenvalues of the same order. In fact in the case of systems of eigenvalues of different range, we have to establish which are essentially the dominant ones and therefore to decide if the dominant subsystem is in low or high frequency range. This can be established also by evaluating the residuals. A high order residual can be associated at low frequency poles and vice versa.

Therefore in the first case, relation (1) guarantees a good agreement between the low and high order models in the other case it is not true. In the latter case conjecture 1 can also be valid changing σ_i^2 with the values of dominant gains of the system. In fact in order for a good agreement between the original model and the approximated one to be correct , expression (1) cannot be generally applied because it does not guarantee the L^2 optimal or suboptimal performances [4] .

Conjecture 2

The following consideration takes into account what has been previously established.

The reduced model S $(\tilde{A}_{11}, \tilde{B}_1, \tilde{C}_1)$ (if $\rho \geq 50$) is "near" to the optimal reduced model obtained from the minimization of a quadratic error function J. To make the optimization routines faster we can start with S $(\tilde{A}_{11}, \tilde{B}_1, \tilde{C}_1)$.

Since in MIMO systems condition (1) is not verified, the optimization procedure to give the reduced model can fail. In fact it is possible to obtain an optimal approximated model while the value of J is unacceptable for a good agreement between the original and the reduced model.

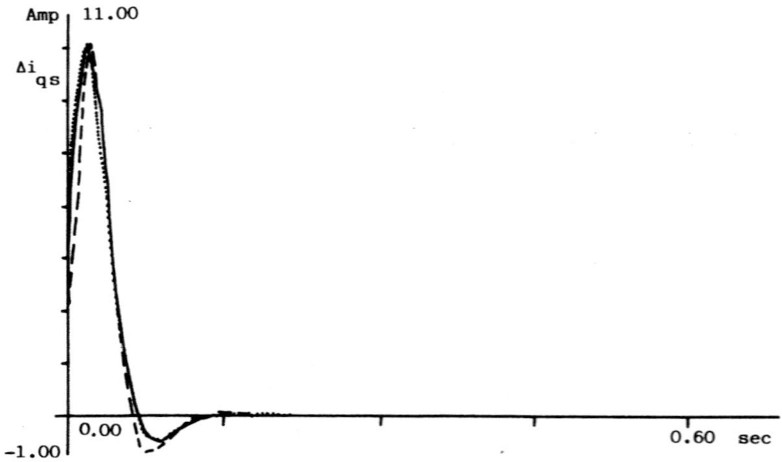

Fig. 1   Impulse response of the quadratic current
in a linearized induction motor system (torque input).

An example of approximation is here reported [1] where, as previously outlined,
it can appear that from simulation as from optimal approximation, the model of a
linearized induction motor system (torque input) is near (Fig. 1) the balanced
one ($\rho \simeq 60$).

Conjecture 3

The optimization routine to obtain reduced models from the original one consists
effectively of two iterative loops: one to obtain the optimal B_r and C_r for a fixed A_r,
and the other to update A_r in order to converge towards the optimal conditions.

From a numerical point of view, the first loop is critical as regards the convergence
of matrix B_r towards the correct value.

It can be shown that by using in the optimization algorithm balanced schemes, both
for the system S (A, B, C) and for the intial choice of S_r (A_r, B_r, C_r), fast convergence can be obtained.

In Figs 2 and 3 the convergence rate of the algorithm adopting a norm criterium
has been shown. The results are referred to reduction problem from a five order
inputs three outputs system. In Fig. 2 the convergence rate to the correct B_r value
has been shown when an initial choice of the reduced system has been made assuming
only its asymptotic stability , while in Fig. 3 the improvement is evident when
a balanced reduced scheme is used.

A further consideration arises: while the convergence inprovement is related to
any balanced scheme,   optimality is reached only with a suitable balanced representation of the original model and of the first step reduced order model, thus either referring to the scheme reported in [6] , in accordance with Conjecture 1, or
to the scheme reported in [4] in the hypothesis of balanced gain representation.

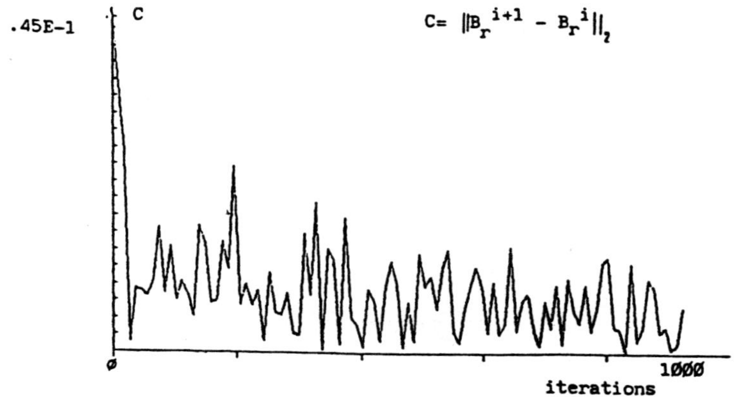

Fig. 2 Convergence rate for generical system representation.

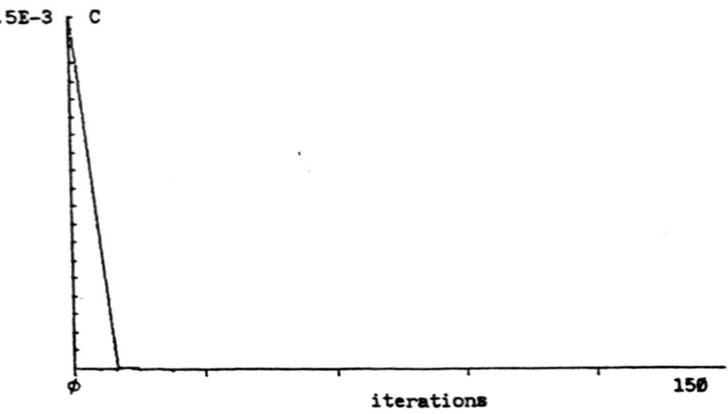

Fig. 3   Convergence rate for balanced system
representation.

5. Conclusions

Generally some useful evident properties of model reduction schemes are emphasized
by using interactive trial-simulation approaches. Some conjectures can be established
and further reworked for future theoretical research.   Model approximation requires high theoretical background, but also a long amount of practice, in
order to get useful results in brief time, and suitable simulations in accordance with
numerical precision bounds. Indeed the reduction algorithm without a powerful numerical
support is inapplicable. The designer must have some criteria and some tools to decide
what is useful. Further considerations can be derived directly from experience of
using interactive reduction programs as UMLLSR. There is not a universal model reduction algorithm: studying the interactions among the various model reduction procedures,
the user can hope to reach   good approximation efficiency.

References

[1] Consoli A., L. Fortuna, and A. Gallo: Induction Motor Identification by a Micro-
     computer-Based Structure. IEEE Trans. on Industrial Electronics, 34 (1987) 4, 422-
     428.

[2] Fortuna L., and A. Gallo: An Interactive Program Package for Linear System Reduc-
     tion. 9th IFAC World Congress, Budapest 1984.

[3] Gray W.S., and E. Verriest: Optimality Properties of Balanced Realizations: Minimum
     Sensitivity. Proc. 26th Conf. on Decision and Control, Los Angeles CA 1987.

[4] Kabamba P.T.: Balanced Forms: Canonicity and Parametrization. IEEE Trans. Autom.
     Contr., 30 (1985), 1106-1109.

[5] Mishra R.N., and D.A. Wilson: A New Algorithm for Optimal Reduction of Multivaria-
     ble Systems. Int. J. of Control, 31 (1980) 3, 443-466.

[6] Moore B.C.: Principal Component Analysis in Linear systems: Controllability,
     Observability, and Model Reduction. IEEE Trans. Autom. Contr., 26 (1981) 1, 17-32.

[7] Paraskevopoulos P.N.: Advances in Control Theory. Academic Press, New York 1986.

Determination of Simplified Models
by means of Chebyshev Polynomials

Janusz Halawa[1], Anna Trzmielak-Stanisławska[2]

A method of simplifying the mathematical models described by high-order transfer functions is presented. The simplified transfer function is found by expanding the initial transfer function into a Chebyshev series and composing a rational approximation of this series. The time and frequency responses of the simplified model should ensure a good, in a sense of the norm assumed, approximation of these for the higher order models. The method presented provides better results than these in continued fraction method.

METHOD DESCRIPTION

The transfer function

$$G(s) = \frac{b_m s^m + a_{m-1} s^{m-1} + \ldots + b_0}{a_n s^n + a_{n-1} s^{n-1} + \ldots + a_0} \tag{1}$$

is approximated with N-order interpolation polynomial having the knots

$$u_{N-1,k} = \frac{\cos k\pi}{N} . \tag{2}$$

This polynomial is given by the formula

$$G_N(s) = \frac{2}{N} \sum_{i=0}^{N}{}'' \left(\sum_{k=0}^{N}{}'' G(u_{N-1,k}) T_i(u_{N-1,k}) \right) T_i(s) = \sum_{j=0}^{N}{}'' c_i T_i(s). \tag{3}$$

Each interpolation polynomial is a projection discrete operator. From among the projection discrete operators, the interpolation polynomial (3) with the knots (2) has the minimal norm. If the function $F(s) = \sum_{i=0}^{\infty} c_i[f] T_i(s)$, and the interpolation polynomial for the $F(s)$ function is given by the formula (3), then the coefficients of this polynomial are

$$c_k[f_N] = c_k[f] + \sum_{i=1}^{\infty} (c_{2iN-k}[f] + c_{2iN+k}[f]), \quad k = 0,1,2,\ldots,N-1$$

and

$$c_N[f_N] = c_N[f] + \sum_{i=1}^{\infty} c_{(2i+1)N}[f] .$$

The series (3) is approximated by a rational function of the form

$$G_{m_1,n_1}(s) = \frac{\displaystyle\sum_{i=0}^{m_1} b_i^* T_i(s)}{\displaystyle\sum_{i=0}^{n_1} a_i^* T_i(s)} . \tag{4}$$

[1] Institute of Technical Cybernetic of Technical University of Wrocław, ul. Janiszewskiego 11-17, 50-372 Wrocław, Poland

[2] Institute of Computer Science of University of Wrocław, ul. Przesmyckiego 20, 50-151 Wrocław, Poland.

because a simplified transfer function for (1) is to be found out. The approximation is composed in a similar way as Pade approximation for Maclaurin expansion, i.e. the coefficients b_i^* and a_i^* are determined so that for the expression

$$G(s)-G_{m_1,m_1}(s) = \frac{\frac{1}{2}c_0 + \sum_{i=0}^{\infty} c_i T_i(s)\quad \sum_{i=0}^{n_1} a_i^* T_i(s)\quad \sum_{i=0}^{m_1} b_i^* T_i(s)}{\sum_{i=0}^{n_i} a_i^* T_i(s)} \tag{5}$$

the coefficients for $T_i(s)$ $(i = 0,1,\ldots, N)$ in the numerator of (5) would vanish, thus

$$(\tfrac{1}{2}c_0 + \sum_{i=0}^{\infty} c_i T_i(s)) \sum_{i=0}^{n_1} a_i^* T_i(s) - \sum_{i=0}^{n_1} b_i^* T_i(s) = \sum_{i=N+1}^{\infty} c_i T_i(s). \tag{6}$$

Because of the identity

$$T_{i+j}(s) + T_{|i-j|}(s) = 2T_i(s)T_j(s)$$

the equation (6) may be written

$$\tfrac{1}{2}c_0 \sum_{i=0}^{n_1} a_i^* T_i(s) + \tfrac{1}{2}\sum_{j=1}^{\infty}\sum_{i=0}^{n_1} a_i^* c_j (T_{i+j}(s)+T_{|i-j|}(s)) - \sum_{i=0}^{m_1} b_i^* T_i(s) = \sum_{j=N+1}^{\infty} c_i T_i(s). \tag{7}$$

The following set of equations for b_i^* and a_i^* results from the equation (7)

$$b_0^* = \frac{1}{2}\sum_{i=0}^{n_1} a_i^* c_i \tag{8}$$

$$b_r^* = \frac{1}{2}\sum_{i=0}^{n_1} a_i^* (c_{|r-i|} + c_{r+1}) \qquad r = 1,2,\ldots,N, \qquad b_i^* = 0 \text{ if } r > m_1$$

We have there N+1 equations with n_1+m_1+2 unknowns b_i^* and a_i^*. We can assume $a_{n_1}^*$ is equal to 1. Hence, we have n_1+m_1+1 unknown coefficients. Taking $N = n_1 + m_1$, we get system of linear equations. Solving the system (7), we obtain the transfer function of lower order for (4).

EXAMPLE

According to the method outlined above, we shall find a third-order simplified transfer function $G_{23}(s)$ for the sixth-order initial transfer function of the form:

$$G_{56}(s) = \frac{y(s)}{x(s)} = \frac{L(s)}{M(s)}, \tag{9}$$

where

$$L(s) = 0.04375s^5 + 0.36500s^4 + 0.82439s^3 + 0.27563s^2 + 0.07500s + 0.01000$$

and

$$M(s) = 0.31250s^6 + 2.65000s^5 + 6.26750s^4 + 2.97081s^3 + 1.21893s^2 + 0.27184s + 0.02040.$$

The transfer function (9) has been used as an example above because its response to the step function is oscillatory function. For such a system it is more difficult to determine the simplified transfer function providing good approximation of the initial system than in case of systems which responses are not the oscillatory functions. As a result of the discussed method we obtain the simplified transfer function $G_{23}(s)$ as follows

$$G_{23}(s) = \frac{y_{23}(s)}{x(s)} = \frac{0.140152s^2 + 0.019943s + 0.009475}{s^3 + 0.283023s^2 + 0.156599s + 0.019295} . \qquad (10)$$

The results achieved with the method presented will be compared with these resulting from the method of continued fractions [2, 3, 4]. With the method of continued fractions the following simplified transfer function has been determined

$$G_{23}^{*}(s) = \frac{y_{23}^{*}(s)}{x(s)} = \frac{0.140152s^2 + 0.019141s + 0.009475}{s^3 + 0.283023s^2 + 0.156599s + 0.019294} . \qquad (11)$$

The graphs of $y(t)$, $y_{23}(t)$ and $y_{23}^{*}(t)$ responses are illustrated in Fig. 1. These graphs overlap each other in this figure. The approximation errors for the method presented are

$$\|y - y_{23}\|_{L^2_{\langle 0,150\rangle}} = 0.0007296, \quad \|y - y_{23}\|_{C_{\langle 0,150\rangle}} = 0.008584$$

and for the simplified model determined with the continued fractions method these are

$$\|y - y_{23}^{*}\|_{L^2_{\langle 0,150\rangle}} = 0.002922, \quad \|y - y_{23}^{*}\|_{C_{\langle 0,150\rangle}} = 0.01433 .$$

The approximation obtained with the method presented, in a sense of these norms, has been proved to be better than that obtained with the known and good method of continued fractions. Tha gain-phase characteristics of systems described with the transfer function (9), (10) and (11) are shown in Fig. 2. As it can be seen, the method presented in this paper provides also better approximation.

CONCLUSIONS

The method presented ensures better approximation of initial system response by the simplified system response than that attainable by the method of continued fractions. The approximation errors were lower both for time responses and frequency responses. It should be noted that one of simplified models may appear to be unstable even so the initial model is stable, the situation having been also met for the method of continued fractions.

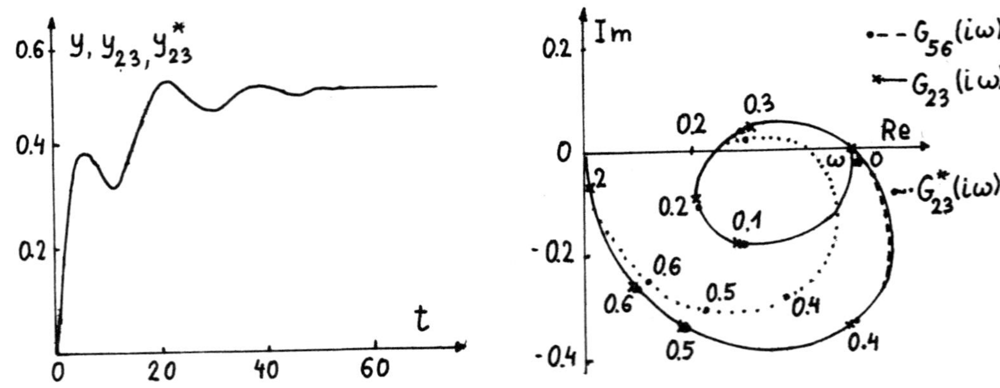

Fig. 1. The time responses of the systems (9), (10) and (11)

Fig. 2. The frequency responses of the systems (9),(10) and (11)

REFERENCES

|1| Bistritz, V., Langholz, G.: Model reduction by best Chebyshev rational approximations in the complex plane. Int. J. Control. 30 (1979) 2, 277-289.
|2| Chen, C.F., Shieh, L.S.: A novel approach to linear model simplification, Int. J. Control. 8 (1968) 6, 561-570.
|3| Halawa, J., Halawa T.: Simulation investigations of interconnected power systems. Reduction of mathematical model. System Analysis and Simulation. t. II. Academie-Verlag, Berlin (1985), 399-402.
|4| Halawa, J., Trzmielak-Stanisławska, A.: Properties of Simplified Transfer Functions as Acquired with a Method of Continued Fractions, Archiwum Automatyki i Telemechaniki, Tom XXXI, Zeszyt 1-2 (1986), 117-132.
|5| Paszkowski, S.: Zastosowania numeryczne wielomianów Czebyszewa, PWN, Warszawa (1975).

A Note on Simplification of Large Dynamic Systems Using a Moment Technique

Janusz Halawa[*)]

A method of finding the transfer functions of the form

$$G_i^*(s) = \frac{y_i^*(t)}{x(t)} = \frac{k\, a_i^i\, \exp\,(-t_{0i}s)}{(s + a_i)^i} \,, \quad i = 3,4 \tag{1}$$

is given. Parameters of the transfer functions are derived from a step response using the moments of the steady-state value, $k = \lim\limits_{t \to \infty} y(t)$ and the response of the original system $y(t)$. The method presented has been compared with two known identifications methods employing the pulse response moments. The results achieved by the method presented are better.

PRESENTED METHOD

Now, we introduce a method of finding the simplified mathematical models using the moments of original system

$$M_{0w} = \int_0^\infty [\,k - y(t)\,]\,dt \quad \text{and} \quad M_{1w} = \int_0^\infty t\,[\,k - y(t)\,]\,dt,$$

where $y(t)$ is a response of the original system to the step function $x(t) = 1(t)$, and $k = \lim\limits_{t \to \infty} y(t)$. Transfer functions (1) are approximated by the ones in which the functions $e^{-st_{0i}}$ are replaced with the first terms of its expansion into a power series. Then we have

$$G_i^*(s) \quad G_i(s) = \frac{k\, a_i^i \sum\limits_{j=0}^{i-1} \dfrac{1}{j!}\,(-st_{0i})^j}{(s + a_i)^i} \,.$$

Let consider the third-order transfer function

$$G_3(s) = \frac{y_3(s)}{x(s)} = k\, a_3^3\, \frac{\dfrac{t_{03}^2}{2}\,s^2 - t_{03}\,s + 1}{(s + a_3)^3}$$

The moments, M_{03} and M_{13} of the difference $R_3(t) = k - y_3(t)$ are

$$M_{03} = \int_0^\infty R_3(t)\,dt = k\,\left(\frac{3}{a_3} + t_{03}\right),$$

$$M_{13} = \int_0^\infty t\,R_3(t)\,dt = k\,\frac{t_{03}^2}{2a_3^2}\,\left(a_3^2 + \frac{6}{t_{03}}\,a_3 + \frac{12}{t_{03}^2}\right) \,.$$

[*)] Institute of Technical Cybernetic of Technical University of Wrocław, ul. Janiszewskiego 11-17, 50-372 Wrocław, Poland

When comparing the moments $M_{03} = M_{0w}$ and $M_{13} = M_{1w}$ we have

$$a_3 = \sqrt{\frac{3}{2\dfrac{M_{1w}}{k} - \left(\dfrac{M_{0w}}{k}\right)^2}} \qquad \text{and} \qquad t_{03} = \frac{M_{0w}}{k} - \frac{3}{a_3} . \tag{2}$$

As it results from the above formulae, the inequalities $t_{03} > 0$ and $a_3 > 0$ imply the following inequalities $\dfrac{a_3}{3} < \dfrac{M_{0w}}{k}$ and $2\dfrac{M_{1w}}{k} > \left(\dfrac{M_{0w}}{k}\right)^2$. The parameters of the fourth-order model

$$G_4(s) = \frac{y_4(s)}{x(s)} = k\, a_4^4 \frac{e^{-st_{04}}}{(s+a_4)^4} \qquad k\, a_4^4 \frac{-\dfrac{t_{04}^3 s^3}{6} + \dfrac{t_{04}^2 s^2}{2} - t_{04}s + 1}{(s + a_4)^4}$$

my be found from relations

$$a_4 = \sqrt{\frac{4}{\dfrac{2M_{1w}}{k} - \left(\dfrac{M_{0w}}{k}\right)^2}} \qquad \text{and} \qquad t_{04} = \frac{M_{0w}}{k} - \frac{4}{a_4} , \tag{3}$$

where

$$\frac{2M_{1w}}{k} > \left(\frac{M_{0w}}{k}\right)^2 \qquad \text{and} \qquad \frac{4}{a_4} < \frac{M_{0w}}{k} .$$

EXAMPLE

For the initial system described with the transfer function

$$G(s) = \frac{y(s)}{x(s)} = \frac{10.11 \ldots 19}{(s+10)(s+11) \ldots (s+19)} \tag{4}$$

in case of presented method we obtain from computer simulation $y_u = 1$, $M_{0w} = 0.7188$, $M_{1w} = 0.2853$. When substituting the values into equations (2) and (3) we get

$$G_3(s) = \frac{y_3(s)}{x(s)} = 7.4586^3 \frac{e^{-0.3166s}}{(s+7.4586)^3} \tag{5}$$

and

$$G_4(s) = \frac{y_4(s)}{x(s)} = 8.6125^4 \frac{e^{-0.2853s}}{(s+8.6125)^4} . \tag{6}$$

The known method of moments $|2|$ is based on the equations

$$G(s) = \int_0^\infty f(t)e^{-st}dt = k\frac{b_p s^p + b_{p-1}s^{p-1} + \ldots + b_1 s+1}{a_n s^n + a_{n-1}s^{n-1} + \ldots + a_1 s+1} := m_0 - m_1 s + m_2 \frac{s^2}{2!} - m_3 \frac{s^3}{3!} \ldots ,$$

where

$$m_i = \int_0^\infty t^i \, f(t)dt, \quad i=0,1,2,\ldots$$

and $p < n$. In these formulae, $f(t)$ is the pulse response of original system. According in this method for original system described by the transfer function (1), we obtain

$$G(s) = \frac{y_{23k}(s)}{x(s)} = \frac{0.013565s - 0.201326s+1}{0.008214s^3 + 0.103210s^2 + 0.521170s+1} . \qquad (7)$$

The second known method of moments proposed by Gilibaro and Lees in work $|1|$, has not given any reduced models in this case.

The responses of the systems described by the transfer functions (4),(5) and (6) are given in figure 1

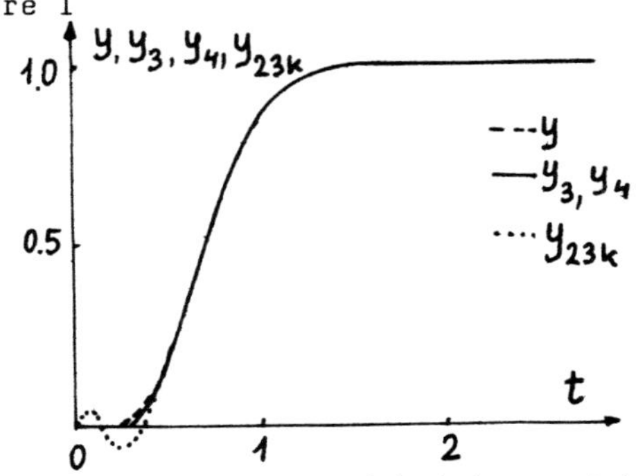

Fig .1. The time responses of systems (4),(5) and (6)

CONCLUSIONS

The assumption made for the simplified transfer function denominator has single roots enables to express the coefficients of the simplified transfer functions with simple formulae. As results from the examples provided, the simplified models given by the presented method have the step function responses which provide better approximation of the initial system response to the same input function than these from simplified models obtainable from the well-known method of moments $|1,2|$ using the pulse response of the initial system. Not out of place to add that the method of Gilibaro and Lees has furnished no **positive** results in this example. The step input function used for identification purposes with the presented method is more convenient in identification of industrial plants than the pulse input function used in the methods $|1,2|$.

REFERENCES

[1] Gilibaro, J.G., Less, F.P.: The reduction of complex transfer function models to simple models using the method of moments. Chem. Eng. Sci. 24 (1969), 85-93.

[2] Górecki, H.: Analiza i synteza układów regulacji z opóźnieniem. WNT, Warszawa 1971.

[3] Halawa, J.: A method using moments of step-function response to reduce of order of mathematical models. Instytut Cybernetyki Technicznej Politechniki Wrocławskiej, Preprint.

Model Reduction and Stability of Nonlinear Dynamical Systems by means of Centre Manifold Theory

Reinhard Boettner[1]

__Abstract__: The application of centre manifold theory to reduction, simplification and stability analysis of nonlinear systems is described. Numerical simulation results demonstrate a reasonable correspondence between the original and reduced system. The utilisation of computer algebra systems as a means of determination the simplified model equations and the essential parameters near a stationary point is outlined.

__Keywords__: NONLINEAR DYNAMICAL SYSTEMS, REDUCTION, SIMPLIFICATION STABILITY ANALYSIS, CENTRE MANIFOLD THEORY, COMPUTER ALGEBRA

INTRODUCTION

In any modeling task very often two conflicting factors play an important role: simplicity and accuracy. This leads us to two desirable properties of a system model: reduced computation and simplified structure. To reach this goal often the nonlinear system is linearized around a stationary operating point. Then the well developed methods of linear model reduction [3,7] are applied and the stability of the nonlinear system near the operating point is determined by the method of first approximation of Ljapunov [8].

This approach has some drawbacks: the nonlinear parts of the system are entirely neglected and this procedure is not practicable if some eigenvalues of the linearized system have zero real parts.

Centre manifold theory [2,4,5,6], in some sense an abstraction of the idea of uncoupled equations leading in linear case to the diagonalization of matrices, provides a promising way out. By means of centre manifold theory it is possible to single out few essential equations and parameters which govern the nonlinear behaviour(e.g. stability and bifurcation) near a stationary point or stationary motion(e.g. limit cycle). As a rule one or two essential equations arise. This corresponds to stability change(a single eigenvalue with zero real part) or to the excitation of sustained oscillation(Hopf-Bifurcation, a conjugate complex pair of eigenvalues crossing the imaginary axis) in dependence upon some critical parameters. The remaining state variables will be computed by a nonlinear coordinate transformation(the centre manifold), the determination of which is the hardest work one has to perform in order to simplify and analyse a concrete nonlinear dynamical system.

[1] Central Institute for Cybernetics and Information Processes of the Academy of Sciences of the G.D.R., DDR 1086 Berlin, Kurstrasse 33

An invariant manifold S for the equation

$$\dot{x} = N(x) \qquad x \in R^n \qquad (1)$$

is said to be a _local_ _invariant_ _manifold_ if for $x(0) \in S$ the solution $x(t)$ is in S for $|t| < T$, $T > 0$. We say that S is an _invariant_ _manifold_ if it is always possible to choose $T = \infty$.

For simplicity and because it is the most interesting and desirable case for practical systems , we assume that the system has no unstable part and that the linear part of the system is in block diagonal form already:

$$x \in R^n \qquad \dot{x} = Ax + f(x,y) \qquad (2)$$
$$y \in R^m \qquad \dot{y} = By + g(x,y) \qquad (3)$$

$x = y = 0$ is a stationary solution and A and B are constant matrices whose eigenvalues have zero real parts and negative real parts, respectively. f and g vanish at the origin along with their first partial derivatives. If $y = h(x)$ is an invariant manifold for (2,3) then we call it a _centre_ _manifold_ if $h(0) = 0$, $h'(0) = 0$.

The center manifold theorem [2,5] states :

1) that there exists a centre manifold for (2,3)

$$y = h(x) \qquad (4)$$

$|x| < \delta$, $h \in C^2$ and that the solution on the centre manifold is governed by the n equations

$$\dot{u} = Au + f(u, h(u)) \qquad (5)$$

2) a) Suppose that the zero solution of (5) is stable(asymptotic stable ,unstable) then the zero solution of (2,3) is stable(asymptotic stable, unstable)

b) If the zero solution of (5) is stable and $(x(t), y(t))$ is a solution of (2,3) with $(x(0), y(0))$ sufficiently small then there exists a solution $u(t)$ of (5) with

$$x(t) = u(t) + O(e^{-\gamma t}) \qquad (6)$$
$$y(t) = h(u(t)) + O(e^{-\gamma t}) \qquad (7)$$

for $t \to \infty$, $\gamma > 0$.

3) The centre manifold $h(x)$ can be approximated by a function $F(x)$, using the identity

$$\dot{y} = h'(x)[Ax + f(x, h(x))] = Bh(x) + g(x. h(x)) \qquad (8)$$

$$N(h(x)) = h'(x)[Ax + g(x, h(x))] - Bh(x) - g(x, h(x)) = 0 \qquad (9)$$

we obtain by substitution of $h(x)$ in (3) and applying the chain rule. If a function $F(x)$ with $F(0) = F'(0) = 0$ can be found that

$$N(F(x)) = O(|x|^q)$$

$q > 1$, $x \to 0$, then it follows that

$$h(x) = F(x) + O(|x|^q) \text{ as } x \to 0.$$

<u>ILLUSTRATIVE EXAMPLE AND SIMULATION RESULTS</u>

To illustrate the application of the above results we consider the system

$$\dot{x} = c_1 x + c_2 xy + c_3 x^3 + c_4 x^2 y^2 \qquad (10)$$

$$\dot{y} = -y + c_5 x^2 + c_6 xy^2 \qquad (11)$$

The parameter(eigenvalue) c_1 is at $c_1 = 0$ a bifurcation point for the system. If $c_1 < 0$ it is asymptotically stable and for $c_1 > 0$ it is asymptotically unstable. When $c_1 = 0$ the linearized system is weakly stable and we cannot say anything about the stability of the nonlinear system without further investigation.

Taking the ansatz

$$F(x) = O(x^2) = ax^2$$

in equation (9) with (10,11) we get

$$N(F(x)) = ax^2 - c_5 x^2 + O(x^4) \qquad (12)$$

Let $a = c_5$, that is $F(x) = c_5 x^2$ it follows $N(F(x)) = O(x^4)$ and

$$h(x) = c_5 x^2 + O(x^4) \qquad (13)$$

If we substitute (13) in (10) we get the equation which determines the behaviour (including the stability) of the nonlinear system near the stationary point on the centre manifold :

$$\dot{u} = (c_2 c_5 + c_3) u^3 + O(u^6) \qquad (14)$$

Thus the zero solution of (10,11) with $c_1 = 0$ is asymptotically stable if $c_2 c_5 + c_3 < 0$ and unstable if $c_2 c_5 + c_3 > 0$. In case $c_2 c_5 + c_3 = 0$ one has to find a better approximating Taylor series $F(x)$ at $x = 0$ and to repeat the described procedure to find out the stability determining equations and parameters.

At this place it is visible that centre manifold theory does not only leads to simplified models but at the same time gives hints which parameters are critical with respect to nonlinear system behaviour(e.g. stability) and therefore should be estimated with special care.

Numerical simulation studies of the system (10,11) confirm the conclusions about the system stability drawn from equation (14). They also show that the reduced system(consisting of only one differential equation and one algebraic equation)

$$\dot{x}_r = (c_2 c_5 + c_3) x_r^3 \qquad (15)$$

$$y_r = h(x_r) = c_5 x_r^2 \qquad (16)$$

provides a good approximate solution to the original system (10,11).This

is also true when the parameter(real part of eigenvalue) c_1 is not zero
but has a small negative value. The linear term then has to be included
into equation (15) also.

The numerical simulation results demonstrate the fulfillment of the essential demands for acceptable model reduction methods[3] :

-satisfactory correspondence of transient behaviour in the neighbourhood
 of the equilibrium point

-stationary exactness

-correspondence of stability properties

-few computational expense.

The last point deserves some comments. To find out the simplifying nonlinear coordinate transformation, the centre manifold (4) and the reduced model equations (5) one has to perform substitutions of functions in
other functions and to calculate indeterminate coefficients. This are
more algebraic manipulations, that is manipulations of symbols than numerical calculations. Thus for treatment of higher dimensional systems we
prepare the automatic generation of approximations to the nonlinear coordinate transformation (4) and the reduced model equations (5) using a
computer algebra system [1], which takes over the necessary algebraic
calculations.

REFERENCES

[1] BUCHBERGER,B.,COLLINS,G.E.,LOOS,R.;Computer Algebra. Symbolic and
     Algebraic Computation.;Springer Verlag Berlin 1983

[2] CARR,I.;Applications of Centre Manifold Theory; Springer Verlag
     Berlin 1981

[3] FÖLLINGER,O.; Reduktion der Systemordnung; Regelungstechnik <u>30</u>
     (1982), S. 367-377

[4] GUCKENHEIMER,J.,HOLMES,P.; Nonlinear Oscillations, Dynamical Sys-
     tems, and Bifurcation of Vektor Fields; Springer Verlag Berlin 1983

[5] HENRY,D.; Geometric Theory of Semilinear Parabolic Equations;
     Springer Verlag Berlin 1981

[6] KELLEY,A.; The stable, center stable, center, center unstable and
     unstable manifolds.; J.Diff.Equs.<u>3</u> (1967),546-570

[7] MAHMOUD,M.S.,SINGH,M.G.; Large Systems Modelling ;Pergamon Press
     Oxford 1981

[8] WUNSCH,G.; Handbuch der Systemtheory ; Akademie Verlag Berlin 1986

Method of Automated Construction
of System Dynamics Models (ACM)

Michal Kejak, Petr Javorský[1]

High complexity of forms and behaviour of social-economical systems as well as the level of knowledge about them, do not permit even to exclude the formulation of a mathematical model by transferring the valid laws into the mathematical language, as it is in the case of the technical systems in the mathematical-physical analysis. Problematical gaining of necessary data and mostly the impossibility to repeat these experiments hinders the application of empirical identification methods very often used in complex technical systems. One of the ways how to gain such a model of complex social-economical reality can be system dynamics (SD) established by Prof. J. Forrester [1].

The methodology of SD suffers from considerable lack of rigour. It obstructs more effective utilization of SD for modelling of complex problems of societal reality. Our method is based on the formalization of "rules" of SD and their algorithmatization. Such computer-aided methodology removes tedious routine and noncreative work, increases credibility of models, lowers cost and shortens the time needed for the development of the model. In addition, the algorithmical procedure leads the modeller to systematical work and observance of the basic rules of SD without often requiring his knowledge of SD. It thus enables even laymen in the field of SD to construct rigorous SD models.

The phases of the SD model construction process can briefly described as follows:

- conceptualization

- formulation

- testing

- implementation.

For its creative and the least structurized content, the first phase, *conceptualization*, gives the least preconditions for algorithmatization. The causal-loop diagram is the final product of this part of the modelling process.

The core of our method is automation of the second phase of the model building - *formulation* - consisting of two steps:

- *converting causal-loop diagram to SD schematics*

- *converting SD schematics to differential equations.*

The algorithm used for the first step is an improved version of Burns's algorithm [2][5]. This new algorithm for automatical classification of quantities (ACQ) has been developed on the basis of the formulation of axioms applied in SD. the properties of digraphs, cycles in digraphs and information of the dimensions of quantities. The input of the algorithm,

---

[1] Michal Kejak, Petr Javorský, Institute for Application of Computing Technique in Control, Revolucni 24, Praha 1, Czechoslovakia

and consequently of the whole ACM method,   is causal-loop diagram given
in the form of a signed digraph.

Automation of the second step, the converting SD schematics to
differential equations, is no less important for the ACM method.
Therefore has been suggested an original algorithm for automatical
derivation of equations (ADE).

The software system of the ACM method contains further algorithms
for *automatized simulation,* for *automatized analysis* (e.g. experimental
sensitivity analysis) and for *optimal control* [3]. Recently the ACM
method has been extended by a new approach to SD enabling to handle
uncertainty in modelled systems by means of the concept of *fuzzy set*
[4].

References

[1]   J.W.Forrester: Principles of Systems. Cambridge, MA: Wright-Allen,
      1968
[2]   J.R.Burns,O.Ulgen,H.W.Beights: An algorithm for converting signed
      digraphs to Forrester schematics, IEEE Trans. on Systems, Man, and
      Cybern., Vol. SMC-9, No. 3, 1979
[3]   P.Javorský: An algorithm for optimal control of nonlinear systems.
      In: this proceedings
[4]   M.Kejak: Simulation of fuzzy system dynamics models. In: this
      proceedings
[5]   M.Kejak: Automated Construction of System Dynamics Models.
      Dissertation, Prague, 1988 (will be published) (in Czech)

A Modular Computer-Aided Modelling and Simulation System in Chemical Engineering

Siegfried Krüger, Winfried Mylius[1]

Summary: The paper presents a computer-aided modelling and simulation system for problems of chemical engineering described by ordinary and/or partial differential equations.
Keywords: modelling and simulation, chemical engineering, model base, method base, simulation language.

1. Introduction

In recent years simulation as a method for solving problems has developed into one of the most important means of the modern engineer. At the Köthen Engineering College a modular simulation system including methodological and system theoretical aspects is developed. Simulation package will be used for computer-aided modelling (CAM) and simulation of chemical engineering processes. Chemical engineering processes are described by systems of ordinary and/or partial differential equations. The simulation package contains the following components

- model base
- method base
- data base
- graphic tools
- simulation language.

2. The Model- and Method Bases

The efficiency of a simulation system first of all depends on the efficiency of the numerical algorithms. There is no algorithm convinient to all types of differential equations. A well posed mathematical model is necessary. Tools for the modelling, simulation and result description are helpful for the engineer using the simulation package.
Let us consider the model structure and the numerical methods depending upon. The main cases of interest will be
- systems of ordinary differential equations with initial conditions,
- time independent elliptic partial differential equations with boundary conditions,
- time dependent parabolic partial differential equations with initial and boundary conditions.

---

1) Department of Mathematics and Computer Engineering
Engineering College Köthen, Bernburger Str. 52-57
Köthen, 4370, GDR

Furthermore the model structure depends on the dimension of the region,
the order of the differential equations, the type of the boundary conditions, the symmetry properties and the independent variables of the
coefficients.
For every special case of the model strucutre special numerical methods
stored in the method base are needed.
The method base contains three classes of numerical algorithms:
elementary modules, base modules and complex modules.
Elementary modules are such algorithms like algebraic operations based
upon matrices and vectors. Base modules are solvers for well defined
problems like integration methods and solvers for systems of algebraic
equations. Complex modules are programs for special model structures
using elementary and base modules.
The complex modul and the classes of base modules are described by the
model structure. The special base modules are determined automatically
or by the user.
The following table shows some examples of the model structure and the
numerical methods determined:

model structure	numerical methods
initial value problems of ordinary differential equations	solver for systems of ordinary diffe- tial equations
non-stiff-problems	explicit methods of Runge-Kutta-Fehl- berg-type with error estimation
stiff problems	linear implicit method of Rosenbrock- Wanner-type, algorithm of Gear
stiff or non-stiff problems	an integration routine being able to switch between a stiff and non-stiff algorithm
symmetric parabolic partial differential equations in one space variable	mesh generator one-dimensional, run-time-system for a system of ordinary differential equations, linear implicit method with symmetric Jacobian matrix, solver for banded linear algebraic equation system
two-dimensional elliptic problem	mesh generator two-dimensional linear basic functions, solver for linear algebraic equation system

3. The simulation language

The component "dialogue language" of the simulation language being the
base for communicating with the system consists of six parts representing several modes:
- command language for communicating with the system and its components,
- language to describe the "continuous" model (CM),
- language to describe the "discrete" model (DM),
- language to describe the "calculation" model (CALM),
- language for simulation experiments (SEX),
- language for output (OUT).

a) Command language: Inputs in mode of the command language include
commands for reading and writing the model base, the method base and
the data base, the commands to change over to other modes, the commands
to select the natural language (English, German) for setting the statements and commands to call the (parts of) compiler and interpreter.
The implementation is accomplished by a compiler or interpreter.
b) The CM component: It is a descriptive component. We want to describe a class of chemical engineering models given by systems of ordi-
nary/partial differential equations of the general master form:

$$A_0 D_t G = \sum_{k=1}^{3} D_k \left(\sum_{i=1}^{3} A_{ki} D_i G \right) + \sum_{i=1}^{3} A_{i+9} D_i G + A_{13} G + A_{14} \qquad \text{with}$$

$$D_1 := \frac{\partial}{\partial x} \,, \; D_2 := \frac{\partial}{\partial y} \,, \; D_3 := \frac{\partial}{\partial z} \,, \; D_t := \frac{\partial}{\partial t} \quad \text{and} \; A_0 = A_0(x,y,z,t,\underline{G}),$$

$$A_i = A_i(x,y,z,t,G,\hat{\underline{G}},D_1 G,D_2 G,D_3 G,\hat{\underline{G}}') \qquad i = 1, \ldots, 14,$$

$\hat{\underline{G}}$ – vector of the residual functions of the differential
equation system.

In addition to the description of the special differential equations
it is necessary to give attached initial- and boundary conditions and
a description of the region. The implementation of CM is realized by
means of a compiler with the target files:
model structure file, "continuous" model procedures, description of
the technical model and a set of CM-statements.
c) The DM component: Inputs in this mode describe the discrete model,
that means the grid of the region, the basic functions for the FEM etc.
d) The CALM component: In this mode the special base modules are determined and completed by conditions of the convergence for instance.
e) The SEX component: Inputs for the simulation experiment are accomplished by this component for instance simulation parameters and model
parameters.
f) The OUT component: The numerical and graphical outputs are controlled by this component including the selection and manipulation
of the states.

The following figure illustrates the interaction of the several components of the simulation package:

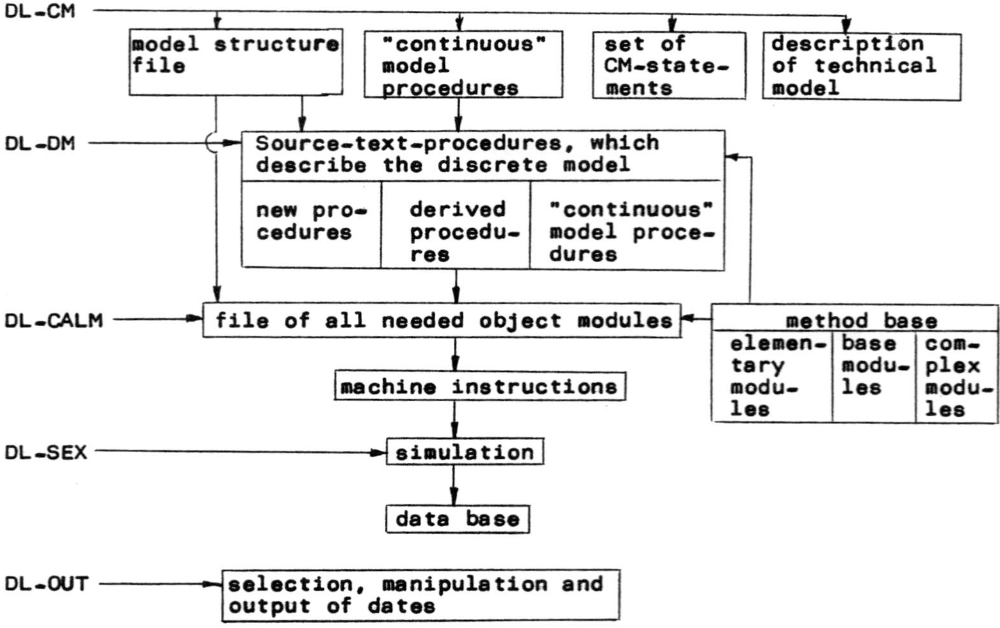

References

/1/ Cellier, F. E.: Simulation Software: Today and Tomorrow, in
    Burger, J., Jarny, Y.(ed.) Simulation in Engineering Sciences,
    Elsevier Science Publishers B.V. (North-Holland) IMACS 1983.

/2/ Gladwell, I.; Wait, R. (ed.)
    A Survey of Numerical Methods for Partial Differential Equations.
    Clarendon Press, Oxford 1979.

/3/ Hake, J.-F.; Homberg, W.: Software-Umgebungen zur Behandlung
    partieller Differentialgleichungen. Angewandte Informatik 5/87,
    Vieweg & Sohn Verlagsgesellschaft m. b. H.

/4/ Machura, M.; Sweet, R. A.: A Survey of Software for Partial
    Differential Equations. ACM Transactions on Mathematical Software.
    Vol. 6, No. 4, Pages 461-488, 1980.

/5/ Mylius, W.: Eine Fachsprache für digitale Simulationssysteme zur
    Lösung partieller Differentialgleichungen aus der Stoffwirtschaft,
    ihre Implementierung in DISIP. Diss. A, Leipzig 1985.

CANDYS/CM – A Dialogue System for Modelling Continuous Dynamical Systems with Chain Structure by Differential Equations

Reinhart Funke[1]

Abstract

For a given set of time series a differential equation system (d.e.s.) has to be performed. A FORTRAN-77 dialogue system has been created for such modelling purposes, in which the special Lotka-Volterra approach to modelling developed in [3] is used.

Introduction

The modeller begins the modelling, previously having a first conception about the structure of the d.e.s., however the structure can be altered during the modelling process. Higher dimensional d.e.s. can be constructed using lower dimensional d.e.s. stored in a model base. The new d.e.s. can be stored again in the model base.

During the modelling process certain operations and algorithms have to be repeated with different data. The codes of the realized main operations are stored in order to generate automatically the right hand side of the d.e.s. It can be either directly calculated interpretatively when the d.e.s. is being numerically integrated by the program system CANDYS, or written out in terms of the programming language used in other programs. Than the r.h.s. can be compiled separately and linked to any object file.

Real systems are not quite deterministic. Stochastic differential equations in Ito's sense are used in order to reflect stochastic fluctuations. A number of one-dimensional stochastic basic processes, mainly diffusion processes, with known distribution, such as the Ornstein-Uhlenbeck process, the linear process and processes with hyper-gamma distribution (see [1,2]) are used, so that confidence intervals ("hoses") for selected components of the d.e.s. with stochastic fluctuations can be stated avoiding thousands of simulation experiments.

---

[1] Akademie der Wissenschaften der DDR, Zentralinstitut für Kybernetik und Informationsprozesse, Kurstrasse 33, DDR-1086 Berlin

The modelling method

The following variables are defined in CANDYS/CM as one-dimensional REAL-arrays of length NT:

```
T                       - for the independent variable (time)
X 1, ...,X NX           - for simulated values
Y 1, ...,Y NY           - for smoothed or performed in other ways time series
Z 1, ...,Z NZ           - for input data
P 1, ...,P NY           - for the time-derivative of the smoothed time series.
```

These variables (time series) can be performed component by component by the following mathematical operations (operations are applied to one or two time series, respectively):

```
+ - addition of two variables
- - subtraction of two variables
* - multiplication of a variable by an other one
/ - division of a variable by an other one
C - copying of a variable to an other one
F - function evaluation of a variable;  the following standard func-
    tions are implemented as yet:
    A - arctg, C - cos, E - exp, I - arcsin, L - log, O - arccos
    Q - square, R - square root, S - sin, T - tg;
I - initialisation of a variable by a constant
L - multiple lineare regression
Q - smoothing the data of a variable with the help of a quadratic
    spline function and calculating the time-derivative
V - valuation by means of the linear regression coefficients.
```

The following service operations aid the modelling process:

```
G - display the graphs or the phase portrait of two variables
P - print the values of a variable
R - read data/a model from the data/model base
S - simulate the model  (numerical integration of the initial value
    problem)
T - definition of the time horizon
W - write data/the model to the data/model base.
```

Differential equations (of the first order) are build up trying to knot the time derivative of the smoothed time series with other (smoothed) time series by the mathematical operations stated above. The model is the better the better is the additional knowledge about the system structure. Several systems in nature, such as food chains, enzyme chains or energy cascades, are characterized by a chain structure (see [3]). Therefore it

is justified to consider d.e.s. with chain structure. Let the i-th component of a d.e.s. be given by the differential equation

$$\frac{d\,X_i(t)}{d\,t} = (\alpha_1\,x_i^{1+\gamma}(t) + \alpha_2\,X_i(t))\,X_{i-1}(t), \qquad X_i(0) = x_i, \qquad x_i > 0, \qquad (*)$$

where $X_{i-1}(t)$ is non-negative and it does not depend on $X_j(t), j > i$. Then
it can be suitably modeled by a diffusion process with known probability
distribution. If the distribution is given explicitely, then confidence
intervals (in dependence on time) can be calculated in discrete points,
interpolated and numerically integrated by quadrature rules. The confidence intervals propagate in this way in the higher components.

The stochastic basic processes

1. The Ornstein-Uhlenbeck process
It is given by the stochastic differential equation (s.d.e.)

$$d\xi = -\alpha\,\xi\,dt + \beta\,dw(t), \qquad \xi(0) = x_o, \qquad \alpha,\beta > 0.$$

Its p.d.f. is normal with asymptotically constant parameters [1]. Therefore it is used to model stationary parameters.
2. The linear process
It is given by the linear s.d.e.

$$d\xi = \alpha\,\xi\,dt + \beta\,\xi\,dw(t), \qquad \xi(0) = x_o, \; x_o > 0, \; \beta > 0.$$

The p.d.f. of the process $\eta(t) = \log\xi(t)$ is normal. The mean of $\xi(t)$ is
equal to the solution of the corresponding deterministic equation $(\beta=0)$.
It is used to model the i-th stage of d.e.s. with the following chain
structure

$$\frac{d\,X_i(t)}{d\,t} = \alpha_i\,X_{i-1}(t)\,X_i(t), \qquad X_i(0) = x_i, \; x_i > 0.$$

3. A simple nonlinear process
It is given by the nonlinear s.d.e.

$$d\xi = (\alpha_1\,\xi^{1+\gamma} + \alpha_2\,\xi)\,dt + \beta\,\xi^{1+\gamma/2}\,dw(t), \quad \xi(0) = x_o, \; x_o > 0,$$

$\beta > 0, \; \gamma \neq 0, \; (\beta^2-2\alpha_1)/\gamma \geq 0$. The process $\eta(t) = x_o^\gamma\,\xi^{-\gamma}(t)$ possesses
hyper-gamma distribution [2]. $\xi(t)$ is used to randomize equ. $(*)$.

References

[1] Bharucha-Reid, A.T., Elements of the Theory of Markov Processes and
    Their Applications, McGraw-Hill, 1960.
[2] Funke, R., A Nonlinear Diffusion With Hyper-gamma Distribution, Sto-
    chastic Analysis and Applications, 6 (1988) 3 (in preparation).
[3] Peschel, M., W. Mende, The Predator-Prey Model: Do We Live in a Vol-
    terra-World ?, Akademie-Verlag Berlin and Springer-Verlag Wien, 1986.

Control of the Observation Process by Probability Criterion

M.N.Krasilshchikov, V.I.Karlov [1]

<u>Problem statement</u>. Consider the system model

$$\dot{x} = Ax + F\xi, \quad t \in (0,T);\tag{1}$$

where $x(t)$ is an $n \times 1$ phase vector, $A(t)$ is an $n \times n$ matrix, $\xi(t)$ is an $m \times 1$ white noise with an intensity matrix $D_\xi(t)$. The original conditions for (1) are a Gaussian vector x with characteristics $\hat{x}_0, \hat{P}_0$ The system is observed during the interval $(0,T)$

$$y = \gamma Hx + \eta,\tag{2}$$

where $y(t)$ is an $\ell \times 1$ vector, $H(t)$ is an $\ell \times n$ matrix, $\eta(t)$ is an $\ell \times 1$ white noise with a nonsingular intensity matrix $D_\eta(t)$ which is not correlated with the noise $\xi(t)$. The function $\gamma(t)$ formalizes the observation programme and satisfies the following restrictions:

$$\gamma(t) \in \Gamma, \quad \Gamma = \{0,1\},\tag{3}$$

$$\tau_\Sigma = \int_0^T \gamma(t)\,dt \leqslant \bar{\tau}_\Sigma\tag{4}$$

where $\bar{\tau}_\Sigma$ is the maximum time interval for the observation. The estimate of an $z \times 1$ vector z at the moment $t = T$ is needed: $z = C^T x(T)$, where C is an $n \times z$ matrix, $\operatorname{rang} C = z$, $z < n$. Then the estimate $\tilde{z}$ of the vector z is $\tilde{z} = \Phi(y(\cdot))$, where $y(\cdot)$ is the $y(t)$ realization during the interval $(0,T)$; Φ is the operator. The synthesis of the operator $\Phi(\cdot)$ will be carried out in two stages: a) forming sufficient statistics of $p(z|(\cdot))$; b) solving a probability optimization problem in the space R^z.

The first stage is realized on the basis of the continuous Kalman filter

$$\begin{cases} \dot{x}^* = Ax^* + \gamma P^* H^T D_\eta^{-1}(y - Hx^*), \quad x^*(0) = \hat{x}_0 \\ \dot{P}^* = AP^* + P^*A^T - \gamma P^*H^T D_\eta^{-1} HP^* + FD_\xi F^T, \quad P^*(0) = \hat{P}_0 \end{cases}\tag{5}$$

Thus, if $t = T$, then $M[z|y(\cdot)] = z^* = C^T x^*(T)$, $M[(z-z^*)(z-z^*)^T|y(\cdot)] = P_z = C^T P^*(T)C$ and $P(z/y(\cdot)) = N\{z^*, P_z\}$.

Probability criterion of the quality of the vector z estimate $\tilde{z}$: it is necessary to minimize the characteristic size β of the criterion area $z(\beta)$, which satisfies the condition $\beta_1 \leqslant \beta_2 \Leftrightarrow z(\beta_1) \supset z(\beta_2)$ in the space R^z, i.e. to solve the following problem

$$\tilde{z}_\alpha = \underset{\tilde{z} \in R^z}{\arg\min}\, \Phi_\alpha(\tilde{z}).\tag{6}$$

where $\Phi_\alpha(\tilde{z}) = \tilde{\beta}_\alpha$; $\tilde{\beta}_\alpha$ is a quantile of the function $\Phi(z - \tilde{z})$, where $\Phi(z - \tilde{z})$ is an equation of the area $z(\beta)$ level line; with β_α satisfying the condition $\mathscr{P}\{\Phi(z - \tilde{z}_\alpha) \leqslant \tilde{\beta}_\alpha\}$, where α is a given

---

[1] Moscow Aviation Institute, 4, Volokolamskoye Shosse, 125871, GSP, Moscow, USSR.

guaranteeing probability.

Application of the generalized minimax (GMM) approach [1] . The GMM
approach is based on a transfer from problem (5) to an equivalent minimax one

$$\tilde{z}_d = \arg\min_{E \in \mathcal{E}_d} \ \min_{\tilde{z} \in R^z} \ \max_{z \in E} \ \Phi(z - \tilde{z}), \tag{7}$$

where $\mathcal{E}_d$ is a set E family in the space of the vector $z = N\{z^*, P_z\}$,
having a probability measure d . For the symmetrical area $z(\beta)$ the
following result is true $\tilde{z}_d = z^*$ and $\tilde{\beta}_d = \min\limits_{E \in \mathcal{E}_d} \max\limits_{z \in E} \Phi(z - z^*)$. The
calculation of $\tilde{\beta}_d$ is carried out with respect to the vector $z - z^* = N\{0, P_z\}$,
so that $\tilde{\beta}_d = \mathcal{Y}\{P_z\}$ and planning the observation process can be carried
out a priori solving the problem of controlling Riccati's equation in
(5) with restrictions (2),(3) on the basis of the criterion

$$\mathcal{I} = \tilde{\beta}_d = \mathcal{Y}\{P_z\} = \mathcal{Y}\{C^T P^*(T) C\} \longrightarrow \min_{\gamma \in \Gamma} . \tag{8}$$

The equivalent problem [2] . Consider the system

$$\begin{cases} \dot{S} = -A^T S + \gamma H^T D_\eta^{-1} H Q \\ \dot{Q} = A Q + F D_\xi F^T S, \ t \in (0,T) , \end{cases} \tag{9}$$

where $S(t), Q(t)$ are $n \times z$ matrices with the original condition

$$\hat{P}_0 S(0) - Q(0) = 0, \tag{10}$$

(9),(10) yeild $P^*(t) S(t) = Q(t), \ \forall t \in (0,T)$. As a result, if we assume that
$S(T) = C$, then $\mathcal{I} = \mathcal{Y}\{C^T P^*(T) C\} = \mathcal{Y}\{C^T Q(T)\}$ and the problem of
controlling Riccati's equation is equivalent to the problem of control-
ling system (9), which is linear by the phase variables. The numerical
algorithm of solving the equivalent problem is based on using the
principle of maximum, with the system (9) linearity taken into account.

The numerical solving algorithm. As system (9) is linear with respect
to the variables $S(t)$ and $Q(t)$, the equivalent control problem can be
successfully solved by using Pontryagin's maximum principle. We shall
not consider the technique of using the principle, and shall concentrate on the final result – the numerical algorithm of optimization
problem solving.

1. An initial observation programme $\gamma^\circ(t)$ is defined.

2. In accordance with $\gamma^\circ(t)$ the $X^\circ - n \times z$ matrix is defined:

$$X^\circ = (\Lambda_{SQ}^\circ(T,0) + \Lambda_{QQ}^\circ(T,0) \hat{P}_0)(\Lambda_{SS}^\circ(T,0) + \Lambda_{SQ}^\circ(T,0) \hat{P}_0)^{-1} C, \tag{11}$$

where $\Lambda_{SQ}^\circ, \Lambda_{QQ}^\circ, \Lambda_{SS}^\circ, \Lambda_{SQ}^\circ$ are blocks of the $n \times n$ fundamental matrix of
system (9). The value of the optimality criterion corresponding to the
observation programme $\gamma^\circ(t)$ is calculated as a result of performing

the operation $\mathcal{J}^\circ = \mathcal{Y}\{C^\mathsf{T}x^\circ\}$

3. A new observation programme is sought for

$$\gamma^1(t) = \begin{cases} 1, & \text{if } M^\circ(t) \geqslant d, \\ 0, & \text{if } M^\circ(t) < d \end{cases} \tag{12}$$

where $M^\circ(t) = t_\ell\{Q^{\circ\mathsf{T}}H^\mathsf{T}D_\eta^{-1}HQ^\circ[\mathcal{Y}_P^\circ]\}$ is a programme function; $[\mathcal{Y}_P^\circ]$ is an $\ell \times \ell$ matrix of partial derivatives of the criterion $\mathcal{J}^\circ = \mathcal{Y}\{P_z^\circ\}$ with respect to $P_z^\circ = C^\mathsf{T}x^\circ$; d is the Lagrange multiplier conditioned by the restriction $\int\limits_0^T \gamma^1(t)\,dt = \bar{\tau}_\Sigma$. The matrix function $Q^\circ(t)$ present in (12) is defined as a result of integrating system (9) from right to left with the restricting conditions $\varsigma^\circ(T) = C,\ Q^\circ(T) = x^\circ$.

Operations 2-3 are repeated with substituting iteration number 1 for number 0, iteration number 2 for number 1, etc.

Guaranteed convergence of the algorithm suggested can be achieved if in operation 3 the control $\gamma^{1'}(t)$ is used instead of the control $\gamma^1(t)$, the former being obtained by combining the controls $\gamma^\circ(t)$ and $\gamma^1(t)$.

References

[1] Kibzun, A.I., V.V.Malyshev: Technical cybernetics, no.1 (1984), 99–106.
[2] Zverev, A.I., V.I.Karlov, M.N.Krasilshchikov: Technical cybernetics no.4 (1984), 102–110.

Simulation Analysis of a Nonparametric Algorithm
for Identification of Discrete-Time Hammerstein System

Jakub Markowski and Maciej Popkiewicz [1]

1. Introduction.

The paper deals with the discrete-time Hammerstein system shown in Fig. 1. The system is described by the equation $y_n = \sum_{i=1}^{n} k_{n-i} m(u_i)$, $i \in C$. Greblicki and Pawlak [1,2,3] presented a new approach for identification of this system based on nonparametric estimate of regression function. For recovering the characteristic of the nonlinear subsystem the Watson — Nadaraya nonparametric kernel estimate is applied. The weighting function of the dynamical subsystem is recovered by the correlation method. In the pre-cited papers a pointwise consistency of the estimate is prooved, the rate of convergency is analised, and convergency in the global sense (mean integrated square error — MISE) is studied.

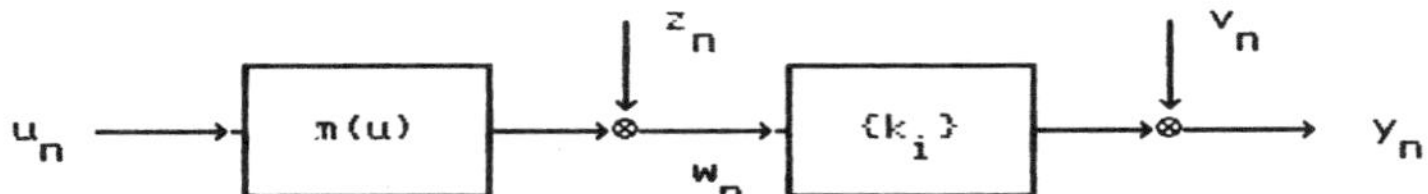

Fig. 1 The identified Hammerstein system.

In this paper the results of simulation analysis of the identification algorithm are presented. The main results concern :accuracy of identification of nonlinear subsystem for finite number of observations, and sensitivity of MISE to the number of observations, type and the parameters of the transfer function of the dynamical subsystem, type of the kernel and its bandwith parameter.

2. Identification algorithm and its properties.

In the system shown in' Fig. 1 an output of the nonlinear subsystem is disturbed by an additive noise z_n, an output of the linear dynamical subsystem having weighting function $\{k_i\}$ is disturbed by an additive noise v_n. Both, signal z_n and signal v_n are independent zero mean, stationary white noises with finite variance. The signals z_n and v_n are independent on input white noise u_n and a signal w_n respectively. The signal w_n interconecting both subsystems is inaccesible to measurements. Because both m and $\{k_i\}$ are recovered from observations (u_i, y_i), $i=1,2,\ldots$ we can estimate them only within to some unknown

---

[1] Institute of Engineering Cybernetics, Technical University of Wrocław, Wybrzeże Wyspiańskiego 27, 50-370 Wrocław, Poland.

factors. For estimation of $m(\cdot)$ we apply

$$\hat{m}(u) = \left[\sum_{i=1}^{n-j} Y_{i+j} \, K\!\left((u-U_i)/h_n\right) \right] \Big/ \sum_{i=1}^{n-j} K\!\left((u-U_i)/h_n\right), \qquad (1)$$

where K is a suitably selected kernel, $\{h_n\}$ is a sequence of positive numbers (bandwidth parameter).

Under assumptions : $m(u)=-m(-u)$, and $|m(u)|\le c_1+c_2|u|$, where c_1, c_2 are positive constans; U_n's are symmetrically distributed i.e. $f(u)=f(-u)$, variance of U_n is finite ; the kernel K fullfilled the restrictions $|u|K(u)\to 0$ as $|u|\to \infty$, $\int K(u)du < \infty$, and the linear dynamical subsystem is asymptotically stable the following theorems have been obtained:

Theorem 1 [1]: Let $h_n\to 0$ as $n\to \infty$ and $n\,h_n\to\infty$ as $n\to \infty$ then $\hat{m}(u)\to \beta\, m(u)$ as $n\to \infty$ in probability at all points $u\in C(f,m)$ at which $f(u)> 0$. ∎

Theorem 2 [1]: Let both m and f have bounded derivatives up to the second order in some neighborhood of $u\in R$ and let $\int xK(x)dx=0$, $\int x^2K(x)dx <\infty$. If $h_n \sim n^{-1/5}$ then $|\hat{m}(u)-\beta m(u)|=O(n^{-2/5})$ in probability. ∎ (Remark: For independent observations $h_n \sim n^{-1/5}$ is asymptotically optimal.)

Theorem 3 [3] (Global estimation error): Let the assumptions of *Theorem 2* are satisfied and $\sup|m(u)|<\infty$. Then $E \int \left(\hat{m}(u)-\beta m(u)\right)^2 du \to 0$. ∎

In the numerical experiments described below as a quality index of estimation an empirical value of MISE was used.

3. Range of numerical investigation.

The simulations were performed on an IBM PC XT digital computer with mathematical co-processor 8087. The PASCAL programming language was used. The signals u_n, z_n and v_n were generated using the Bays and Durham [4] algorithm for uniformly distributed random variables and the Box and Muller [4] algorithm for normal r.v.. The maximum number of observations (U_i, Y_i) was choosen to be 1500. Four types of nonlinearity of memoryless subsystems were investigated: $m(u)=(u$ for $|u|\le 1$ and $\mathrm{sign}(u)$ otherwise) — saturation; $m(u)=(0$ for $|u|\le 1$ and $\mathrm{sign}(u)$ otherwise); $m(u)=u\,\mathrm{sign}(u)$ and $m(u)=(2/\pi)\mathrm{arctg}(u)$. The following dynamical subsystems were investigated A: $y_n=u_n$, B: $y_n-0.5y_{n-1}=u_n$, C: $y_n-0.95y_{n-1}=u_n$, D: $y_n=u_n-0.5u_{n-1}+0.1u_{n-2}$, E: $y_n=u_n-5u_{n-1}+0.2u_{n-2}$. The systems B and C are recursive filters. An impuls response of the system C tends to zero very slowly. E and D are simple filters (with finite memory) and their equations parameters were choosen tangibly different. In the estimator (1) the Gauss kernel $K(u)=(2\pi)^{-1/2}\exp(-u^2/2)$ and the window kernel $K(u)=(1/2$ for $|u|\le1$ and 0 otherwise) were applied. The bandwidth parameter was selected as $h_n=cn^{-\alpha}$, where $c= 0.25, 0.5, 1, 2, 4$ and $\alpha= 0.025, 0.1, 0.2, 0.4$. The empirical value of MISE was calculated from

$$Q_n=(1/10) \sum_{i=1}^{10} \int_{-2}^{2} \left(\hat{m}_{ni}(u)-m(u)\right)^2 du, \qquad (2)$$

where $\hat{m}_{ni}(u)$ denotes $\hat{m}(u)$ for the i-th sequence $\{(u_k, y_k)\}_{k=1}^{n}$.

4. Results of simulation.

The main goal of simulation was investigation of the estimator behavior for finite number of observations. Selected computational results are depicted in figures and tables. In all the cases, if not stated otherwise, $U_n \sim$ uniform$(-3,3)$, $Z_n, V_n \sim$ normal$(0, 0.1)$, $K(u)$ – window kernel, nonlinear subsystem– saturation. The assumptions of *Theorem 3* [2] were satisfied in each case, but decreasing of performance index (2) was relatively slow for n>1000. From the other side, application of the estimator (1) for n<100 gave unsatisfactory results . In Fig. 2. composite results of simulation for the linear objects A ÷ E are available. The worst results of identification were obtained for the objects C and E. In the first case the reason was very slow decrease of weighting function and consequently strong dependence of the pairs (U_i, Y_i). In the second case decided a big value of the ratio k_1/k_0. The last reason may be eliminated by seting j>0 in (1). For the linear object E and for j≠0 we obtained $Q_{200}=3.71$, $Q_{500}=1.56$ but for j=1 $Q_{200}=0.012$, $Q_{500}=0.007$. Dependency of identification accuracy on the offset parameter j is also shown for the linear object B in Table 1 . Conclusions of these results are presented in Sec. 5.

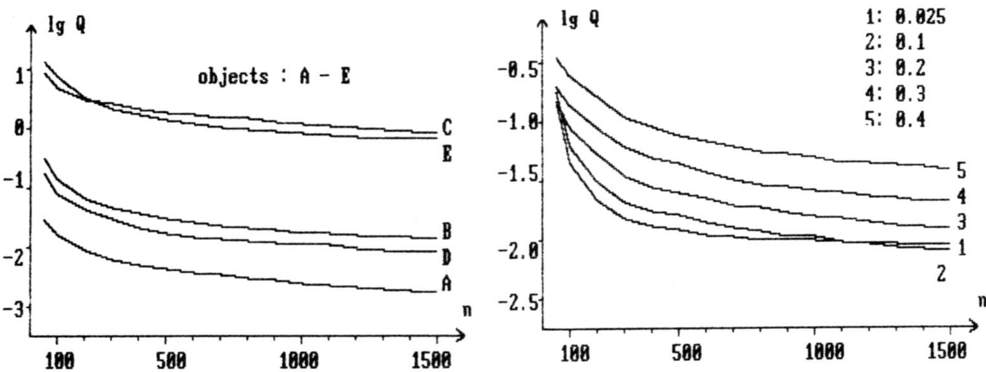

Fig. 2. Identification error Q_n vs. number of observations for
    a) selected objects,   b) different parameters α of h_n.

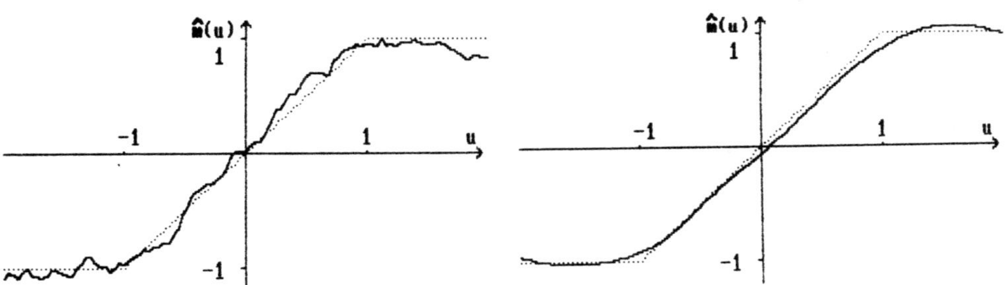

Fig. 3. Estimate $\hat{m}_{200}(u)$ for   a) the window kernel,   b) the Gauss kernel.

Computational experiments did not show any effect of a kernel type on identification accuracy but the Gauss kernel gives more smooth $\hat{m}(u)$ then the window kernel what is shown for the dynamical object B in Fig. 3.

Selected results showing an effect of bandwidh parameter $h_n = cn^{-\alpha}$ on identification accuracy are summarized in Table 2 and Fig. 2b. In this case nonlinear characteristic was choosen as $m(u) = (2/\Pi) \arctg(u)$ and $U_n \sim$ normal$(0,3)$ hence the assumptions of *Theorem 2* [1] were satisfied. A parameter α of 0.1 was found to yield the best identification accuracy for $n > 1000$. An influence of a parameter c on Q_n (2) for $\alpha = 0.2$ is shown in Table 2.

Table 1. Identification error Q_n vs. offset parameter j.

$n \backslash j$	0	1	2	3	4
200	0.066	0.584	3.031	14.54	51.95
500	0.033	0.347	1.523	6.61	30.86

Table 2. Identification error $Q_n \cdot 10^2$ vs. parameter c of h_n.

$n \backslash c$	0.25	0.50	1.00	2.00	4.00
500	9.509	4.718	2.512	1.725	3.163
1500	3.843	2.127	1.220	0.864	1.453

5. Final remarks. Modification of the algorithm.

Generally the results of simulation show consistency of the properties of the algorithm for finite number of observations with the asymptotical properties described in recently published papers. However the results show also that the problem of optimal choise of the bandwidth parameter for finite number of observations is still open. The results shown in Fig.2b. are rather unexpected in comparison with *Theorem 2* [1].

The strong dependency of (1) on parameter j has been observed (Tab.1, Q_{200}, Q_{500} for system B).Parameter j related to the maximum value k_j of a weighting function of the dynamical subsystem gives the best accuracy of estimation. This leads to the following modification of the estimation algorithm: *Step 1:* estimation of the weighting function $\hat{k}_i$, $i = 1, \ldots, i_{max}$; *Step 2:* Calculating j^*: $|k_{j^*}| \geq |k_i|$, $i = 1, \ldots, i_{max}$; *Step 3:* Application of algorithm (1) with $j = j^*$ (instead of $j = 0$).

References

[1] Greblicki, W.,Pawlak, M.: Identification of Discrete Hammerstein System Using Kernel Regression Estimates, IEEE Trans. Automat.Contr., Vol. AC-31 (1986),1,74-77.

[2] Greblicki, W.,Pawlak, M.: Nonparametric Identification of Two-Channel Nonlinear Systems, Proc. 25th Conf. on Decision and Control, Athena (1986), 2012-2015.

[3] Greblicki, W.,Pawlak, M.: Hammerstein System Identification by Nonparametric Regression Estimation, Int. J. Control, Vol.45, (1987) ,1, 343-345.

[4] Press, W.H.,Flanney, B.P.,Teukolsky, S.A.,Vetterling, W.T.: Numerical Recipes, Cambridge University Press, 1986.

Remarks on Pole Assignment by Constant Output Feedback

Edward Jezierski [1]

1. INTRODUCTION

Consider the linear dynamical system described by the equations

$$\dot{x}(t) = Ax(t) + Bu(t) \qquad (1)$$

$$y(t) = Cx(t) \qquad (2)$$

where $x(t) \in R^n$, $u(t) \in R^m$, $y(t) \in R^r$ and A, B, C are constant matrices of appropriate dimensions. The Wonham's result states that if the system is controllable then it is also pole-assignable using a constant state feedback. Unfortunately, in practical situations usually the state vector is not accessible but only the output vector. In that case the problem of pole assignment consists in finding the constant output feedback

$$u(t) = Ky(t) \qquad (3)$$

to assign the prespecified location of the closed-loop system poles

$$\sigma(A + BKC) = \{\hat{\lambda}_1, \hat{\lambda}_2, \ldots, \hat{\lambda}_n\} \qquad (4)$$

which are real valued or complex valued in conjugate pairs.

In this problem there arise two main questions:

1° What conditions must be satisfied by the matrices A, B, C to assure an existence of a real valued K for a freely chosen spectrum (4);

2° How to find the desired matrix K in the simplest way.

So far there are known partialy answers to the first question presented in [1-4,7-9]. By counting dimensions, it is clear that a necessary condition for pole placement using output feedback is $mr \geq n$ but it is not a sufficient condition. The sharpest result to date, due to Kimura (1975) asserts that if the system is controllable and observable and if $n \leq m+r-1$ an almost arbitrary set of distinct closed-loop poles is assignable. Recently Fletcher and Magni (1987) examined in detail exact pole assignment under the same condition and pointed some pathological cases. Herman and Martin (1981) proved that if $mr > n$ and the matrices $CB, CAB, \ldots, CA^{n-1}B$ are lineary independent then generic pole assignment is possible provided K is allowed to be complex-valued. Morse,

[1] Technical University of Łódź, Żwirki 36, 90-924 Łódź, Poland

Wolovich and Anderson (1983) analysed detaily the problem of output assignability of 2-output, m-input and 2m-dimensional systems, for the cases m=2 and m=3. They showed that the nature of the problem is nonlinear and they characterized the classes of generically assignable and completely assignable systems for this cases.

The main purpose of this paper is to show a constructive procedure which reduces the problem of output pole placement to finding a solution of the set of nonlinear scalar algebraic equations. This is very useful in particular cases to discuss the problem of output assignability and to obtain the solution for K in an effective manner.

2.   RELATIONS BETWEEN OPEN- AND CLOSED-LOOP CHARACTERISTIC POLYNOMIALS AND A FEEDBACK MATRIX

Let us assume that the system described by (1) possesses a characteristic polynomial of the form

$$d(\lambda) = \det(\lambda 1_n - A) = \lambda^n + \sum_{i=0}^{n-1} a_i \lambda^i, \tag{5}$$

and after introducing the feedback (3), the characteristic polynomial of the closed-loop system is

$$\hat{d}(\lambda) = \prod_{i=0}^{n} (\lambda - \hat{\lambda}_i) = \lambda^n + \sum_{i=0}^{n-1} \hat{a}_i \lambda^i \tag{6}$$

Jezierski and Kuźmiński (1987) proved that for the coefficients of the polynomials $d(\lambda)$ and $\hat{d}(\lambda)$ the following identity holds:

$$\begin{bmatrix} a_0 - \hat{a}_0 \\ a_1 - \hat{a}_1 \\ \vdots \\ a_{n-1} - \hat{a}_{n-1} \end{bmatrix} = \begin{bmatrix} a_1 & a_2 & \cdots & a_{n-1} & 1 \\ a_2 & a_3 & \cdots & 1 & 0 \\ \vdots & \vdots & & \vdots & \vdots \\ 1 & 0 & \cdots & 0 & 0 \end{bmatrix} \begin{bmatrix} w_1 \\ w_2 \\ \vdots \\ w_n \end{bmatrix} \tag{7}$$

where the scalar elements w_α for $\alpha = 1, 2, \ldots, n$ are given by

$$w_\alpha(K) = \sum_{P=1}^{\min(\alpha, m)} (-1)^{P-1} \sum_{\substack{\langle j_1, \ldots, j_P \rangle \\ \langle q_{j_1}, \ldots, q_{j_P} \rangle}} \left\| \begin{bmatrix} k_{j_1} \\ \vdots \\ k_{j_P} \end{bmatrix} \begin{bmatrix} CA^{q_{j_1}} b_{j_1} \cdots CA^{q_{j_P}} b_{j_P} \end{bmatrix} \right\| \tag{8}$$

k_{ji} denotes the ji-th row of K and b_{ji} is the ji-th column of B. The summation is carried out for all sets of natural numbers $j_1,\ldots,j_p$ which satisfy the condition $1 \le j_1 < j_2 < \ldots < j_p \le m$ and for all sets of non-negative integers $q_{j_1},\ldots,q_{j_p}$ for which $q_{j_1} + q_{j_2} + \ldots + q_{j_p} = \alpha - p$.

From (7) it follows that the vector w of all coefficients $w_\alpha(K)$ is uniquely determined by the spectrum (4). Let the mr-vector k be formed from rows of the feedback matrix in the following way $k = [k_1\ k_2\ \ldots\ k_m]^T$ then the vector w takes the form

$$w = E_1 k + E_2(k \otimes k) + \ldots + E_m(k \otimes k \otimes \ldots \otimes k) \tag{9}$$

where $\otimes$ denotes a tensorial product of vectors and E_1, $E_2,\ldots,E_m$ are the matrices which uniquely depend on A, B, C. The matrix E_1 represents linear dependencies between k and w, and from (8) it follows that $E_1 = H^T$, where $mr \times n$ matrix H is equal to

$$H = \begin{bmatrix} Cb_1 & CAb_1 & \ldots & CA^{n-1}b_1 \\ Cb_2 & CAb_2 & \ldots & CA^{n-1}b_2 \\ & & \ldots & \\ Cb_m & CAb_m & \ldots & CA^{n-1}b_m \end{bmatrix} \tag{10}$$

Unfortunately forms of the matrices $E_2,\ldots,E_m$ are not as clear as the form of E_1. Additionaly it is worth noting that in a general case of m-input system calculation of the vector k from the equation (9) is rather difficult.

3. THE CASE $n=4$, $m=2$, $r=2$

In this case the condition $n \le m+r-1$ is not satysfied but it is the first nontrivial case when the relation $n \le mr$ is fulfilled, so we can expect existing the matrix K which assign the spectrum (4). From (8) it follows that the vector w is of the form

$$w = H^T k - g|K| \tag{11}$$

where

$$g = \begin{bmatrix} 0 \\ |Cb_1\ Cb_2| \\ |CAb_1\ Cb_2| + |Cb_1\ CAb_2| \\ |CA^2 b_1\ Cb_2| + |CAb_1\ CAb_2| + |Cb_1\ CA^2 b_2| \end{bmatrix} \tag{12}$$

If H is invertible then it is possible to find the solution of (11) for the vector k as a function of the determinant $|K|$, namely

$$k = H^{-T}(w + g|K|),\qquad(13)$$

what is equivalent to

$$K = \begin{bmatrix} \tilde{h}_1(w + g|K|) & \tilde{h}_2(w + g|K|) \\[2mm] \tilde{h}_3(w + g|K|) & \tilde{h}_4(w + g|K|) \end{bmatrix}\qquad(14)$$

where $\tilde{h}_1$, $\tilde{h}_2$, $\tilde{h}_3$, and $\tilde{h}_4$ are rows of H^{-T}. By comparison of determinants of both sides of the above matrix the following scalar quadratic equation with only one unknown $|K|$ is obtained

$$g^T M g |K|^2 + (w^T M g + g^T M w - 1)|K| + w^T M w = 0\qquad(15)$$

where $M = (\tilde{h}_1)^T \tilde{h}_4 - (\tilde{h}_3)^T \tilde{h}_2$. The last equation gives direct possibility to discuss the problem of existing the real-valued solution for K.

4.   REFERENCES

[1] Brockett R.W., Byrnes Ch.,I.: Multivariable Nyquist criteria,root loci, and pole placement: a geometric viewpoint. IEEE Trans. Autom. Control AC-26(1981)1, 271-283.

[2] Fletcher L.R.:   Exact pole assignment by output feedback - part 2. Int. J. Control 45(1987)6, 2009-2019.

[3] Fletcher L.R., Magni J.F.: Exact pole assignment by output feedback - part 1. Int. J. Control 45(1987)6, 1995-2007.

[4] Hermann R., Martin C.: Applications of algebraic geometry to systems theory - part 1. IEEE Trans. Autom. Control AC-22(1977)1, 19-25.

[5] Jezierski E., Kuźmiński K.: New results in modal control synthesis. Systems Science 7(1981)1, 42-54.

[6] Jezierski E., Kuźmiński K.: Comments on relations between coefficients of the closed-loop characteristic polynomial and output feedback gains. Int. J. Control 46(1987)4, 1487-1489.

[7] Kimura H.: Pole assignment by gain output feedback. IEEE Trans. Autom. Control AC-20(1975)4, 509-516.

[8] Magni J.F.: Exact pole assignment by output feedback - part 3. Int. J. Control 45(1987)6, 2021-2033.

[9] Morse A.S., Wolovich W.A., Anderson B.D.O.: Generic pole assignment: preliminary results. IEEE Trans. Autom. Control AC-28(1983)4, 503-506.

On Computational Solution of Differential Equations with Delay

P.S.Szczepaniak[1] and A.Małolepszy[1]

A computational method of solving nonlinear differential-difference equations is described. In the method, the initial function is determined approximately with the use of the orthogonal series expansion technique and the final problem is in the form of a sequence of initial value problems for ordinary differential equations. A numerical example is given to illustrate accuracy of the method.

1. Introduction

The differential-difference equations with retarded argument

$$\dot{x}(t) = f[t,x(t),x(t-h)] , \quad t > 0 \tag{1}$$

$$x(t) = \eta(t), \quad -h \leq t \leq 0 , \quad h > 0 \tag{2}$$

where η is a given initial function, are often met in both theory and practice. For the theory and numerical methods of solving differential equations of this kind see [2,7] and [5,6], respectively.

In the present paper, a method of finding numerical solution of equation (1) is presented and for convenience, the one-dimensional case is considered. The extension to the multidimensional case is not difficult. The method uses Bellman's idea [1] but the initial function is generated here in a different way. Bellman's approach was also applied in [3,4] and [12].

2. Computational solution
2.1. Reduction to a system of ordinary differential equations

Following Bellman [1] , let the sequence of functions

$$x_k(s) = x([s+(k-1)]h) , \tag{3}$$

where $0 \leq s \leq 1$ and $k=0,1,2,\ldots$, be introduced. Equation (1) may now be written for $k=1,2,\ldots$ as the following system of equations

$$\dot{x}_k(s) = h\, f[(s+(k-1))\,h,\, x_k(s),\, x_{k-1}(s)] , \quad 0 \leq s \leq 1 \tag{4}$$

with the initial conditions

$$x_k(0) = x_{k-1}(1). \tag{5}$$

To determine the solution of equation (1) from the system (4) one needs to know the function $x_o(s)=\eta[(s-1)\,h]$ and the values $x_k(0)$ for $k=1,2,\ldots$. The initial function $x_o(s)$ may be stored (which is of no interest here) or generated in any possible way, while system (4) is integrated.

---

[1] Technical University of Łódź, Dept of Technical Physics and Applied Mathematics /I-1/, Piotrkowska 220, 90-369 Łódź, Poland

2.2. Bellman s approach

In his approach, if the initial function $x_0(s)$ is a polynomial, a trigonometric polynomial or a sum of exponentials, it can be generated by means of the equation

$$x_0^{(n)}(s) + b_1 x_0^{(n-1)}(s) + \ldots + b_n x_0(s) = 0 \tag{6}$$

with the initial values

$$x_0(0)=c_0 \; , \quad x_0(0)=c_1 \; , \quad \ldots \; , \quad x_0^{(n-1)}(0)=c_{n-1} \; . \tag{7}$$

In each k-th step ($k=1,2,\ldots$), the system involving equation (6) and k equations of the form

$$\dot{x}_i(s) = h \, f[(s+(i-1))h, \, x_i(s), \, x_{i-1}(s)], \; i=1,2,\ldots,k \tag{8}$$

with known $x_i(0)=d_i$ is solved in the interval $[0,1]$. The values d_i become known in the previous steps from (5), i.e. $d_i=x_{i-1}(1)$ with $i=1,2,\ldots,k$. As seen Bellman's approach avoids both storage of the values of the previous segment of solution as well as a possible interpolation of the past values which were not computed by the applied method of numerical integration.

The alternative method proposed below retains the same advantages.

2.3. An alternative method

Consider a two-argument function $x(t,\theta)= x(t+\theta)$, $t \geqslant 0$, $-h \leqslant \theta \leqslant 0$, which represents a segment of the function subject to pure delay effect. A model of the delay may be given in the form of the following partial differential equation [10]

$$\frac{\partial x(t,\theta)}{\partial t} = \frac{\partial x(t,\theta)}{\partial \theta} \tag{9}$$

with the boundary conditions

$$x(t,0)= x(t) \text{ and } x(0,\theta)=\eta(\theta) \tag{10}$$

assuming that the function $x(t,\theta)$ is differentiable with respect to t and θ in its domain. The substitutions

$$t=[s+(k-1)]\,h \; , \quad \theta=(\tau-1)\,h$$

where $0 \leqslant s \leqslant 1$, $0 \leqslant \tau \leqslant 1$ and $k=1,2,\ldots$, allow to write (9) in the form

$$\frac{\partial x_k(s,\tau)}{\partial s} = \frac{\partial x_k(s,\tau)}{\partial \tau} \tag{11}$$

with (after an additional use of (3)) conditions (10) transformed to

$$x_k(s,1)= x_k(s) \text{ and } x_0(1,\tau)= x_1(0,\tau)= x_0(\tau)=\eta[(\tau-1)h]. \tag{12}$$

If, for $k=1$, the considered function can be expanded into an orthogonal series, then

$$x_1(s,\tau)=\sum_{i=0}^{\infty} p_i(\tau)\,a_i(s) \tag{13}$$

a.e. in the interval $0 \leqslant \tau \leqslant 1$. The coefficients of the expansion are calculated by the formula

$$a_i(s) = \int_0^1 w(\tau)p_i(\tau)x_1(s,\tau)\,d\tau \tag{14}$$

where $w(\tau)$ denotes the weighting function and $p_i(\tau)$ are the basis functions orthogonal on the interval $[0,1]$ - for a detailed theory of orthogonal functions see e.g. $[8,9,11,13]$. Let the functions $w(\tau)$ and $p_i(\tau)$ be once differentiable and the function $x(t+\theta)$ have the properties of the solution of differential-difference equations $([7],\text{chap.}1)$, i.e. the first derivatives of the functions $x_0(\tau)$ and $x_1(s)$ be continuous for $0\leq\tau\leq1$ and $0\leq s\leq1$ respectively, but $\dot{x}_0(1)\neq\dot{x}_1(0)$. Then the equation

$$\dot{a}_i(s) = \int_0^1 w(\tau)p_i(\tau)\frac{\partial x_1(s,\tau)}{\partial s}\,d\tau \tag{15}$$

holds everywhere in $0<s\leq1$ since for $k=1$ equation (11) holds everywhere except the point $\tau=1-s$. After transformations like those done in $[10]$ but with additional consideration of the particular properties of the function $x_1(s,\tau)$ one obtains for $0<s\leq1$

$$\dot{a}_i(s) = -\sum_{j=0}^{m}\left\{\int_0^1\frac{d}{d\tau}[w(\tau)p_i(\tau)]\,p_j(\tau)\,d\tau + w(0)p_i(0)p_j(0)\right\}a_j(s) +$$
$$+ w(1)p_i(1)x_1(s)\,, \qquad i=0,1,2,\ldots,m \tag{16}$$

or equivalently, after the integration by parts

$$\dot{a}_i(s) = \sum_{j=0}^{m}\left\{\int_0^1 w(\tau)p_i(\tau)\frac{dp_j(\tau)}{d\tau}\,d\tau - w(1)p_j(1)p_i(1)\right\}a_j(s) +$$
$$+ w(1)p_i(1)x_1(s)$$

where for practical reasons, the range of summation is limited to finite m. The initial conditions are

$$a_i(0) = \int_0^1 w(\tau)p_i(\tau)x_0(\tau)\,d\tau \tag{17}$$

The function $\tilde{x}_0(s)$ approximating $x_1(s,0)=x_0(s)$ is then

$$\tilde{x}_0(s) = \sum_{j=0}^{m}p_j(0)a_j(s) \tag{18}$$

Note that the final form of equations (16) obtained for particular orthogonal basis functions p_i may be surprisingly simple. For example, the use of the Legendre polynomials orthogonal in the interval $[0,1]$ yields the following system of equations

$$\dot{a}_i(s) = -2\sqrt{2i+1}\sum_{j=0}^{i}\left\{\sqrt{(2j+1)}\,[(i+j)\bmod 2]\,a_j(s)\right\} +$$
$$+ \sqrt{2i+1}\,[x_1(s)-(-1)^i\,\tilde{x}_0(s)] \tag{19}$$

Consider now equation (4) for $k=1$ and use (18) to approximate the last argument of the function f. Then, the function f depends on s, $x_1(s)$ and $a_i(s)$ $(i=0,1,2,\ldots,m)$ and for the approximate solution of (4) for k=1 we can use the equation

$$\dot{x}_1(s) = h\, f[sh,\ x_1(s),\ \tilde{x}_0(s)],$$
$$x_1(0) = x_0(1) \tag{20}$$

combined with (16) and (18). The initial values for (16) should be previously calculated by formula (17) using the known initial function. The system of ordinary differential equations so obtained can now be solved in a routine way. If the value $x_1(1)$ has been computed, then the next equation of (4)

$$\dot{x}_2(s) = h\, f[(s+1)h,\ x_2(s),\ x_1(s)],$$

with
$$x_2(0) = x_1(1) \tag{21}$$

may be added and the system consisting of equations (16),(20),(21) is solved. This procedure is further continued as in the Bellman approach as far as necessary, i.e. in each k-th step, the system involving the particular form of equations (16), equation (20) and the k-1 equations (4) with the initial values given by (17) and (5) is solved. The above results can be summarized in the following algorithm (N - a given number determining the length of the interval in which equation (1) should be solved):

1: set k=0, compute $a_i(0)$ from (17);
2: k=k+1;
3: solve simultaneously system (16) and the k equations (4) with initial conditions (17),(5); save $x_k(1)$;
4: if k < N then go to step 2, else stop.

2.4. Numerical example

t	x(t)		x(t+1)		x(t+2)	
	B-K	O	B-K	O	B-K	O
0.0000	1.000000	1.000000	0.306853	0.306271	0.048608	0.046246
0.0625	0.939375	0.939284	0.277019	0.274586	0.042593	0.039884
0.1250	0.882217	0.881835	0.249832	0.247407	0.037274	0.034614
0.1875	0.828150	0.827554	0.225046	0.222636	0.032575	0.029963
0.2500	0.776856	0.776362	0.202442	0.200048	0.028427	0.025862
0.3125	0.728066	0.728003	0.181827	0.179442	0.024770	0.022250
0.3750	0.681546	0.682083	0.163033	0.160642	0.021549	0.019074
0.4375	0.637095	0.638169	0.145909	0.143497	0.018716	0.016285
0.5000	0.594535	0.595879	0.130320	0.127876	0.016230	0.013842
0.5625	0.553713	0.554944	0.116144	0.113668	0.014050	0.011706
0.6250	0.514492	0.515227	0.103275	0.100775	0.012144	0.009843
0.6875	0.476752	0.476712	0.091614	0.089105	0.010479	0.008222
0.7500	0.440384	0.439478	0.081072	0.078575	0.009029	0.006814
0.8125	0.405293	0.403663	0.071571	0.069102	0.007769	0.005595
0.8750	0.371391	0.369425	0.063036	0.060608	0.006676	0.004541
0.9375	0.338602	0.336919	0.055402	0.053015	0.005729	0.003633

Table. Comparison of the solutions (B-K - Bellman and Kotkin,
O - results of the presented method for m=4)

The accuracy of the method is shown on the equation considered in [4] by Bellman and Kotkin, namely

$$\dot{x}(t) = - x(t-1) \cdot (t+1)^{-1}, \quad t > 0$$

with $x(t) = 1$ for $-1 \leqslant t \leqslant 0$. In the table, the solution obtained via the presented method using the Legendre polynomials - equations (19) - is compared with the results given in [4]. The fourth-ordor Runge-Kutta method was used and the step of integration was the same in both cases, $h=2^{-8}$. Greater values of m give better results but the accuracy becomes worse when N increases.

3. Conclusions

A method similar to the one derived by Bellman [1] for computational solution of delayed equations is presented. It differs from the latter in the way of approximate determination of the initial conditions, i.e. in Bellman's approach the initial function was assumed to be a polynomial, a trigonometric polynomial or a sum of exponentials and it was generated from an ordinary differential equation of order n, whereas in the present method the requirements imposed on the initial function are substantially weaker since it is determined by an orthogonal expansion. The coefficients of the expansion are the solutions of a system of simple ordinary differential equations, which are solved simultaneously with the integration of the analysed equation. The final problem is computationaly simple and standard subroutines may be used in a straightforward manner to obtain the numerical solution. The method is obviously time-consuming if the number N is large.

4. References

[1] Bellman R.: On the computational solution of differential-difference\equations. J.Math.Anal.Appl.,$\underline{2}$(1961),108-110

[2] Bellman R.,Cooke K.L.: Differential-difference equations. Academic Press, New York, London 1963

[3] Bellman R.,Cooke K.L.: On the computational solution of a class of functional differential equations. J.Math.Anal.Appl.,$\underline{12}$(1965),495--500

[4] Bellman R.,Kotkin B.: On the numerical solution of a differential--difference equations arising in analytic number theory. Math.Comp. $\underline{16}$(1962),473-475

[5] Cryer C.W.: Numerical methods for functional differential equations. In: Klaus Schmitt editor : Delay and functional differential equations and their applications. Academic Press, New York and London 1972, 17-101

[6] Elsgolc L.E.: Približonnye metody integrirovanija differencjalno-raznostnych uravnenij. Uspechi mat.nauk, $\underline{8}$(1953)4,81-93

[7] Elsgolc L.E.,Norkin S.B.: Vvedenije v teoriju differencjalnych uravnenij s otklaniajuščimsa argumentom. Nauka, Moskva 1971

[8] Jackson D.: Fourier series and orthogonal polynomials. Mathematical Association of America, 1941

[9] Kaczmarz S.,Steinhaus H.: Theorie der Orthogonalreihen. Fundusz
     Kultury Narodowej, Warszawa-Lwów 1935

[10] Reeve P.J.: A method of approximating to pure time delay. Int.J.
     Control, $\underline{8}$(1968)1,53-63

[11] Sansone G.: Orthogonal functions. Interscience, New York 1959

[12] Szczepaniak P.S.,Szczepański J.: A method using orthogonal functions
     for analysis of delayed-differential systems. In: Proc.of the 5th
     Int.Symp."System-Modelling-Control".Zakopane, Poland, 1986, vol.2,
     175-180

[13] Szegö G.: Orthogonal polynomials. American Mathematical Society,
     New York 1939

A Theory of Elementary Social Systems as a Basis for the Analysis and Modelling of Decision Situations

Christian Dahme [1]

A theory of elementary social systems which makes it possible to analyse activities and social systems is described. Within this theory every activity can be described as an interaction of elementary social systems. With the aid of such interaction one can build up a field of possibilities of the behaviour, i.e., all possibilities that an individual has to achieve the aim through an (elementary, composed, complex or composed complex) activity. Also through interaction a composed social system can arise.

Keywords: systems theory for social systems, elementary social systems, activity, aim, interaction

Why do we need a theory of elementary social systems for the analysis and model ling of decision situations and what does this theory contain and what novel fetures does it have ? At present the interpenetration of all social processes is increasing and with it the complexity within or between social systems. The analysis and the mastery of such complexity (especially for every decision, in which this complexity has to be considered) have become an essential need. On the other hand the mastery of such complex situations depends on the methods and instruments developed for this purpose. In the last few years different approaches (with regard to methods and instruments) have been developed for the support of the mastering of decision situations that are based on the gametheory, the systems theory, the operation research, the cybernetics and the like. These approaches were characterised by the fact that a methodical-instrumental part relativly well developed is taken as a basis for these approaches whereas a contextual-conceptional part was scarcely or not developed. This deficit has lead to problems, difficulties and failures.

The theory of elementary social systems which is outlined in the following with regard to the essential pre-requisites and results is an attempt at overcoming this situation.

1st <u>methodological pre-requisites</u>

   a) In the development of a scientific discipline there are different phases with regard to the mastery and the understanding of the object:

      1st the descriptive or on the whole oriented phase

      2nd the phase of the orientation on the parts of the whole or the analytical phase

      3rd the phase of the re-orientation on the whole or the synthetical phase.

      In order to be able to explain and to gain mastery of an object in its wholeness, it is nessecery to carry out all phases. [1], [2]

   b) In order to build a model you need

      an aim of the modelling,

---

1 Humboldt-Universität zu Berlin, Sektion WTO, A.-Bebel-Pl., DDR-1086 Berlin

- an object of the modelling and.
- a theory or theoretical knowledge of this object (or the field from which this object comes). [3]

The method of modelling can be used with regard to a) in the 2nd phase (to gain scientific findings about the part of the whole) and in the 3rd phase (for the synthesis).

c) With regard to the systems analysis of social objects GWISCHIANI has formulated principles of systems approach [4].

2nd theoretical pre-requisites and positions

a) There are different kinds of movement or of organization of matter (physical, biological, social systems etc.). Every kind of organization is characterised by its typical laws. There are a connection between the different kinds of organization.

b) In our opinion, a social system is a system with its exterior and interior objective and subjective conditions that consists of one or more individuals which has emerged under concrete social (historical) and natural conditions to achieve an aim and is characterized by typical activities for achieving that aim. Therefore, the basis for a theoretical concept with regard to social system as an object being investigated is the so-called Marxist principle of activity [5]. This principle was used as a basis of psychology by LEONTIEV [6] and the following makes use of this principle.

c) The second essential idea in understanding social objects is the aim. Therefore, it was nessecery to claryfy what an aim (goul, taget) is. A phylogenetical access was selected [7]. In this connection a system with an aim is a self-referable system and the aim of the system is to maintain its own system. In the evolution there comes into being additional possibilities for self-reference. In my opinion, aims occur on the biological organization-level and all the aims of human actions (actings) have a specification over them.

   Arising from a need the _aim_ of human actions and with that the aim of the social system belonging to it, is to achieve a situation which makes it possible to overcome (serve; cancel, neutralize) the need.

Aims can spring up on three different levels:

1st An aim from an (individual) need

2nd The aim of the maintanance of an individual
   (this can be a special aim of the 1st kind and it is more complex as an aim of the 1st kind in general)

3rd Social aims.

d) For the creation of aims the following principles were developed:
- the union of aim and social system
- the differenc between the idea of an aim and an aim
- the differenc between aim and criterion (of the evaluation)
- the relativity between aim and criterion

e) <u>Elementary social systems</u> are such social systems that can not be decomposed in social subsystems and consist of
- one individual with
- an aim (based on an need) that carries out
- one (internal or external) activity to achieve this aim.
Interior conditions of an elementary social system may be:
1st a need (an aim)
2nd possibilities to a valuation
   - of such a need,
   - of the possibilities and of the conditions for the carrind-out of such an activity that makes possible a satisfaction (overcoming, canceling ...) of this need,
   - of the result of the activity which will be expected.
3rd the abilities and accomplishments for such a valuation and with that for a decision on the carrying-out of such an activity
4th an individual who carries out this valuation and decision
5th the abilities and accomplishments for the carrying-out of this activity
6th the object of this activity
7th the means for the carrying-out of this activity
8th an individual who carries out this activity
An elementary social system for which the conditions 1 to 8 are right and which going out from a need creates an aim and carries out the activiety itself to achieve this aim is called an <u>a - system</u>.
An elementary social system for which the conditions 1 to 4 are right and the conditions 5 to 8 are partly or not right and which going out from a need creates an aim and does not carry out the activiety itself to achieve this aim is called an <u>a - p- system</u>.
If an individual is an object of the activity which serves the achievment of an aim of an a- or a-p-system then such an object is called a <u>"p-system"</u> (a "p-system" is not a social system !) [7]

3rd <u>Results</u>
a) Every activity can be described as an interaction of elementary social systems.
b) The following kinds of interaction exist:
   - interaction between an a-system and its environment
   - interaction of diffenrent a-systems with regard to a common resource
   - interaction between one a-p-system and one a-system
   - interaction between some a-p-systems and one a-system
   - interaction between some a-p-systems and some a-systems
   - interaction between one a-p-system and some a-systems
   - interaction between some a-p-systems to achieve together the aim (to carry out together a complex activity)
With the aid of this interaction one can build up a (relatively closed) field of possibilities of the behaviour, i.e., all possibilities that an

individual has to achieve the aim through an (elementary, composed, complex or composed complex) activity.

  c) It was not possible to build up the field of possibilities on deductive kind alone. A deductive approach leads to too many possibilities, a completeness was impossible and the ideas "complex" and "composed" activity can not be found in this way.
  With the aid of this approach it was possible to distinguishe between elementary, composed, complex or composed complex activity and between activities of the 1st or 2nd kind. It could be pointed out under which conditions (and causes) a composed or complex activity and under which conditions through interaction a composed social system arise.

  d) The question "Under which condition is an individual willing to carry out the activity for an a-p-system (i.e., to be service-rendering a-system for such an a-p-system) ?" leads to the problem of co-operation.

  f) In this theory an individual is an integration of elementary social systems and the aim of the maintenance require a composed activity.

Decision situations spring up in connection with needs (with the creation of aims) or within an actictvity, i.e., within elementary social systems and their interaction. The analysis of decision situations supposes the analysis of elementary social systems belonging to it.

With the aid of this theoretical approach one are able to build up models of (micro-sociological) objects and this theory can be a basis for an advirory system. (This theory of elementary social system is described in [2].)

<u>References:</u>

[1] Fuchs-Kittowski, K.; Hager, Th.; Dahme, Ch.: Zum Gegenstand der Medizin aus wissenschaftstheoretischer Sicht; in: DDR-Med.-Rep.; 12 (1983) 6; S. 489-495

[2] Dahme , Ch.: Methodologische und theoretische Voraussetzungen für die Analyse komplexer Entscheidungssituationen, Diss. B, eingereicht 1988 an der Humboldt-Universität

[3] Dahme, Ch.: Einige methodologische Bemerkungen zur Modellbildung; Deutsche Zeitschrift fär Philosophie; Heft 4 1987, S. 358-364

[4] Gwischiani, D.: Die philosophische Grundlage der Systemforschung, Gesellschaftswissenschaften 31 (1982) 3 pp. 65-79

[5] Marx, K.; Engels, F.: Die deutsche Ideologie; MEW Bd. 3; Berlin 1981; S. 29f.

[6] Leontjew, A. N.: Tätigkeit Bewußtsein Persönlichkeit; Berlin 1982

[7] Dahme, Ch.: Ziele und Zielvorstellungen; Deutsche Zeitschrift für Philosophie (in print)

[8] Dahme, Ch.: On the Particularity of the Model-Building of Elementary Social Systems - A Draf for a Systems Theory of Elementary Social Systems, in: Sydow, A., Thoma, M. and Vichnevetsky, R., (eds.), Systems Analysis and Simulation 1985 - Volume I (Akademie-Verlag, Berlin 1985) pp. 324-327

An Approach to the Development of Supporting Systems for Analysis and Construction/Influencing Social Systems

Th. Hager, Ch. Dahme [1]

With regard to the building up of supporting systems two tasks of modelling can be distinguished (i.e., 1st the modelling of the object or 2nd the modelling of the mastering of problem-situation). The 2nd task is considered. A framework for computer aided analysis and mastering a situation is outlined .

Keywords: supporting systems, mastering of situations, social systems, framework for supporting systems

1. The Problem

The growing complexity and dynamics of social development and the growing availability of modern information technologies induce the need for supporting systems (advirory systems) in domains of human activity without experience in application of models, too.

The building-up and the application of supporting systems bring to light the existence of contradictions within the problem solving resp. the decision behaviour, which is discussed by some authors in connection with solving multi-criteria-optimization-tasks [1, 2].

Obviously, the subject (that means, the "user" of a supporting system/computer aided decision support system) needs not only the discharge of formal tasks, but also an extensive support of the treatment of nonformal tasks to avoid inadmissible reductions (for instance of criteria and of constraints).

This way, in the case of building up supporting systems two tasks of modelling emerge [3]:

a) modelling the object of the analysis (spreading the field of possibilities of human activity concerning a given system)

b) modelling the mastering of problem-, and especially of decision situations (analysis of the typical behaviour/of typical microstrategies under given constraints)

2. Theoretical basis

Our approach is founded on the concept of human activity developed by LEONT'EV. Within this concept a man's activity is considered not only under the aspect of treating/changing the object as a means of satifying human needs and of reaching human goals but also under the aspect of the learning process within mastering a situation [4].

---

1 Humboldt-Universität zu Berlin, Sektion WTO, A.-Bebel-Pl., DDR-1086 Berlin

Mastering a situation, the subject reflects the goals, the criteria of the evaluation of the results of his activity, his knowledge and possibilities.
This concept leads to such theoretical constructions (ideas) as "social micro-object/ social system", "situation", "micro-strategies", and "situation mastering" and is discussed in [3], [5] and [6] with regard to the modelling of social systems and some applications.
In this work we consider social micro-objects under the aspect of the second task of modelling (see above).
Such social micro-objects/social systems represent a connection of individuals (and their working means) under given social and natural conditions. Such systems emerge and/or reproduce in the case of a common goal and are characterized by typical activities for approaching the goal [3, 5].

3. A framework for a supporting system

It is obvious to interpret and to model a system, which consists of a "user" (who has the aim to master a complex situation in the sense of a prospective orientation) and his working means (the computer aided working place), as a social system corresponding to the above-mentioned definition.
Based on a corresponding model of such a system it is possible to construct a framework for computer aided analysis and mastering a situation, which is to be outlined in the following:

The proposed framework consists of two phases, a phase of orientation, and a phase of problem treatment.

A) Phase of orientation/orientational phase:
   Determination of the system and the actual and/or future demand situations which have to be mastered as a prerequisite to future successfull reproduction. After these starting activities the "user" has to define the main deficiencies which hinder mastering of typical situations:
   objective conditions (e.g. resources) and/or knowledge:
   a) knowledge concerning future behaviour of the system
   b) knowledge concerning methods for removing the deficiencies
   c) knowledge concerning the mastering of a decision situation
      (like variants of system's development, alternatives, criteria, methods
      for selecting alternatives)
   d) knowledge concerning the realization of a decision
   The orientational phase ends with the determination of a task and the transition to the second phase or the termination of the activity.

B) Phase of problem treatment:

   in this phase the "user" attempts to remove the deficiencies of knowledge
   corresponding to the above-defined tasks. After solving the tasks he has the
   possibiliy to return to the orientational phase and to define the next task
   or to terminate his activities.

Obviously, such kind of a supporting system (of an advisory system) has to
correspond to the very dynamic and lateral behaviour of the human problem
solver in a complex cognitive situation. In the sense of LEONT'EV it is the
main task in building-up a supporting system to offer the "user" the possibility to elaborate a subject-related system of tools. As a result of the subject's
activities a kind of scientific instrument emerges which integrates the human
activitiy.

References

[1] Laritcev, O.I.: Ob'ektivnye modeli i sub'ektivnye resenija.- Moskva 1987

[2] Wittmüß, A.; Straubel, R.; Rosenmüller, R.: Interactive Multi Criteria
    Decision Procedure for Macroeconomic Planning.- in: Systems Analysis Model-
    ling Simulation.-Berlin 1(1984)5.,S.411ff.

[3] Hager, Th.: Zur Entwicklung und Erprobung systemanalytischer Instrumenta-
    rien für die Leitung gesellschaftlicher Prozesse. In: Wiss. Zeitschr. der
    Humboldt-Universität zu Berlin,- R.Ges.Wiss. 37(1988)1. - S.87-92

[4] Leont'ev, A.N.: Tätigkeit-Bewußtsein-Persönlichkeit; Berlin 1982

[5] Dahme, Ch.; Hager, Th.: Intentional and Operational Aspects of Decision
    Behaviour and their Modeling. In: Proceedings of the IFIP TC/WG9.1 Con-
    ference on System Design for Human Development and Participation and Be-
    yond, Berlin, GDR,12-15 May 1986. Edited by: P.Docherty, K.Fuchs-Kittowski,
    P.Kolm, and L.Matthiassen, North-Holland Amsterdam 1987, S. 359 - 370

[6] Dahme, Ch.: A Theorie of Elementary Social Systems as a Basis for the Ana-
    lysis and Modelling of Decision Situations; in this volume

Distributed Message Exchange System Modelling

Zenon Mital[1]

1. Introduction

In the paper some aspects of the process cooperation in a distributed
operating system are discussed. The concept of a communication protocol
between the nodes of the system is presented. The algorithm of the
cooperation between the system nodes is modelled by means of Petri nets.
As a hardware facility supporting an inter-node communication, standard
RS adapters are assumed.

2. A functional description of the system

Each node ND of the system is implemented as a separate microcomputer
node with adequate hardware configuration. It can be regarded as a
subsystem of parallely executed processes, composing an application level
software dedicated to particular node appropriation.

The operating system kernel of each subsystem ND creates an
environment for the concurrent process subset operation. The kernel
offers the media for process creation, termination and cooperation. The
process cooperation is performed on the base of the message exchange
[1,4]. The kernel provides the following basic functions [2,3] for
process manipulation and communication:

- CREATE_a_process,
- TERMINATE_a_process,
- START_a_process,
- SEND_a_message   (if the receiving port is full, a
                         process-sender is suspended),
- WAIT_for_a_message.

To distribute a system onto several nodes, the following functions
available in the node have been expanded: CREATE_a_process,
START_a_process, SEND_a_message. These functions use two-level process
addresses. They are called inter-node functions. Besides the name of a
message receiving process (for SEND function) and the name of a process
which has to be created or started, there is also the node address
within a full process address. So, it is possible to send a message to
any process operating in any other node. We can also create and start
processes outside the "home-node".

As a hardware facility supporting inter-node communication, standard
RS adapters are used.

The messages between two processes operating within the same node are

----

[1] Institute of Computer Science, Gdansk University of Technology,
     ul. Majakowskiego 11/12, 80-952 Gdansk, Poland

sent directly. Also CREATE_a_process and START_a_process functions within the same node are performed normally, without special problems. A specific procedure is applied while transmitting a message to a process which operates in the other node, or while creating and starting processes there. To fulfil an inter-node cooperation task, the operating system level process INCP has been constructed. Process INCP receives special kernel messages which can be considered as requests of inter-node communication. Then, it initiates a transmission to the other node via RS adapter. The identical process INCP in the receiving node cooperates in the transmission.

If the execution of a kernel function requires an inter-node communication (inter-node type function), a process-function caller can not be continued until it receives thediagnostic reply information.

To suspend process while waiting for the completion of inter-node communication, the process state space has been expanded: the state "Wait for the completion of inter-node communication" (WCC) has been introduced.

A process assumes the WCC state during CREATE, START and SEND functions execution, if they refer to process in other node.

3. An inter-node communication protocol

The communication procedure includes the following steps:
- the kernel sends a special kernel message to the INCP process. The message contains the parameters of the function (and the text of a message for SEND function),
- the process INCP initiates the transmission of the function parameters via RS adapters to the receiving node. The identical process INCP in the receiving node cooperates in the transmission. The function parameters are buffered in the private memory of the INCP receiving process,
- the INCP process in the receiving node, using functional interface to the kernel in the receiving node, calls the function and receives a diagnostic reply,
- the INCP process in the receiving node transmits the diagnostic data via RS adapter to the INCP process in the sending node,
- the INCP process in the sending node, using the functional interface to the kernel, informs the kernel that the communication protocol has been completed (providing the diagnostic reply received from the receiving node).

We assume that SEND_a_message function is executed in a bit different way depending on the place, where the message receiving process is located:
- If the receiving process is located in tne same node, where a process-sender operates, the last-mentioned one can be suspended

during the SEND function execution (if the receiving port is full);
- If a message is sent to the other-node-located process, a process-
  sender can not be suspended even if the receiving port is full.   In
  this case the process-sender receives a reply diagnostic "a message
  has not been provided to addressee". If we had suspended the
  processes in this case, we would have had to distribute a queue of
  the processes suspended during SEND function execution among a few
  different nodes. This would have resulted in serious problems while
  SEND and WAIT functions executing.

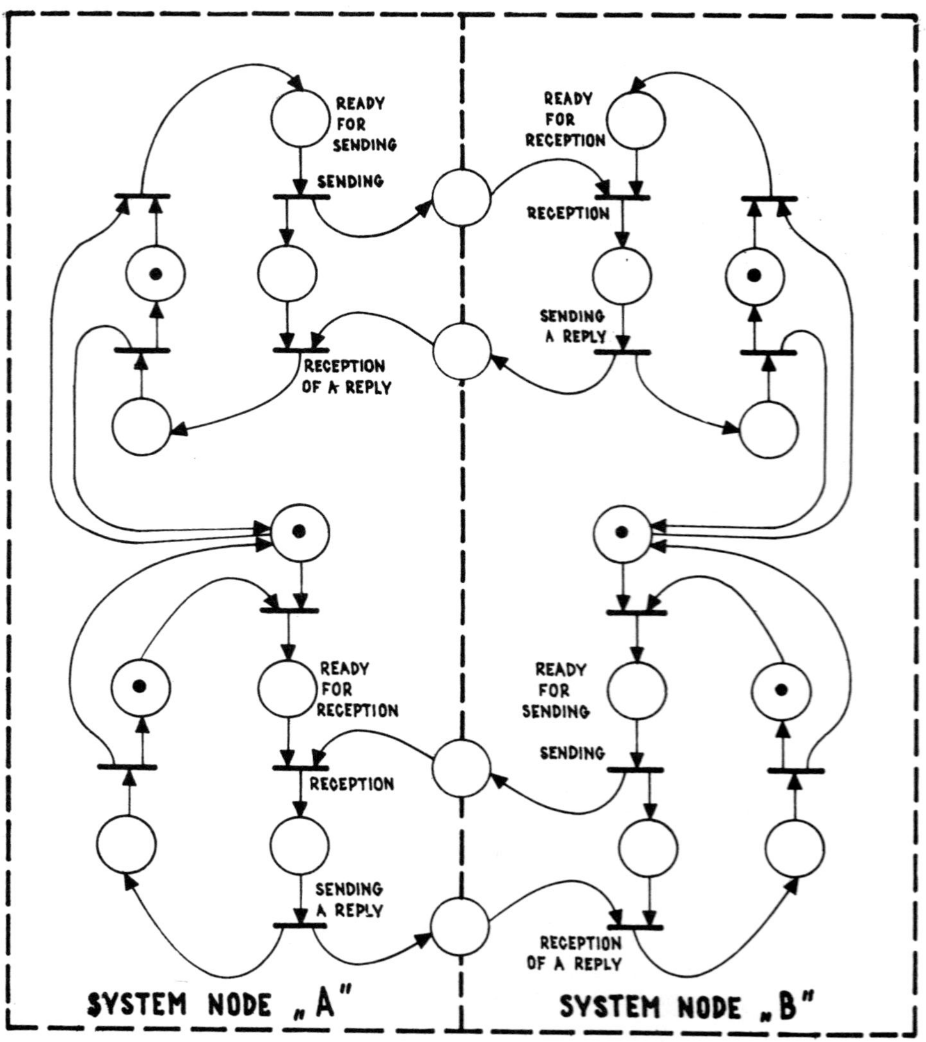

Fig. 1. Petri net model of an inter-node communication process.

4. Petri net model of inter-node communication process functioning

As it was mentioned in section 3, the INCP processes control an inter-node communication via RS adapters. The execution of any inter-node function includes the following two phases:
- transmission of a function code and its parameters from a sender node to an addressee node,
- transmission of a reply message from the addressee node to the sender node.

A process which calls the inter-node function, is switched to WCC state. It will be suspended in this state until the INCP process in the same node receives a reply message. Then, the reply information is transferred by the INCP process to the kernel (via functional interface). At that time the kernel can change the WCC state and return a diagnostic reply to the process-caller.

In Fig. 1 the functioning of the INCP processes, which control the communication between the system nodes, is shown. This is modelled by means of Petri nets.

5. Final remarks

Although the execution of the inter-node function consumes much more of total time, it is not necessary to wait actively for the completion of the inter-node communication process. The processes, which have to wait until the inter-node communication is finished, are switched to the WCC state. So, a processor can be assigned to the other processes.

There are not special restrictions for application level processes to use the inter-node functions. It is only required to establish an appropriate capacity of the port of the INCP process (where the special kernel messages are sent) during the system generation phase. It depends on the frequency of using of inter-node functions and on the RS data transmission rate. Simulation methods can be used in order to evaluate an appropriate port capacity.

Literature

[1] Brinch Hansen P.: Operating System Principles (Polish edition). WNT, Warszawa, 1979.

[2] Gorski J., Napiatek J.: Operating System ECTTOS — Kernel Functions. Technical Report (in Polish). Institute of Computer Science, Gdansk, 1985.

[3] Mital Z.: CDDT — Concurrent System Operation Display and Debugging Tool. 13-th Symposium on Microprocessing and Microprogramming EUROMICRO'87. Short Note Session volume. Portsmouth, UK, 1987.

[4] Shaw Alan C.: The Logical Design of Operating Systems (Polish edition). WNT, Warszawa, 1980.

Analysis and Optimization by means of Estimations in Measurement

Eugen-Georg Woschni[1]

ABSTRACT

In the paper beyond the exact solutions using system-theoretical formulations methods of estimations and approximations are treated. Attention is given to the importance of a-priori information as a supposition for the efficiency of these methods.

Especially the paper deals with examples important both for demonstration the method as for practice: The selection of measuring systems based on the approximation of the measuring errors caused by these systems; optimization and optimal filtering; estimation of the memory capacity necessary to store the information gained by measurement and of the channel-capacity.

INTRODUCTION

In comparision with other fields of technical cybernetics in measurement the output signal is given and the input is searched for. From this difference follows that estimations and approximations especially in measurement play an important role, because before solving tasks of system analysis and optimization model-signals for the - before measuring unknown - input singals are to be supposed using the a-priori information of the process to be measured.

In the paper some important examples are treated showing both the method and the results with relevance to practice.

SELECTION OF MEASURING SYSTEMS TAKING INTO CONSIDERATION DYNAMIC BEHAVIOUR

Instead of the exact solution

$$h(t) = L^{-1} \left\{ G(p)/p \right\} \tag{1}$$

the transient response $h(t)$ (response to a step function) may be approximated by a linear increasing function with the transient time T_T as Figure 1 shows.

Assuming a pulse-shaped input Figure 1 shows three typical cases:

[1] Technische Universität Karl-Marx-Stadt, PSF 964, DDR-Karl-Marx-Stadt 9010

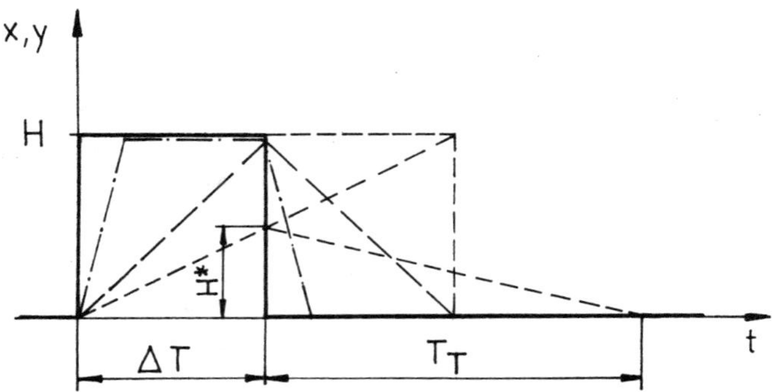

Fig. 1. Approximation to select Measuring Systems

Long-dashed – the transient time is equal to the pulse width, $T_T = \Delta T$. Now the pulse height is still correctly indicated, whereas the pulse shape is strongly distorted. On the other hand too long a transient time leads to an error of the pulse height (short dashes). To permit proper determination of the pulse shape, the transient time has to be substantially shorter than the pulse width (chain curve). The same considerations may be applied to analyse measuring errors: If the response has a heavily prolonged trailing edge, an error will be present. Furthermore it is possible to estimate the error, using the relation of flank sides of the output. The real height H of the input is, using the symbols of Figure 1,

$$H \approx H^* T_T / \Delta T. \qquad (2)$$

It is noteworthy that minor and medium errors in amplitude measurement often produce more detrimental effects in practice than very large errors. For errors in the order of some 10 % often cause no failure due to the safety margins, in particular the elements will withstand the initial tests. Later on, however, all elements will fail at the same time when the machines have been produced in series. On the other hand, major measuring errors mostly become evident during the testing period.

OPTIMIZATION AND OPTIMAL FILTERING

Theory of optimal filtering leads as well-known considering the criterion of minimal mean-square error to the optimal frequency response /4/

$$G_{opt}(j\omega) = G_{id}(j\omega) \frac{S_{xx}(\omega)}{S_{xx}(\omega) + S_{zz}(\omega)} \qquad (3)$$

In practice only the frequency response $G_{id}(j\omega)$ of the ideal systems is given against what the frequency densities of signal $S_{xx}(\omega)$ or noise $S_{zz}(\omega)$ often are unknown. Therefore in these cases approximations are used describing the original system by means of an ideal low-pass with the limiting frequency ω_{co} and the corrected system with ω_c i.e. the factor of correction is $a = \omega_c/\omega_{co}$.

Assuming the system to be linear and white noise $S_{zz\,o}$ not to be correlated to the band-limited signal

$$S_{xx}(\omega) = \frac{S_{xx\,o}}{1 + (\omega/\omega_o)^2} \qquad (4)$$

we get an extreme value of the degree of correction /2/

$$a_{extr.} = \frac{\omega_o}{\omega_{co}} \sqrt{\frac{S_{xx\,o}}{S_{zz\,o}} - 1} \qquad (5)$$

and the optimal error reduction

$$\left.\frac{\overline{\varepsilon^2}}{\overline{\varepsilon_0^2}}\right|_{min} = \frac{\dfrac{\omega_o}{\omega_{co}}\left\{\sqrt{\dfrac{S_{xxo}}{S_{zzo}} - 1} + \dfrac{S_{xxo}}{S_{zzo}}\left[\dfrac{\pi}{2} - \arctan\sqrt{\dfrac{S_{xxo}}{S_{zzo}} - 1}\right]\right\}}{1 + \dfrac{\omega_o}{\omega_{co}}\dfrac{S_{xxo}}{S_{zzo}}\left[\dfrac{\pi}{2} - \arctan\dfrac{\omega_{co}}{\omega_o}\right]} . \qquad (6)$$

The results are made up in the Figures 2 and 3.

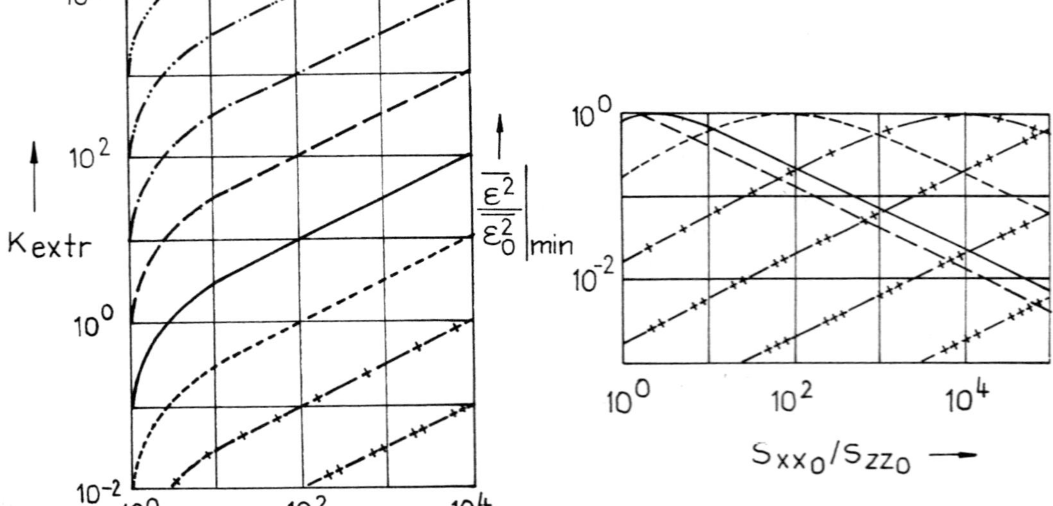

Fig. 2. Extreme values of the degree of correction K_{extr}
$\omega_o/\omega_{co} = 10^4 (-\cdots-)$; $10^3 (-\cdot\cdot-)$;
$= 10^2 (-\cdot-)$; $10^1 (---)$;
$= 10^0 (\text{———})$; $10^{-1} (----)$;
$= 10^{-2} (-+-)$; $10^{-3} (-++-)$

Fig. 3. Values of the best attainable error-reduction
$\omega_o/\omega_{co} \geq 10^1 (---)$; $=10^0 (\text{———})$;
$= 10^{-1} (----)$;
$= 10^{-2} (-+-)$; $10^{-3} (-++-)$;
$= 10^{-4} (-+++-)$; $10^{-5} (-++++-)$

The investigations show that the larger the "dynamic" or "statistic redundancy" ω_{co}/ω_o or S_{xxo}/S_{zzo} is, the more effective the error reduction will be and that for systems with statistic redundancy a > 1, for systems with dynamic redundancy a < 1 is to be chosen in conformity with physical considerations. The possibility of error reduction increases the demands on the quality of the original system.

MEMORY AND CHANNEL CAPACITY

As a measure of the quality of systems the maximum number of amplitude steps m_a or of power steps m_p as a function of the static error Δy or of the mean-square error $\overline{\varepsilon^2}$

$$m_a = 1 + \hat{y}/\Delta y; \quad m_p = 1 + P_y/\overline{\varepsilon^2} \tag{7a,b}$$

and the limiting frequency f_c may be used /4/.
Assuming the general quality criterion

$$Q.C. = \sum_r \lambda_r (c_r) \tag{8a}$$

with

$$\lambda_1 = {}^{10}\log (2 f_c)$$

$$\lambda_2 = {}^{10}\log ({}^2\log m_a) = {}^{10}\log (\tfrac{1}{2} {}^2\log m_p) \tag{8b,c}$$

one gains the information flow I or the channel capacity C_t of Shannon's information theory /1/ as a criterion considering both the static and dynamic behaviour

$$I = 10^{Q.C.} = 2f_c \, {}^2\log m_a = f_c \, {}^2\log m_p. \tag{9}$$

An interpretation by means of physical perception uses the memory capacity to store m_a equally probable measuring values s or the signal to noise ratic P_s/P_n

$$s/\text{bit} = {}^2\log m_a = {}^2\log \frac{\hat{y}+\Delta y}{\Delta y} = {}^2\log \sqrt{m_p} = \tfrac{1}{2} {}^2\log \frac{P_s+P_n}{P_n} \tag{10}$$

leading with $1/T_T = 2f_c$ values per second to the well-known equation for the channel capacity

$$C_t = f_c \, {}^2\log \frac{P_s+P_n}{P_n} \approx 0{,}33 \, f_c \left. \frac{P_s}{P_n} \right|_{dB}. \tag{11}$$

CONCLUSION

The examples given in the paper are showing that especially in measurement approximations play an important role, because before solving tasks of system analysis and optimization one has to suppose model-signals for the - before measuring unknown - input signals.

LITERATURE

/1/ Shannon, C. E., "A Mathematical Theory of Communication",
    Bell System Techn. J., 28, 1948, pp. 379 - 423 and 623 - 656.

/2/ Woschni, Eugen-Georg, "Reduction of Either Statistical or
    Dynamic Errors in Measurement by Means of Correction Programs
    or Networks and Demands to the Quality of the System",
    Z. elektr. Inform.- u. Energietechnik, Leipzig, 10, 1980,
    pp. 435 - 444.

/3/ Woschni, Eugen-Georg, "Signals and Systems in the Time and
    Frequency Domain", in Sydenham, Peter H. (Ed), HANDBOOK OF
    MEASUREMENT SCIENCE, Vol. 1, John Wiley & Sons, Chichester,
    New York, Brisbane, Toronto, Singapore, 1982, pp. 95 - 201.

/4/ Woschni, Eugen-Georg, INFORMATIONSTECHNIK, 3rd. Ed. VEB Ver-
    lag Technik, Berlin and A. Hüthig Verlag, Heidelberg, 1988

/5/ Zadeh, Lotfi A. and Desoer, C. A., LINEAR SYSTEM THEORY,
    McGraw-Hill, New York, 1963.

Simulation of Fuzzy System Dynamics Models

Michal Kejak[1]

1. Introduction

This paper describes a new approach to handle with vagueness in deterministic simulation dynamical models via *fuzzification* [2]. The main attention is devoted to *simulation of fuzzy (or fuzzified) system dynamics models*. Principal theorems on which such simulation is based are given. For lack of space proofs of the theorems are omitted. In conclusion a class of algorithms for simulation of FSD models is outlined.

2. Basic concepts

Theorem 1.: Each fuzzy set A can be represented as a *family of α-cuts* A_α = { A_α ; $\alpha \in (0,1\rangle$} if we define $\mu_A(x)$ = sup min(α, $\mu_{A\alpha}(x)$),
$$\alpha \in (0,1\rangle$$
where $\mu_{A\alpha}(x)$ = 1 iff $x \in A_\alpha$ and $\mu_{A\alpha}(x)$ = 0 otherwise.

Remark 1.: α-cuts are ordinary intervals in the case of fuzzy numbers. We shall denote them as A_α = $\langle a_\alpha^L, a_\alpha^R \rangle$. Fuzzy number (FN) is a convex normalized fuzzy set of the real line R. Let A, B be FNs and * be an algebraical operation then for A * B very simple rules hold [1] and for unary operation too [2].

3. Algebraical equations with fuzzy numbers

Definition 1.: An equation y = $g(z_1, z_2, \ldots, z_m)$ = $g(z_i)$, $i \in [1,m]$ with the function $g : R^m \longrightarrow R$ which can be defined in the recursive way $g = u$ (or $g = u*v$), where u and/or v is either (i) z_i, $i \in [1,m]$; or (ii) $w * s$; or (iii) $\xi(w); w$ and s again being defined as u and/or v and * is an algebraical operation and $\xi(.)$ is a unary operation (see [1], [2]), we shall call an *algebraical equation*.

Theorem 2.: Let Z_i = { $Z_{i\alpha}$; $\alpha \in (0,1\rangle$ } be fuzzy numbers of variables z_i, $i \in [1,m]$. If the equation y = $g(z_1, z_2, \ldots, z_m)$ is an algebraical equation and each variable z_i, $i \in [1,m]$ occurs only once in the function g then Y = $f(Z_1, \ldots, Z_m)$ = { Y_α ; $\alpha \in (0,1\rangle$ }, where Y_α = $\langle y_\alpha^L, y_\alpha^R \rangle$ and y_α^L = $g(z_{i\alpha}^L, z_{j\alpha}^R)$, $i \in I^L$, $j \in J^L$ and y_α^R = $g(z_{i\alpha}^L, z_{j\alpha}^R)$, $i \in I^R$, $j \in J^R$ moreover $I^L, J^L, I^R, J^R \subseteq [1,m]$, $I^L \cap J^L \cap I^R \cap J^R = \emptyset$ and $I^L \cup J^L \cup I^R \cup J^R = [1,m]$.

---

[1] Michal Kejak, Institute for Application of Computing Technique in Control, Revolucni 24, Praha 1, Czechoslovakia

Theorem 3.: Let $Z_i = \{ Z_{i\alpha} ; \alpha \in (0,1> \}$ be FNs of variables z_i, $i \in [1,m]$. If the equation $y = g(z_1,\ldots,z_m)$ is an algebraical equation and the index set $K \subseteq [1,m]$ contains indices of variables which occur in the function g more than once then $Y = g(Z_1,\ldots,Z_m) = \{Y_\alpha ; \alpha \in (0,1>\}$, where $Y_\alpha = < y_\alpha^L, y_\alpha^R >$ and $y_\alpha^L = \inf g(z_k, z_{i\alpha}^L, z_{j\alpha}^R)$, $i \in I^L$, $j \in J^L$
$$z_k \in Z_{k\alpha}, \text{ for all } k \in K$$

and $y_\alpha^R = \sup g(z_k, z_{i\alpha}^L, z_{j\alpha}^R)$, $i \in I^R$, $j \in J^R$ moreover I^L, J^L, I^R, $J^R \subseteq$
$$z_k \in Z_{k\alpha}, \text{ for all } k \in K$$

$\subseteq [1,m]$, $K \cap I^L \cap J^L \cap I^R \cap J^R = \emptyset$ and $K \cup I^L \cup J^L \cup I^R \cup J^R = [1,m]$.

Remark 2.: In case of the occurence of unary unimodal operations in the function g, maxima and/or minima of these operations can appear at the enumeration of α-cuts. Note in this theorem we find extremes (inf or sup) only on $|K|$-dimensional region in R^m and the rest of the function enumerates identically with Theorem 2.

Remark 3.: It is very important to stress that all theorems described above hold on the condition that the FNs entering into an equation are independent or noninteractive [1].

4. Differential equations with fuzzy numbers

For our purpose we shall consider an autonomous system of ordinary nonlinear differential equations in the vector form

$$\underline{x}(t) = \underline{f}(\underline{x}(t)), \qquad \underline{x}(0) = \underline{x}_0 \qquad (1)$$

where $\underline{x} = (\underline{x}_1,\ldots\underline{x}_n) \in R^n$ and $\underline{f} : R^n \longrightarrow R^n$. Because this paper is devoted to using fuzzy sets in system dynamics (SD), we shall view the solution of this system from the standpoint of SD, where the solution is gained through simulation by means of the simple Euler integration method. So that we have a system of nonlinear difference equations

$$\underline{x}(t+dt) = \underline{x}(t) + \underline{f}(\underline{x}(t))dt, \quad \underline{x}(0) = \underline{x}_0 \qquad (2)$$

For the simplicity we shall write this equation in the form

$$\underline{x}(T+1) = \underline{x}(T) + \underline{f}(\underline{x}(T)), \qquad \underline{x}(0) = \underline{x}_0 \qquad (3)$$

In case of one-dimensional discrete dynamical system with control $x(T+1) = x(T) + f(x(T),u(T))$, where $u(T)$ can be also fuzzy, we can normally use the rules for algebraical operations on FNs together with Theorems 2. and 3.

However many problems arise when we begin to concern with a multi-dimensional discrete dynamical system with FNs. First we shall consider initial fuzzy conditions of the system (3) without fuzzy control. Though FNs in the components of the initial state vector are mutually independent or noninteractive (see Remark 3.),the gained new state fuzzy vector generally will not have the noninteractive components vector. This is caused by the fact that minimally one variable occurs on the right-hand side of (3) more than once.

Theorem 4.: Let $\underline{X}_\rho = (\underline{X}_{\rho 1}, \ldots, \underline{X}_{\rho n}) = \{ \underline{X}_{\rho \alpha} \; ; \; \alpha \in (0,1\rangle \}$ be a n-dimensional vector of FNs representing an initial fuzzy condition of discrete dynamical system (3) then $\underline{X}(T) = \{ \underline{X}_\alpha(T) \; ; \; \alpha \in (0,1\rangle \}$, where

$$x_\alpha^L(1) = \inf_{\underline{x} \in \underline{X}_\alpha(0)} (\underline{x} + \underline{f}(\underline{x})), \quad x_\alpha^L(2) = \inf_{\underline{x} \in \underline{X}_\alpha(0)} (\underline{x} + \underline{f}(\underline{x}) + \underline{f}(\underline{x} + \underline{f}(\underline{x}))) \quad \ldots$$

and similarly for $x_\alpha^R(T)$.

Remark 4.: Note $\underline{x} \in \underline{X}_\alpha$ is the same as $\underline{x} \in X_{1\alpha} \times X_{2\alpha} \times \ldots \times X_{n\alpha}$, that is a n-dimensional interval for fixed $\alpha \in (0,1\rangle$. In case that variables x_i are interactive the whole n-dimensional interval cannot be admissible but only a region $H_\alpha \subseteq X_{1\alpha} \times \ldots \times X_{n\alpha}$. If we had available information about boundary of this region, we would be able to continue simulation from the last state instead of starting simulation again from initial state, that means $x_\alpha^L(T+1) = \inf_{\underline{x} \in H_\alpha(T) \subseteq \underline{X}_\alpha(T)} (\underline{x} + \underline{f}(\underline{x}))$ where $H_\alpha(T)$ is admissible region at the time T and $\underline{X}_\alpha(T)$ is the least n-dimensional interval of state containing $H_\alpha(T)$. In principal $X_{1\alpha}(T)$ is a projection of $H_\alpha(T)$ on x_i, in other words $X_i(T) = \{ X_{i\alpha}(T) \; ; \; \alpha \in (0,1\rangle \}$ is a marginal possibility distribution. It can be sometimes sufficient to approximate $H_\alpha(T)$ by the "deformed" n-dimensional interval for the description of which 2^n of points suffice.

Remark 5.: When we shall involve control and/or parameters to the system (3), we shall get $\underline{x}(T+1) = \underline{x}(T) + \underline{f}(\underline{x}(T), \underline{u}(T))$ and the enumeration of such an equation is also based on theorems given above.

Let us now concern with α-cuts for which obviously holds $\underline{X} = \{ \underline{X}_\alpha \; ; \; \alpha \in (0,1\rangle \} = \bigcup_{\alpha=0}^{1} \alpha \cdot \underline{X}_\alpha$, where $\alpha \cdot \underline{X}_\alpha$ is a subnormal fuzzy set whose membership function we shall obtain by multiplying characteristic function of α-cut $\underline{X}_\alpha \subseteq R^n$ by the number α.

Theorem 6.: Let us be to solve a problem of finding infimum of the function $g : R^n \longrightarrow R$ on the vector of FNs $\underline{X}$ represented by the vector of families of α-cuts $\underline{X} = \{ \underline{X}_\alpha \; ; \; \alpha \in (0,1\rangle \}$ then from $y_\alpha^L{}' = \inf_{\underline{x} \in \underline{X}_{\alpha'}} g(\underline{x})$ for any $\alpha' \in (0,1\rangle$ follows

$$y_\alpha^L = \inf_{\underline{x} \in \underline{X}_\alpha} g(\underline{x}) = \inf (\inf_{\underline{x} \in \underline{X}_\alpha - \underline{X}_{\alpha'}} g(\underline{x}), \; y_\alpha^L{}') \quad \text{for each} \quad \alpha < \alpha'.$$

Remark 6.: Given theorem tells if we start "from the top" of FNs, that means from $\alpha = 1$, we can embed the gained solution into a family of related problems; the solution for any $\alpha' \in (0,1\rangle$ is useful for all solution, where $\alpha < \alpha'$.

5. Some notes on algorithms for simulation of FSD

As is known, a SD model can be expressed in the form of a system of *algebraical equation* $\underline{y} = \underline{h}(\underline{x}, \underline{u}, \underline{v})$, where $\underline{x} \in R^n$ is a state, $\underline{u} \in R^r$ is a vector of control and parameter variables, $\underline{v} \in R^p$ is a vector of auxiliary and rate variables, $v_i = h_i(\ldots,v_i,\ldots)$ holding for all $i \in [1,p]$ (there exist no fast loops) and a system of *differential equations* $\underline{\dot{x}} = \underline{f}(\underline{x}, \underline{u}, \underline{v})$ with an initial condition $\underline{x}(0) = \underline{x}_0$ that can be rewritten to a system of *difference equations* $\underline{x}(T+1) = \underline{x}(T) +$ $+ \underline{f}(\underline{x}(T), \underline{u}(T), \underline{v}(T))$.

On the base of theorems described above, we are able to handle with control fuzzy variables, fuzzy parameters and initial fuzzy conditions expressed in the form of fuzzy numbers. An algorithm for simulation of FSD models has generally this structure:

(i) arrangement of the model equations to the *simulation sequence*;

(ii) identification of *parallel paths* in causal-loop diagram (that is graphical expression of the SD model (see [1])); this is important for handling with FNs (see Theorem 3. on multiple occurence of variables in the equation);

(iii) identification of the *strong component* associated to of the state variables in causal diagram;

(iv) *modification* of the model equations on the base of the parallel paths and the strong components to the form suitable for simulation of the FSD model;

(v) algorithms for proper *simulation* of the FSD model.

6. Conclusion

Now concrete algorithms for proper simulation are studied and it is worked on application of fuzzy relations, which are identified on the base of linguistical variables, for describing very vague parts of SD models. Results of this work shall be involved in the *ACM system* (the ACM system is a CAD system for the construction of SD models [2]).

7. References

[1]   D.Dubois, H.Prade: Fuzzy Sets and Systems: Theory and Applications. New York: Academic, 1980

[2]   M.Kejak: Automated Construction of SD Models. Dissertation, Prague, 1988 (will be published) (in Czech)

Multicriteria Fuzzy Decisions

J. Ester, Karl-Marx-Stadt[1]

Recently, an increasing number of papers, the topic of which is the introduction of
FUZZY THEORY /1, 2/ into MCDM, have been published /3, 4, 5/.
In MCDM-problems fuzziness (or uncertainty, pertubations etc.) can be located in three
places ($z(.)$ - membership function):

a) The set X of all feasible decisions is a fuzzy set:

$$X = \left\{ x \mid z_X(x) \right\}, \; x \in R^n \tag{1}$$

b) The mapping from the decision space into the criteria space is a fuzzy set:

$$\left\{ (x,q) \right\} = \left\{ (x,q) \mid z_{XQ}(x,q) \right\}, \; (x,q) \in R^n \times R^m \tag{2}$$

c) The rule of the choice of the final decision from the set of the best solutions
   (e.g. the Pareto set) is fuzzy.
In these equations the symbols mean:

        x - vector of decision variables,

        q - vector of criteria.

Problems a) and b) are typical of decision making in general.
Problem c) arises especially in multicriteria decision making.
In this paper we do not discuss the mathematical operations in detail, in particular
the aggregation rules, which are needed for fuzzy sets.

1. Basic methods for fuzzy-modelling

1.1. Fuzzy set of feasible decisions

The first case is given by the use of fuzzy numbers.

$$X = \left\{ x \mid g'(d',x) \geq b' \right\}, \tag{A.1.}$$

where d', b' are vectors of fuzzy numbers and, consequently, g' is also a vector of
fuzzy numbers.
The next possibility is the modelling with so-called fuzzy relations.

$$X = \left\{ x \mid g(x) \gtrsim b \right\}, \tag{A.2.}$$

---

[1] Technische Universität Karl-Marx-Stadt, Sektion Automatisierungstechnik
DDR-9010 Karl-Marx-Stadt PSF 964

where "$\gtrsim$" is a fuzzy relation and means "almost equal or greater than".

1.2. Fuzzy mapping

In this case we can use fuzzy numbers or fuzzy relations, too.

$$q' = f(c',x) \, , \, (B.1.) \; ; \quad q \approx f(x) \, , \, (B.2.),$$

$$c' - \text{vector of fuzzy parameters,}$$
$$q' - \text{vector of fuzzy criteria,}$$
$$"\approx" \quad \text{means "almost equal".}$$

1.3. Fuzzy decision rule

The first method is the definition of fuzzy goals.

$$z(q' \geq u') \quad \text{or} \quad z(q \gtrsim u) \qquad \text{with } z=\max! \, , \hspace{3cm} (C.1.)$$

where q' and "$\gtrsim$" are as defined above, u are lower bounds of the criteria and u' are fuzzy lower bounds. With the decision rule z=max! we get the so-called optimal fuzzy decision. The second possibility is the use of fuzzy preference relations.

$$q \gtrsim q^0 \Longleftrightarrow q - q^0 \in P(q^0) \, ,$$

where "$\gtrsim$" is the "weak" dominance relation and $P(q^0)$ is the dominating set of the reference solution q^0. For fuzzy preference relations this set is fuzzy:

$$z_{P(q^0)} = f(q,q^0). \hspace{4cm} (C.2.)$$

Quite often we obtain a third possibility , the use of so-called substitute problems with fuzzy numbers, e.g.

$$\sum_{i=1}^{m} g_i' \, q_i(x) = \max! \hspace{4cm} (C.3.)$$

(where g_i' are fuzzy weights, q_i, are the single criteria), fuzzy trade-offs, fuzzy rank-orders or other fuzzy parameters, which characterize the decision behaviour of the decision maker.

Summerizing, we can say that most of the papers, dealing with fuzzy multicriteria decision making, use the type of fuzzy modelling, given by (A.2.),(B.2.) and (C.1.).

2. Aggregation rules

A good analysis of desirable properties of aggregation rules used in fuzzy multicriteria decision making is given in /3/.
These properties are:

        a) commutative,

b) associative,

c) inductive,

d) strongly monotonically increasing,

e) idempotent,

f) compensatory,

g) stable,

h) continuous with respect to each component,

i) continuously differentiable,

j) continuously controllable between intersection and union.

The usual aggregation rules (e.g. MIN/MAX-operator, product/alg.sum, γ-operator, "and"/"or"-operators etc.) cannot satisfy all these properties. In /5/ the L_p-operator is described, which has all these desirable properties /6/

3. Fuzzy decision rules

Let us consider problem c) in more detail. We assume problems a) and b) to be deter-mininistic. In classical MCDM, we obtain two different approaches to solving the decision making problem.

1. Comparison of the objectives among one another.

Determination of a so-called "substitute optimization problem", which must be solved. The solution of this problem is the decision proposal.

2. Comparison of the alternatives among one another.

This can be carried out with a preference relation. Then we can compute the set of the nondominated alternatives. Both methods are also used in multicriteria fuzzy decision making. Let us start with an example of the first approach:

° For each criterion we get upper and lower bounds from the decision maker. Then we define membership functions z_i:

$$
z_i(x) = \begin{cases} 1 & \text{for } q_i(x) \geq c_{io} \\[2mm] \dfrac{q_i(x) - c_{iu}}{c_{io} - c_{iu}} & \text{for } c_{iu} \leq q_i(x) \leq c_{io} \\[2mm] 0 & \text{for } q_i(x) \leq c_{iu} \end{cases} \quad , \quad (3.1.)
$$

where c_{iu} are the lower bounds and c_{io} are the upper bounds, $i = 1(1)m$.

° The next step is the aggregation of the fuzzy sets, defined by (3.1.). This can be carried out in different ways.

For example

$$
z(x) = \min_i (z_i(x)), \qquad\qquad\qquad \text{Slowinski}
$$

$$
z(x) = R(\min(z_i(x)) + (1-R)\, \frac{1}{m} \sum_1^m z_i(x) \, , \qquad \text{Werners}
$$

$$
z(x) = \prod_1^m z_i(x) \, , \qquad\qquad\qquad \text{Hannan}
$$

$$z(x) = \left(\sum_1^m g_i(z_i(x))^p \right)^{1/p} , \qquad\qquad \text{Ester}$$

where $z(x)$ is the membership function of the fuzzy decision set.

° With $z(x)=\max!$ we get the so-called optimal fuzzy decision, which is proposed to the decision maker.

Let us continue the consideration with an <u>example of the second approach:</u>

° For each alternative q^i we define the fuzzy set of dominating solutions $P(q^i)$ with the membership function $z_{P(q^i)} = z(q,q^i)$.

° The fuzzy set of decisions D, which dominate the whole feasible set Q in the criteria space, is given by its membership function $z_D(q(x))$:

$$z_D(q(x)) = \operatorname*{aggr}_{q^i \in Q} (z(q,q^i)) = z_D(x)$$

In order to do this aggregation, we can use different rules. We propose the L_p-operator. In discrete MADM-problems we have

$$z_D(q) = \left(\sum_1^m g_i(z(q,q^i)^p \right)^{1/p}.$$

In continuous problems we obtain

$$z_D(q) = \left\{ \iint_Q g(q^i)(z(q,q^i))^p \, d(q^i) \right\}^{1/p}.$$

° With $z(x)=\max!$ we get the optimal fuzzy decision.

The comparison between the first and the second approach shows that the 2nd approach is more complicated, and it seems that it is more suitable for discrete MADM-problems /7,8/.

<u>Literature:</u>

/1/ Zadeh, L.A., Fuzzy sets, Inform. and Control 8 (1965) 338-353

/2/ Bellmann, R.E.; Zadeh, L.A., Decision-making in a fuzzy enviroment, Management sciences 17 (1970) 141-164

/3/ Werners, B., Interaktive Entscheidungsunterstützung durch ein flexibles mathematisches Programmiersystem, Minerva, München, 1984

/4/ Slowinski, R., A multicriteria fuzzy linear programming method for water supply system development planning, Fuzzy sets and systems 19 (1986) 217-237

/5/ Ester, J., Concepts of efficiency and fuzzy aggregation rules, in: Large-scale modelling and interactive decision analysis (ed. by Fandel, G. u.a.), Springer, 1986

/6/ Ester, J., Einbeziehung der Theorie der Unscharfen Mengen in die Methoden der mehrkriteriellen Entscheidung, WZ der TH Ilmenau, 33 (1987) Heft 6

/7/ Ester, J., Eine Methode zum mehrkriterialen Variantenvergleich bei Unsicherheit, WZ der TH Ilmenau 31 (1985) Heft 2

/8/ Wäscher, G., Innerbetriebliche Standortplanung, Betriebswirtschaftlicher Verlag Gabler, Wiesbaden, 1982

A Method for Multihuman and Multi-Criteria Decision Making

Anlan Song   and   Wei—Min Cheng

School of Economics and Management
Tsinghua University
Beijing, 100084
PR of China

ABSTRACT

In many multiple—criteria decision analysis problems, there are often several decision makers involved. To copy with this kind of problems, we propose a multiple aspiration levels method which is an extension of the reference point method by Wierzbicki. This interactive method can be used to derive satisfing solutions for the decision problems. It is implemented in a decision support and analysis system EXDASS and successfully applied to agricultural planning.

I. INTRODUCTION

The reference point method is a very useful tool for multiple criteria decision making (Wierzbicki, 1981). This method makes it possible to take into account the desires of a decision maker directly. It assumes that the decision maker can expresses his preference in terms of aspiration levels, i.e., the required value of individual objective. Practical applications have shown that it is more convenient for the decision maker to think in these terms than to estimate the trade—off coefficients or utilities required by other methods. The reference point method was widely used in many IIASA's decision support systems and is recently being used in China (Fedra, 1987).

In many practical problems, however, the decision may involve several decision makers in partly conflicting, partly cooperative situations. This is essentially a problem of multihuman decision making (Arrow, 1963). The existing reference point method is not so satisfactory to copy with this kind of problems.

To solved above problems, we propose a multi—aspiration levels method which is a generalization of the reference point method. This method is implemented in an Expert knowledge—based Decision Analysis and Support System———EXDASS, and successfully applied to regional agricultural planning.

II. BASIC IDEA

Assume that there are a group of k decision makers (DMs) involved in decision making. Each of them has a degree of importance or veto power which can be determined by discussion or assigned by a super decision maker. To make the problem simpler, we make following assumptions:

First, it is assumed that for each of the objectives, the decision makers have a common understanding about what means to improve it or not. This dose not mean that all of the DMs have the same preferences on various objectives. Each DM can express his preference by his own aspiration levels of objectives. Second, it is assumed that the DMs are rational, i.e., Von Neumann—Morgenstern's axioms of utilities are satisfied. And furthermore, it is assumed that if all DMs in the group consider two alternatives as indifferent, then the group's preference on these two alternatives is indifferent.

We now introduce a interactive method which is similar to that of Wierzbicki's but extended to multi—objective and multihuman decision making. In the process of man— machine interaction of our method, the DMs give out targets or aspiration levels which may be modified in the process, and the system gives out solutions that satisfy or come close to these targets.

In one interaction, if the aspiration levels given by one of the DMs are not attainable, he demands to get as close to these levels as possible according to his power. If his aspiration levels are attainable with some surplus, then the surplus is allocated to satisfy other DMs' aspiration levels. If the aspiration levels of all DMs are attainable with some surplus, then the surplus is allocated to get results which are close to certain ideal objective levels. In each interaction, the system gives only Pareto optimal solutions to the problem. If the DMs are not satisfied with resulted solutions, they can modify their aspiration levels in the next interaction. The process continues until some satisfactory solutions are resulted.

III. MAIN RESULTS

Mathematical Basis

Let D be dicision space, $D_0 \subset D$ be admissible decisions. Let G be a space of objectives. Given the mapping $Q:D_0 \to G$. Let $Q_0 = Q(D_0)$ be the attainable objectives set. To simplify discussion, assume G is linear topological space and all the objectives are to be maximized. Given a positive cone $\Lambda \subsetneq G$. We have following partial preordering:

$$q^1, q^2 \in G \qquad q^1 \leq q^2 \iff q^2 - q^1 \in \Lambda \qquad\qquad (1)$$

or strong partial preordering:

$$q^1, q^2 \in G \qquad q^1 \lneq q^2 \iff q^2 - q^1 \in \Lambda_P \triangleq \Lambda \setminus (\Lambda \cap -\Lambda) \qquad (2)$$

or strict partial preordering:

$$q^1, q^2 \in G \qquad q^1 < q^2 \iff q^2 - q^1 \in \text{int}(\Lambda) \qquad\qquad (3)$$

where $\text{int}(\Lambda)$ is the interior of Λ. Now we can express Pareto optimal objective:

$$q \in Q_0 \text{ is Pareto optimal} \iff Q_0 \cap (q + \Lambda_P) = \phi \qquad\qquad (4)$$

and weakly Pareto optimal objective:

$$q \in Q_0 \text{ is weakly Pareto optimal} \iff Q_0 \cap (q + \text{int}(\Lambda)) = \phi \qquad (5)$$

For a function $s:G \to R$, we have:

1) $s(q)$ is order preserving if
$$q_1 \leq q_2 \implies s(q_1) \leq s(q_2) \qquad\qquad (6)$$

2) $s(q)$ is strictly order preserving if
$$q_1 < q_2 \implies s(q_1) < s(q_2) \qquad\qquad (7)$$

3) $s(q)$ is strongly order preserving if
$$q_1 \lneq q_2 \implies s(q_1) < s(q_2) \qquad\qquad (8)$$

M—Scalarizing Function

Now let's construct a multi—aspiration levels achievement scalarizing function.

Definition A multi—aspiration levels achievement scalarizing function (M—scalarizing function for short) is a function $s:G \to R$ with:

$$s(q, q^0, q^1, ..., q^k) \triangleq \sum_{i=0}^{k} \rho_i s^i(q - q^i) \qquad\qquad (9)$$

where

1) $q^0 \geq q^i$ $\qquad\qquad\qquad\qquad\qquad\qquad\qquad\qquad (10)$

2) $\rho_0 > 0; \qquad \rho_i \geq 0, \ i=1,2,...,k$ $\qquad\qquad\qquad (11)$

ρ_i represents the degree of importance of the ith decision maker.

3) $s^0(q - q^0)$ should be strictly order preserving or strongly order preserving, and:
$$S^0 \triangleq \{ q \in G \mid s^0(q - q^0) \geq 0 \} = q + \Lambda; \quad s^0(0) = 0 \qquad (12)$$

4) Each $s^i(q - q^i)$ $(i \geq 1)$ should be order preserving and
$$s^i(q - q^i) \leq 0 \qquad\qquad \text{and} \qquad\qquad\qquad (13)$$
$$S^i \triangleq \{ q \in G \mid s^i(q - q^i) = 0 \} = q^i + \Lambda \qquad\qquad (14)$$

Obviously, $s(q, q^0, q^1, ..., q^k)$ is strictly (or strongly) order preserving. This guarantees that the optimal solution of s is Pareto optimal in Q_0.

Similar to Wierzbicki's results, we have following result:

Theorem 1   If $q^0 = \hat{q}$ is Pareto optimal, then

$$Q_0 \cap (\hat{q}+\Lambda_P) = \phi \implies \hat{q} \in \text{Arg max } s(q,\hat{q},q^1,...,q^k)$$

$$\max s(q,\hat{q},q^1,...,q^k) = 0 \tag{15}$$

We can find many M-scalarizing functions. Following function is very useful in practice:

maxmize $s(q,q^0,q^1,...,q^k)$

$$\triangleq \rho_0 \min (q_i-q^0_i) + \Sigma \rho_j \min \{ (q_i-q^j_i) , 0 \} + \varepsilon q \tag{16}$$

or equivalently

minimize $s(q,q^0,q^1,...,q^k)$

$$\triangleq -\rho_0 \min (q_i-q^0_i) - \Sigma \rho_j \min \{ (q_i-q^j_i) , 0 \} - \varepsilon q \tag{17}$$

where

$$\rho_0 > 0, \ \rho_j \geq 0, \ j=1,2,...,k \tag{18}$$

$$\varepsilon = (\varepsilon_1,\varepsilon_2,...,\varepsilon_P) > 0 \tag{19}$$

p is the number of the objectives.

Linear Case

In the case of multi-objective linear programming, the computational model can be easily derived.

Given (MLP)

max cx

s.t.

$Ax = b$

$x \geq 0$

$c \in R^{P\times N}, \ A \in R^{M\times N}, \ b \in R^M, \ x \in R^N$

From (9) we can get following computational model (EMLP):

min $\Sigma z_i - \varepsilon cx$

s.t.

$z_0 I + \rho_0 cx \geq \rho_0 q^0$ (α_0)

$z_1 I + \rho_1 cx \geq \rho_1 q^1$ (α_1)

$\cdots\cdots\cdots\cdots$

$z_k I + \rho_k cx \geq \rho_k q^k$ (α_k)

$Ax = b$ (π)

$x \geq 0$

$z_1,z_2,...,z_k \geq 0$

Given a real number t, the set $S_t \triangleq \{ q \mid s(q,q^0,...,q^k) = t \}$ is called a level set. Assume (EMLP) has optimal solution $(\hat{z}_0,\hat{z}_1,...,\hat{z}_k,\hat{x})$ and corresponding dual vector $(\alpha_0,\alpha_1,...,\alpha_k, \pi)$. Denote:

$\hat{q} = c\hat{x}$ be corresponding objective vector.

$\hat{s} = \Sigma\hat{z}_i - \varepsilon c\hat{x}$ be the optimal value of M-scalarizing function.

We have following result:

Theorem 2   The hyperplane

$$\Gamma = \{ q \mid (\varepsilon+\Sigma\rho_i\alpha^T_i)(q-\hat{q}) = 0 \} \text{ separates } Q_0 \text{ and } S_t, \text{with}$$

$$\hat{q} \in Q_0 \cap S_t$$

Let $w=(\varepsilon+\Sigma\rho_i\alpha^T_i)$, we can get trade-off rates of the objectives from theorem 2:

$$T_{i,j} \triangleq \partial q_i(x) / \partial q_j(x) \mid_{x=\hat{x}} = -w_i / w_j \tag{20}$$

IV. APPLICATION

Above method is implemented in a software module IMDAS, and built into an Expert knowledge-based Decision Analysis and Support System--EXDASS. We now present a regional agricultural decision analysis problem as an example.

Consider a simplified multi—objective agricultural planning model:

Objectives

1. Livestock products
   $Livpr = 1.4x_6 + 0.59x_7$ ==>   max
2. Investment
   $Inves = 10x_2 + 0.5x_6 + 5x_7 + 4x_9 + 0.13x_{10} + 2.7x_{13}$ ==>   min
3. Grain yield
   $Grain = x_{12}$ ==>   max
4. Income per person
   $Incom = x_{15}$ ==>   max

Decision variables

$x_1, x_2, x_3, ..., x_{15}$ (the meaning of them are not given here)

Constraints (omitted)

The aspiration levels of the objectives given by local authority   can be divided into two groups, one are optimistic:

Livpr = 58000;   Inves = 20000;   Grain = 5.4;   Incom = 200

another are conservative:

Livpr = 45000;   Inves = 32000;   Grain = 5.4;   Incom = 200.

Following results have been obtained by the process of man—machine interaction.

1. The utopia point and nadir point of the problem are

object	Livpr	Inves	Grain	Incom
utopia	103878	14286	7.85	147.68
nadir	40000	50000	5.00	100.0

2. The solutions corresponding to the optimistic aspiration levels are

object	Livpr	Inves	Grain	Incom
values	40000	42134	5.36	140.00

We can see the solutions are not so satisfactory.

3. Consider both groups of the aspiration levels, we get following satisfactory results:

object	Livpr	Inves	Grain	Incom
values	45000	32000	5.36	126.9

V. CONCLUSION

The Interactive Multi—objectivet Decision Analysis Sub—system IMDAS based on our method is successfully used in agricultural planning and decision analysis. In practical applications, we feel that our method is both flexible and conveniet.

REFERENCES

1. Arrow, K., Social Choice and Individual Values, Yale University, New Haven, 1963
2. Fedra, K. etal, Expert Systems for Integrated Development: A Case Study of Shanxi Province, The People's Republic of China, IIASA, SR—87—1.
3. Kallio, M. etal, An Implementation of the Reference Point Approach for Multiobjective Optimization, IIASA, WP—80—35.
4. Tanino, T. etal, On Methodology for Group Decision Making, in J. N. Morse ed, Organizations: Multiple Agents with Multiple Criteria, Springer—Verlag, 1981, p409—423.
5. Wierzbicki, A., A Mathematical Basis for Satisfing Decision Making, ibid, p465—485.
6. Yuan, B., Anlan Song, Min Lan, Socio—Economic System Approach to Helping the Zhouning County in Eradicting Poverty, Proceedings of the 1st International Conference on Agricultural Systems Engineering, Aug. 1987, Changchun, China

Evolutionary Learning Optimum-Seeking
on Parallel Computer Architectures

Hans-Paul Schwefel[1]

Abstract

On the one hand side many people admire the often strikingly efficient results of organic evolution. On the other hand side, however, they presuppose mutation and selection to be a rather prodigal and unefficient trial-and-error strategy. Taking into account the parallel processing of a heterogeneous population and sexual propagation with recombination as well as the endogenous adaptation of strategy characteristics, simulated evolution reveals a couple of interesting, sometimes surprising, properties of nature's learning-by-doing algorithm. 'Survival of the fittest', often taken as Darwin's view, turns out to be a bad advice. Individual death, forgetting, and even regression show up to be necessary ingredients of the life game. Whether the process should be named gradualistic or punctualistic, is a matter of the observer's point of view. He even may observe 'long cycles'.

Introduction

Evolution can be looked at from a large variety of positions. Beginning with the closest physico-analytic viewpoint, one might focus attention to the molecular and cellular processes. A more distant point of view centers on the behaviour of populations and species. Another difference emerges from whether one emphasizes the homeostatic aspect of the adaptation to a given environment, which is more relevant in the short term, or the euphemistic view of development to the more complex, higher, or even better in the long term.

The instruments used here, will be a macroscope and a time accelerator. Moreover, for methodological reasons, an optimistic point of view will be shared by comparing macroevolution with iterative optimization, or, even more adequately, with permanent meliorization techniques, i.e. hill-climbing or ridge-following procedures. By means of a simple algorithmic formulation of the main evolutionary principles, it is possible to reveal some properties of the process which in some cases are striking at the first glance. These findings may not only be helpful for better understanding 'nature's intelligence' but also be

---

[1] University of Dortmund, Department of Computer Science
   P.O.Box 50 05 00, D-4600 Dortmund 50, Fed. Rep. Germany

beneficial for global long-term planning and other groping-in-the-dark situations.

Modelling evolution

Ashby's homeostat [1] was a device which should find back to a feasible state by a sequence of random trials, uniformly distributed over a given parameter space. Many people have made the mistake of thinking of mutations as 'pure' random trials. A couple of them was malignant. They wanted to show that evolution theory never will be able to explain how 'nature' found a way to complex living beings within about 10^{17} seconds - the age of our globe. Montroll's random walk [7] paradigm, on the other side, neglects the selection principle of evolution. Both mutation and selection (chance and necessity) are the first principles, which, of course, have to be programmed properly.

Broadly accepted hereditary evidence has led to the proverb 'The apple does not fall far off of the tree'. A better model of mutations therefore is a normal distribution for parameter changements between generations, its maximum being centered at the respective ancestor's position. The rôle of chance in such a model is not explicative, however, but only descriptive. An important question now is the suitable size of the standard deviation(s) of the changes, which may be addressed as mean step size(s) from one generation to the next. This question arises with all optimization or meliorization schemes.

Modelling the selection principle is far more easy, as it seems first. 'Survival of the fittest' is the maxime which was derived from Darwin's observations. Some evolution programmers have taken it for granted: According to a given selection criterion, a descendant is rejected if its vitality is less than that of its ancestor, the ancestor otherwise. This scheme may be called a (1+1)- or two membered evolution strategy (E.S. in the following), resembling the 'struggle for life' between one ancestor <u>and</u> one descendant. Rechenberg [10] has derived theoretical results for the convergence velocity of that process in an n-dimensional parameter space. Most important was his finding that for an endless ridge following situation as well as for a minimum (or maximum) approaching situation the convergence rate is inversely proportional to the number of parameters. Distances growing with the square root of n, the number of iterations or generations needed to proceed from one to the other arbitrary point in space, increases with $O(n^{1.5})$ only, and not geometrically as in the case of simple Monte-Carlo strategies.

This type of creeping random search strategy (see e.g. Brooks [5], Schumer and Steiglitz [11], or Rastrigin [9] was first devised for

experimental optimization, where measurement inaccuracies drop out one-
variable-at-a-time and gradient-following procedures due to their inabi-
lity of non-local operation.

Bremermann's 'simulated evolution' [4] does not differ so much from
Rechenberg's as e.g. G.E.P. Box's 'Evolutionary Operation' EVOP [2]
does, an experimental design technique, and the so-called Simplex and
Complex strategies of Nelder and Mead [8] and M.J. Box [3] for numerical
optimization. Whereas random trials are vividly rejected by G.E.P. Box,
he centers several experiments (principally at the same time) in a
deterministic way around the position of the current best point in a
low-dimensional parameter space. The best of all then is taken as the
center of the next trial series. Nelder, Mead, and M.J. Box, however,
using a polyhedron for placing the trials, reject the worst position and
find a new one by reflecting the worst with respect to the center of the
remaining points of the simplex or complex.

The first concept may be called a $(1+\delta)-$, the latter a $(\mu+1)-$
evolutionary scheme, μ denoting the number of parents, δ the number of
children within one generation. More general, therefore, is a $(\mu+\delta)-$
scheme with μ ancestors which have δ descendants, the μ best of all
become parents of the next generation. The fact that individual life
times are limited is reflected by the $(\mu,\delta)-$ version, first introduced
by Schwefel [12]. Now the μ parents are no longer included into the
selection, thus δ must be greater than μ. Theoretical results so long
are available for the $(1+\delta)-$ and $(1,\delta)-$ evolution strategies only. All
further observations in the following, therefore, were found by computer
simulation only.

<u>Self-adaptation of strategy parameters</u>

As for all optimization techniques, the appropriate step size
adjustment is of crucial importance. Rechenberg found that there is a
'window' of one decade only within which the $(1+1)-$ evolution strategy
has a reasonable convergence velocity. He devised a simple rule for
exogenously adjusting a near optimum performance of the process, i.e. to
control the success probability which should be in the vicinity of one
success among five trials. This advice is good for many but not all
situations. Moreover, it does not give any hints to adapt the standard
deviations of the parameter changements individually. Some may be too
large, others too small, at the same time. Only within the multimembered
strategy, one can include the step size or even different step sizes
(mutation rates) into the set of the individual's genes and adapt them
endogenously. There is some evidence that by means of repair enzymes the
effective mutation rates are controlled, the rate of premutations due to
environmental conditions being constant over long periods.

Let us think at first, however, of one common step size for all
object parameters. Within a $(1,\delta)$- E.S. the correct step size turns out
to be even more important than within a $(1+\delta)$- version [12]. In the
first case regression takes place instead of progress when the step size
is too large, whereas stagnation is the worst case in the latter. At a
first glance, therefore, 'survival' of an ancestor might be a good
advice. Simulation results, however, show that the opposite is true.
This is the first surprise. Figure 1 demonstrates the difference between
a $(1+10)$- and a $(1,10)$- E.S. when minimizing the function

$$F_1 = \sum_{i=1}^{n} x_i^2 \quad \text{with } n=30.$$

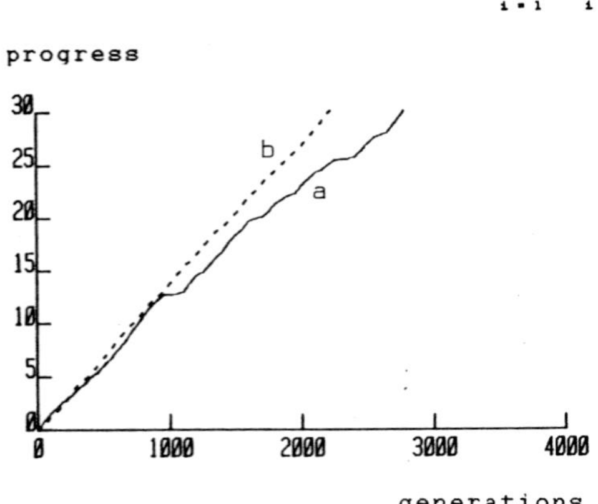

Fig. 1:
Self-learning of the mutation rate
(gross mean step size here) for
function F_1, n=30
a) $(1+10)$- evolution strategy
b) $(1,10)$- E.S.

The 'progress' is measured in terms
of $\log(\sqrt{(F^\circ/F')})$, where F° denotes
the start value, F' the current
value of the objective function.

The number of variables, n, was taken as large as 30 in order to
avoid improper conclusions. In lower dimensional cases nearly every
strategy may achieve good results. One common step size σ (or standard
deviation, more precisely) for all x_i is changed by mutation, i.e. by
multiplying the ancestor's value with a random number, drawn from a
logarithmic normal distribution in order to avoid exogenous drift. The
$(1,10)$- strategy turns out to be superior. An explanation for this
surprising fact is the following: if an ancestor happens to arrive at a
superior position, this might be - by chance - in spite of a non-optimum
step size, or a step size which is not suitable for further generations.
The $(1+\delta)$- scheme preserves the unsuitable step size as long as with it
a further success is placed. This leads to periods of stagnation. Within
a $(1,\delta)$- E.S. the good position, occasionally won with an unsuitable
step size, is lost, together with the latter, during the next generation. This short term regression, however, enhances the long term
velocity of the whole process by a stronger selection with respect to
the suitable step size (strategy parameter). Simply speaking: Forgetting
is as important as learning, the first must be seen as a necessary
integral part of the latter. One might interpret the fact of an inherent
finite life time (preprogrammed maximum number of cell divisions) of
living beings as an appropriate measure of nature to overcome the

difficulties of undeserved success - or, in a changing environment, of forgetting obsolete 'knowledge'.

Collective learning of proper scalings

In most cases it is not sufficient to adapt one common step size for all object parameters. For an objective function like

$$F_2 = \sum_{i=1}^{30} i \cdot x_i^2 \; ,$$

for example, individual standard deviations σ_i, appropriately scaled, could speed up the progress rate considerably. To achieve this kind of flexibility within the multimembered evolution strategy, each individual is characterized by a set of n step sizes in addition to the n object parameters. They are mutated by multiplication with two random factors, one being common for all step sizes as before, the other acting individually, however. Thus general and specific scaling can be learned at the same time. Operating with an $(1,\delta)$- strategy - however large δ may be - leads to a second surprise: this kind of process does not work at all, it gets stuck prematurely by approaching a relative optimum in a lower dimensional space. The reason is rather simple: As said above, the convergence rate is inversely proportional to n, the dimension of the parameter space. Descendants operating in a subspace by sharply reducing some of the step sizes have a short term advantage. Selecting the fittest descendant to become the one and only parent of the next generation, is counterproductive in the long term, as was the survival of the ancestor.

Figure 2 demonstrates how to overcome the difficulty. If more than one, not only the best of the descendants, become parents of the next generation _and_ recombination by sexual propagation takes place, i.e. mixing of the information gathered by different individuals during the course of evolution, then stagnation can be overcome. Now the convergence rate steeply goes up with the population size. On a conventional one-processor computer the parallelism of that scheme cannot be realized, but multi-processor machines are coming. That is why all figures show the progress over the number of generations and not over computing time.

The overwhelming success of recombination demonstrated here, may explain the early appearance of sexual propagation on earth. But it is unprobable that only the additional variability provokes the success. A better explanation might be the following: The typical situation during the meliorization process is ridge following. Within a population some individuals have a position on one side, others on the other side of the

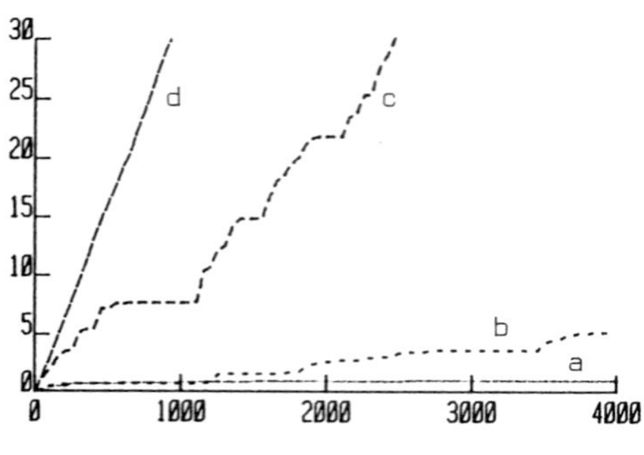

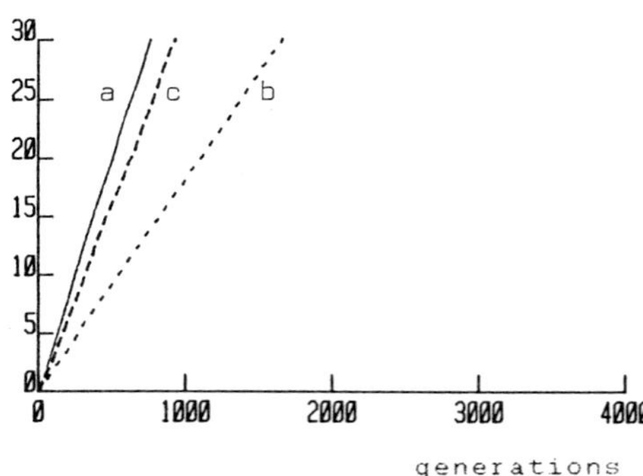

Fig. 2:
Learning of the scaling I
a) (1, 10)- E.S.
b) (3, 10)- E.S. with recombination
c) (6, 30)- E.S. with recombination
d) (15,100)- E.S. with recombination
(2d equals 3c)

Fig. 3:
Learning of the scaling II
a) (1,100)- E.S.
b) (15,100)- E.S. without recomb.
c) (15,100)- E.S. with recombin.
a) & b) with prefixed opt. scaling
c) with scaling as learned

ridge. Mixing genetic information is a means of riding the ridge more
efficiently. A similar argument holds for mixing the step sizes:
Individuals on one side of the ridge have 'internal models' (made up of
the set of step size relations) of the response surface which are
different from those on the other side. Even if both models are wrong,
at least in the long term, since both may be locally adapted only (if
the 'model' learned is not a law of nature), some 'mean' model (or
better: hypothesis) may turn out to be more useful for the future.
Simply speaking, one may say that 'natural intelligence' is distributed.

Now the question of the appropriate selection pressure prevails:
How many of the descendants should be selected as new parents. The
answer - at least for the objective function chosen - is given by figure
4. All other conditions being held constant, including the number of
descendants δ within one generation, only μ, the number of parents, was
changed. Three cases were investigated for function F_1 (see Figure 4).

Whereas in both cases a) and b) $\mu=1$ is the best choice, in a
learning situation (case c), it is better, even necessary, to increase μ
far beyond 1. The diagram moreover demonstrates the effectivity of the
collective learning process. Under proper conditions nearly the same
convergence rate as with total knowledge of the optimum scaling can be
achieved - even with lower selection pressure. This is the third
surprise.

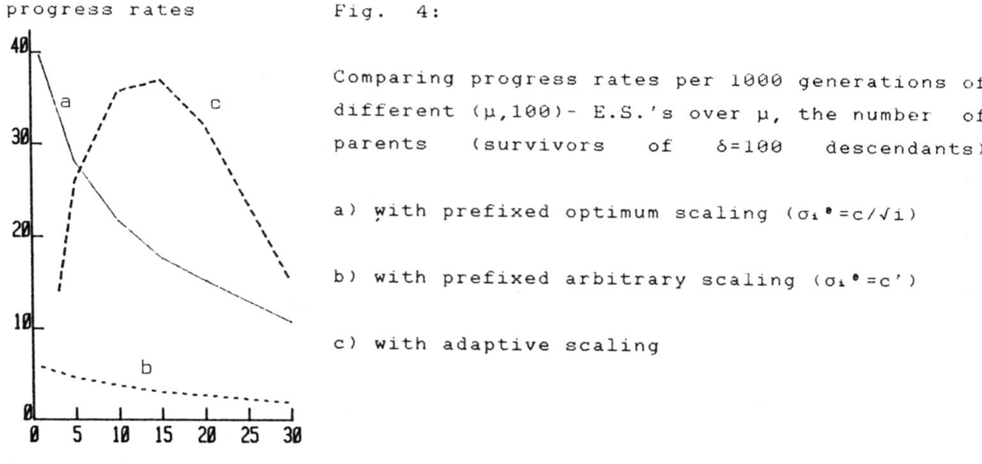

Fig. 4:

Comparing progress rates per 1000 generations of different $(\mu,100)$- E.S.'s over μ, the number of parents (survivors of $\delta=100$ descendants)

a) with prefixed optimum scaling $(\sigma_i{}^* = c/\sqrt{i})$

b) with prefixed arbitrary scaling $(\sigma_i{}^* = c')$

c) with adaptive scaling

Learning of a metric and the epigenetic apparatus

Topologies of vitality response surfaces normally are not as simple as assumed above. The next possible complication is to incline the main axes of the hyper-ellipsoids which form the subspaces F = const. Now scaling alone does not help in achieving optimum performance. What can nature do, what has it done, to overcome the difficulty? In many cases one has found that a single gene influences several phenotypic characteristics (pleiotropy) and vice versa (polygeny). These are the two sides of the same coin, which is correlation, the perhaps best known example of it being allometric growth. The transmission mechanism between

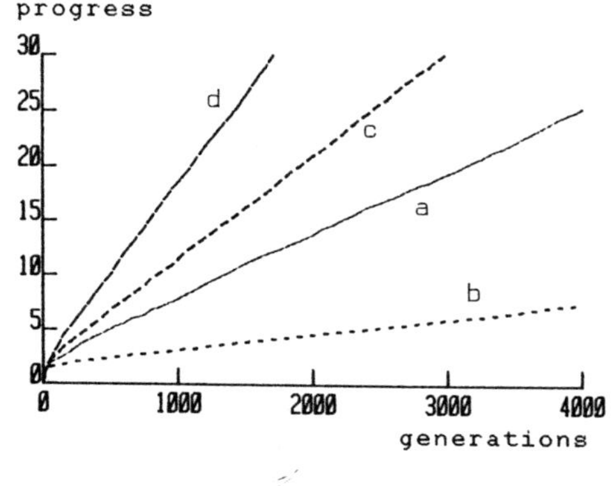

Fig. 5: Learning of a metric
a) (1,100)- E.S. with constant but arbitrary scaling
b) (15,100)- E.S. under the same conditions, i.e. without recombination
c) (15,100)- E.S. with recombination and adapted individual step sizes
d) (15,100)- E.S. as before, with additional learning of linear correlations between the pheno-the phenotypic mutations

genotype and phenotype, called epigenetic apparatus, in a first order may be approximated by linear correlation. In addition to individual step sizes, now correlation coefficients or angles of inclination of the ellipsoid, forming the surface of constant probability density of a mutation, have to be learned. Figure 5 shows first results for the objective function

$$F_3 = \sum_{i=1}^{n} \left(\sum_{j=1}^{i} x_j \right)^2 .$$

Four cases were simulated , three of them corresponding to those of figure 3. In both cases c) and d), recombination, discrete for the step sizes, intermediate for the object parameters, was used. It is obvious that these sampling conditions bear a variety of possibilities with respect to the mutabilities of step sizes and correlation angles so that simulations c) and d) might not yet respresent the best choices. Nevertheless the results show how much may be gained in terms of progress velocity by allowing to learn a simple 'internal model' of the topology of the environment, the 'real world'. Nevertheless, such a model in most cases will not be a correct 'theory', but simply a useful local or temporal hypothesis.

Gradualism or punctualism, and so-called long waves

Up until now all figures showing the evolutionary progress over time or generations were drawn for objective function values only every 50th generation and for the mean of the population moreover. If we have a closer look by picking out one of the decision variables, e.g. x_{15} for function F_2, and look at it at every generation (figure 6) then the picture reveals more details.

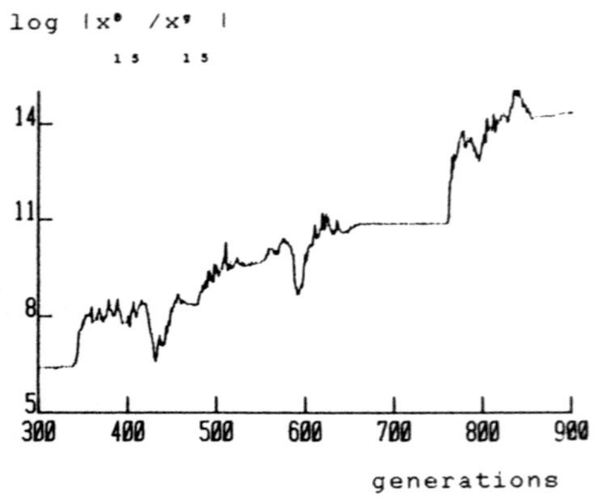

Figure 6: Time cut of one of the variables.

It depends on the density of the historical record whether we may speak of a gradual or a punctualistic process (see e.g. Stanley [13]). Due to the fact that the objective function depends on many parameters, a single one of them must not resemble the progress as a whole. Great success in one direction forgives regression in others - at least in a nonlinear world. And , for sure, looking at some arbitrary cut of the time record, one might get the impression of stochastically disturbed 'long waves' (normally three) with a more or less constant period. Even more aggregated sub-objectives, like the GNP of a nation, could exhibit such behaviour, if the underlying process operates left of the maximum of curve c) in diagram 4.

Conclusions

Many people today, when speaking about long term planning, environmental forecasting, technology assessment etc., are embarrassed by the degree of our ignorance. Very often then they speak of the uncertainties involved. But looking more precisely, isn't it a fact that there are, at the same time and with access to the same data, different certainties, i.e. different interpretations of the past and different aspirations for the future, or, in other words, different 'internal models' of the world? In the light of the simulation results above, one should appreciate, not regret, that. Due to the findings of a rather new field of science, i.e. nonlinear dynamics, we must admit that knowledge about the long-term future is principally unavailable for an open dissipative system operating far off of equilibria. We really are groping in the dark [6]. Therefore we should not try to establish one best model of the world, but make the best of the different individual ones of the ridge we are trying to follow without clearvoyance. Even if all of them were wrong they might as well be worthwhile to be recombined with each other. Instead of relying upon too strong competition, which leads to stagnation, as we have seen, we better should agree upon further experiments in cooperation.

References

[1] Ashby, W.R. (1960) _Design for a brain_. New York: Wiley, 2nd ed. .

[2] Box, G.E.P. and Draper, N.R. (1969) _Evolutionary operation: A statistical method for process improvement_. New York: Wiley.

[3] Box, M.J. (1965) A new method of constrained optimization and a comparison with other methods. _Computer Journal_, _8_, 42-52.

[4] Bremermann, H.J. (1968) _Numerical optimization procedures derived from biological evolution processes_. In: Cybernetic problems in bionics. (H.L. Oestreicher and D.R. Moore, eds.), New York: Gordon and Breach.

[5] Brooks, S.H. (1958) A discussion of random methods for seeking maxima. _Oper. Res._, _6_, 244-251.

[6] Meadows, D., Richardson, J. and Bruckmann, G. (1982) _Groping in the dark - The first decade of Global Modelling_. Chichester: Wiley.

[7] Montroll, E.W. and Shuler, K.E. (1979) Dynamics of technological evolution: random walk model for the research enterprise. _Proc. Natl. Acad. Sci. USA_, _76_, 6030-6034.

[8] Nelder, J.A. and Mead, R. (1965) A simplex method for function minimization. _Computer Journal_, _7_, 308-313.

[9] Rastrigin, L.A. (1965) _Sluchainyi poisk v zadachakh optimisatsii mnogoparametricheskikh sistem_. Riga: Zinatne.

[10] Rechenberg, I. (1973) _Evolutionsstrategie: Optimierung technischer Systeme nach Prinzipien der biologischen Evolution_. Stuttgart: Frommann-Holzboog.

[11] Schumer, M.A. and Steiglitz, K. (1968) Adaptive step size random search. _IEEE Trans._, _AC-13_, 270-276.

[12] Schwefel, H.P. (1981) _Numerical optimization of computer models_. Chichester: Wiley.

[13] Stanley, S.M. (1981) _The new evolutionary timetable_. New York: Basic Books.

Optimization and Simulation in Control Design

Inge Troch, Wien 1)

Abstract. Methods for computing optimal controls are reviewed and compared in view or their applicability for solving real world problems. Further, the importance of a solution-oriented modelling is emphasized.

1. Introduction

Optimization constitutes an important topic in the analysis and design of systems. Optimization is used during the modelling stage e.g.in identification algorithms as well as for the design of open-loop or feedback controls. Being an important and powerful tool for engineers, biologists, economists etc., optimization needs nevertheless the support of various fields of scientific computation, especially the support of simulation. In the following some main aspects occuring in the optimal design of controls shall be emphasized together with views on the role of simulation.

2. Control Design and Optimality

The requirement to design a control being optimal in some sense may occur in various areas of application. It may be connected with passive control of mechanical structures as well as with the dynamic control of discrete or continuous processes occuring in engineering, economics, biology, physiology etc. The mathematical formulation of the control problem may be taken out of quite a large number of possibilites, but the essentials of the methods used for dynamic processes will be quite the same independently of the concrete application. Nevertheless, the concrete realization of one of the possible methods will depend on the actual formulation of the task. Consequently, already the modelling of the system should be done in view of available devices such as methods, computers, program packages, experiences etc. A model which is not alone problem-oriented but also solution-friendly can reduce the effort and the time required considerably. One should always keep in mind that the solution derived will constitute the solution to a certain mathematical problem being a more or less accurate model of the real task. Therefore, any the model of a system will meet the actual requirements only to a certain extent.

Design of control strategies by using optimality concepts is quite frequent and rather well understood. Further, the call for an *optimal solution* is appealing. But, it is important to note that *control objectives should affect modelling decisions*. Otherwise, one might end up with a solution to the mathematical model being not realizable in practice. Each model, even a very sophisticated one, is imperfect. Hence, the resulting solution has to be examined carefully. Simulation provides a powerful and

---

Work supported by Österr. Fonds zur Förderung der wissenschaftl.Forschung

1) Technische Universitaet, Wiedner Hauptstr.8-10, A-1040 Wien

successful tool for these investigations, especially for the necessary
sensitivity studies. For that reason, a design procedure based on optimality considerations will usually follow the lines indicated in Fig.1.

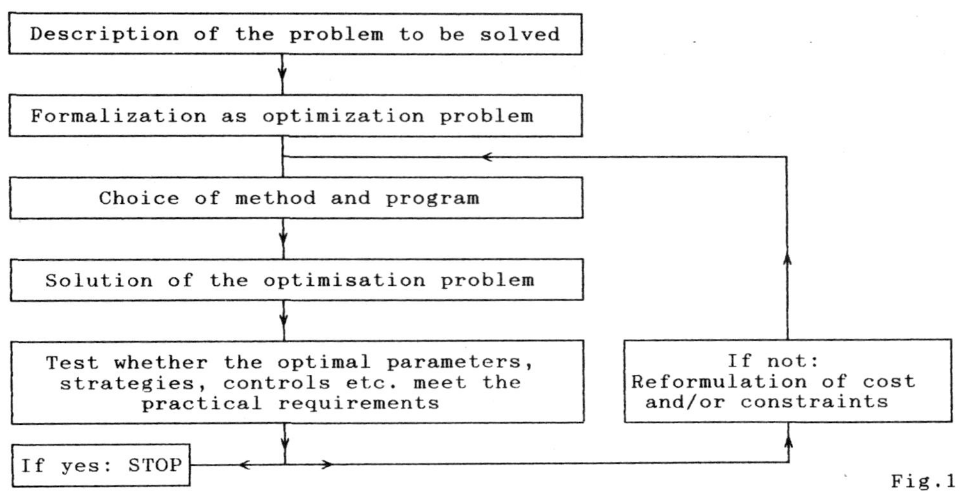

Fig.1

3. Problem Description

The modelling of a control task as optimization problem requires the
description of *the dynamic behaviour* of the system (including *initial and
final conditions*) which must not contain more parameters than can be identified with sufficient accuracy and at acceptable expense. These considerations together with the observation that methods for linear systems are
developed best lead to a preference for linear time-invariant systems.
Such systems are not a reflection of a belief in a linear time-invariant
real world, but instead a reflection of the present state of the art of
describing and analysing the real world, [24].

Further, all *restrictions* have to be formalized no matter whether they
are imposed on the state of the system or on the available controls. Often
such restrictions are rather vague such as 'during normal operation the
speed of a motor must not exceed a certain number of revolutions, but during start-up it can be up to three or four times as much'. Finally, the
optimality requirement is to be formalized. This is by no means trivial
because in most practical applications more than one goal have to be observed which may interfer or even be conflicting. Here, so-called vector-
optimization may help, but it is to be noted that such problems will be
solved by reducing them to one or to a series of normal optimization problems.

Finally, a problem formulation results containing the following
elements:

   * dynamic behaviour of the system depending on controls and

disturbances acting on the system
* initial state
* desired final state or manifold, respectively, and/or
  desired movement
* bounds on the controls
* bounds on the states
* measure of the quality of the controls

These main elements are more or less the same for quite different types
of systems. The following considerations are based on continuous-time
systems with concentrated parameters. For analogous discrete-time systems,
the statements hold also with mild deviations due to the different dynamic
characterization. Things become more involved for systems with time-lags
and/or with distributed parameters. There are some theoretical results
and practical methods available for such systems too. But due to restricted space they shall not be considered further.

4. Methods for Solving Optimal Control Problems

As already stated, we shall concentrate on systems which can be model-
led by ordinary differential equations and which are investigated best.
But it should be noted that especially time-discrete systems can be trea-
ted in a quite similar manner. Theory provides four groups of tools:
* Maximimum principle of Pontryagin
* Dynamic programming by R.Bellman
* Parametrization of controls
* Direct variation of controls

The first three approaches reduce the original infinite-dimensional
problem to a finite-dimensional one, whereas the last approach - although
closely related to the maximum principle - yields an iteration method
allowing the direct computation of the optimal control using the systems
and the adjoint equations, e.g. [18]. The *maximimum principle* transforms
the optimal control problem to a two- or multi-point boundary value problem where the number of points is unknown in advance whenever state constraints are present or singular arcs occur. In applications, *dynamic
programming* applies first appropriate discretization and yields by this a
parameter optimization problem of high dimension which is solved in a
specific way. *Parametrization* results in an approximation of the optimal
control by an element of a finite-dimensional function space. If chosen
appropriately, the true optimal control may result.

The maximum principle has the advantage that it allows the computation
of the exact solution, often it provides at the same time sufficient conditions for the optimality of a solution. Last but not least the maximum
principle provides information about the type of optimal controls for certain types of problems. A rather general result says that optimal controls
can be composed of singular and regular arcs and that the latter are either free arcs (i.e. lie in the interior of the admissible domain in the

state space) or boudary arcs. For certain types of problems the following
results on free, regular optimal trajectories and controls are of rele-
vance for applications (for details see e.g.[1], [8]):
  * Bang-Bang-Controls appear for problems being linear in the controls,
    which are subject to simple bounds.
  * Max-0-Min-Controls are optimal for problems being linear in the abso-
    lute values of the controls and in the controls themselves, again the
    latter have to be constrained by simple bounds (fuel-optimaity)
  * Linear state feedback is optimal for linear-quadratic problems, i.e.
    problems with linear dynamic equations and quadratic cost.
Information like the one just mentioned allows reducing the optimal
control problem to a parameter optimization problem (i.e.to a finite-
dimensional one) and computing the exact solution. Theoretically, dynamic
programming is equivalent to the maximum principle, [10], as be seen by
comparing the Bellman functional equation and the maximality condition for
the Hamiltonian. Yet, it has the advantages, that state constraints can be
observed more easily and that it is better suited for computerization. But
it should be noted also, that in most cases only open-loop controls and no
feedback laws will result.

The maximum principle yields computational difficulties due to the
extreme stiffness of the resulting boundary value problem whereas dynamic
programming requires a large memory and long computation times what can
be overcome to a certain extent by using methods like DDDP [20], or the
adaptive search tube, [27]. It should be mentioned that dynamic programming allows an interpretation of the adjoint variables which in some cases
may be used to facilitate the numerical solution of the boundary value
problem resulting from the application of the maximum principle.

5. Parametrization of Controls

For linear-quadratic problems the optimal control is paramterized by
the maximum principle quite naturally. But also for other types of problems, *parametrization* constitutes a promising and powerful alternative.
State constraints can be considered rather easily. Further, the problem
investigator can introduce his experiences and wishes on the structure of
a good control. Combining these ideas with results derived from the maximum principle allows in some cases the computation of the exact solution
of the formalized problem in this manner. Knowing e.g. that the optimal
control is of bang-bang-type suggests to use the switching times - computationally more appropriate: the lengths of the switching intervals - as
parameters to be optimized. The same holds for fuel-optimal problems. But
an additional difficulty results from the generally unknown number of
switchings. Of course, such a procedure may yield a relative extremum only
and not the absolute one. Normally, the optimal control is approximated by
a linear combination of elements of an appropriately chosen function space
such as e.g. the space of

* piecewise constant functions
* piecewise polynomial functions of a given degree
* splines of a given order
* functions known to be well-suited for practical realization

Especially the last possibility is of interest for applications, [16], because the optimization procedure then results in a control which can be realized without additional difficulties by applying the available means. It does not make much sense to compute with great effort the exact solution of the mathematical model and to effectuate then this solution only in an approximate way and with big efforts for the real system. In such cases the direct computation of such an 'approximation' will be of greater value.

Obviously, the resulting parameter optimization problem is a problem with linear and nonlinear constraints, which can be solved by choosing a method well suited for the concrete type of problem, e.g. [9], [11], [12], [13], [14], [15], [19].

6. Simulation Languages and Optimization

As indicated already in Fig.1, analysing the resulting optimal control by simulation is an appropriate procedure. The optimization as well as the simulation require programming of the model including all constraints. For this reason only one model description should be used both for optimization and for simulation. Simulation languages offer a comfortable possibility for setting up the model and for carrying out experiments. But normally, they dont support the optimization itself, for which rather powerful routines are available in standard software libraries. But, using both possibilities at the same time requires a relatively involved programming work. This leads to reflections about the necessity and the possibility of extending existing simulation languages by appropriate optimization routines, [2], which support at the same time the computation of optimal controls, [6], [28].

Existing simulation languages have to be enlarged by a suitable optimization environment in such a way that the user can either simulate the dynamic behaviour of a system subject to given controls or, that he can optimize parameters and/or the structure of controls etc. according to a given optimization criterion. It is important that these two tasks can be handled independently one from each other because otherwise computation times would become so large that simulation gets exhausting. For more details the reader is refered to [7], [17], [26].

7. Conclusions

An evaluating survey on existing tools for designing optimal controls demonstrates that parametrization is not only one possibility among many others, but indeed an valuable, application-oriented tool for using optimization concepts for controller design. This holds especially in case

results from theory are used already during the modelling stage in such a
way that the maximum principle can be applied without additional difficulties and yields information about the structure of the optimal control.
This applies in particular for systems being linear or for systems being
linear at least in the control and in the absolute values of the controls.

8.References

[1] M.Athans, P.L.Falb, Optimal Control, McGrawHill, New York, 1966.
[2] I.Bausch-Gall (1982), Parameteroptimierung bei technischen Modellen
     mittels einer kontinuierlichen Simulationssprache. In: Simulations-
     technik (Ed.: M.Goller), Springer Verlag, Berlin, 266-273
[3] R.E.Bellman, S.E.Dreyfus, Applied Dynamic Programming, Princeton
     University Press, Princeton, 1962.
[4] W.G.Boltjanski, Mathematische Methoden der optimalen Steuerung,
     Akadem. Verlagsges., Leipzig, 1971.
[5] W.G.Boltjanski, Optimale Steuerung diskreter Systeme, Akad.Verlags-
     ges., Leipzig, 1976.
[6] F.Breitenecker, Optimierung in kontinuierlichen Simulationssprachen -
     Aspekte bei Modellen technischer Systeme. In: Simulationstechnik
     F.Breitenecker, W.Kleinert), Springer Verlag, 1984, 656-660
[7] F.Breitenecker, R.Ruzicka, A.Sauberer, I.Troch: Die ACSL-Optimie-
     rungsumgebung GOMA zur Berechnung optimaler Steuerungen. In "Stand
     und Entwicklungstendenzen der Automatisierungstechnik" (Ed.:
     P.Kopacek), ÖPWZ, Wien, 1988, 61-65
[8] A.E.Bryson, Y.C.Ho, Applied Optimal Control, Wiley, New York, 1975
[9] D.Burley, Studies in Optimization, Intertext, Leighton Buzzard, 1974
[10] S.E.Dreyfus, Dynamic Programming and the Calculus of Variations,
     Academic Press, New York, 1965.
[11] R.Fletcher (Ed.), Optimization, Academic Press, New York, 1969.
[12] R.Fletcher, Practical Methods of Optimization, I,II, Wiley,
     Chichester, 1980.
[13] P.E.Gill, W.Murray, M.Wright, Practical Optimization, Academic Press,
     New York, 1981.
[14] P.E.Gill, W.Murray (Eds.), Practical Methods for Constrained Optimi-
     zation, Academic Press, London, 1974.
[15] C.Großmann, A.A.Kaplan, Strafmethoden und modifizierte Lagrangefunk-
     tionen in der nichtlinearen Optimierung, Teubner, Leipzig, 1979.
[16] M.Gräff, Die Berechnung von optimalen Steuerungen für dynamische
     Prozesse durch Parameteroptimierung, Bericht 2 (1986), Abt.Regelungs-
     math.,Hybridrechen- und Simulationstechnik, TU Wien, 63 pp.
[17] M.Gräff, F.Breitenecker, I.Troch, Praktische Parameteroptimierung in
     ACSL. Interface Nr.23, (1986) HRZ, TU Wien, 9-14.
[18] L.Hasdorff, Gradient Optimization and Nonlinear Control, Wiley, New
     York, 1976.
[19] M.Hestenes, Conjugate Direction Methods in Optimization, Springer,
     New York, 1980.
[20] D.H.Jacobson, D.Q.Mayne, Differential Dynamic Programming, Wiley, New
     York, 1970.
[21] H.W.Knobloch, Das Pontryagin'sche Maximumprinzip für Probleme mit
     Zustandsbeschränkung, I, II, ZAMM 55 (1975), 545-556, 621-634.
[22] J.P.LaSalle, The Bang-Bang-Principle, Proc.1st IFAC Congress (1960),
     493-497.
[23] G.Leitmann, Einführung in die Theorie optimaler Steuerungen,
     Oldenbourg München, 1974.
[24] R.S. Pindyck, An Application of the Linear Quadratic Tracking Problem
     to Economic Stabilization Policy, IEEE-Trans., AC-17 (1972), 287-300
[25] L.S.Pontrjagin et al., Mathematische Theorie optimaler Prozesse,
     Oldenbourg, München, 1964.
[26] A.Sauberer, R.Ruzicka, F.Breitenecker, I.Troch: Implementation der
     Optimierungsumgebung "GOMA" in ACSL. In: Simulationstechnik (Ed.:
     J.Halin), Springer-Verlag, Berlin, 1987, 222-231
[27] H.K.Schulze, Die Methode des adaptiven Suchschlauchs zur Lösung von
     Variationsproblemen mit Dynamic-Programming-Verfahren, EDV 3 (1966)
[28] I.Troch: Simulation and Optimisation. Proc.European Congress on
     Simulation, Prag 1987 (Ed.: V.Hamata), vol.B, 315-321.

System-Engineering Methodology for Simulation and Control of Dynamical Networks

Manfred Peschel, Hans-Michael Voigt
Academy of Sciences of the GDR
Centre for Scientific Instrumentation (ZWG)

Werner Mende
Academy of Sciences of the GDR
Institute for Geography and Geo-Ecology

Felix Breitenecker
Technical University of Vienna
Institute for Technical Mathematics

Introduction

In [1] we developed a system-concept for rate-coupled systems. Based on an appropriate structure-design principle for dynamical systems described by ordinary differential equations we can represent the corresponding models by normal forms, e.g. by Lotka-Volterra equations.

An equivalent concept was introduced by M. Savageau [2].

In [3] we presented the method of "Direct Iteration" for approximating solutions of ordinary differential equations.

In this method the "integration" operator is realized by the following recursion corresponding obviously to an Euler step

$$x(t+D) = x(t)\left(1 + D\frac{y(t)}{x(t)}\right) \quad if \quad \frac{y}{x} \geq 0$$

$$x(t+D) = \frac{x(t)}{1 + D\left|\frac{y(t)}{x(t)}\right|} \quad if \quad \frac{y}{x} < 0.$$

This method is based on the substitution of a given time function by a tangential straight line for every time instant.

More growth-oriented is a local substitution of the given time function by tangential exponential [6,7].

The corresponding simulation scheme we propose to call "Natural Iteration". It is given by the following recursion approach for the simple integrator

$$x(t+D) = x(t)e^{DFx}$$

with the logarithmic derivative

$$Fx = \frac{d\ln x}{dt}.$$

Both approaches for modelling and system simulation will be generalized in this paper referring to [4]. With this respect the goal of this paper is to show the great variety of possible ways for the simulation of ordinary differential equations.

In [5] we derived certain control concepts connected with local models for plant control and proposed some new controller concepts. In this paper we will more completely connect the modelling approach with the design of controllers.

General network concepts for the description and simulation will be outlined with special respect to neural networks.

Local Models as Approximate Description

of System Dynamics

Both methods mentioned in the introduction for the transformation of an ordinary differential equation to a recursion the application of a local model of the following form is common:

$$\frac{x'}{x} = g(D, Fx).$$

Here x' = x(t+D) is the consecutive value to x = x(t), D is the time-step size, and Fx is the logarithmic derivative. In this context a local linear model obviously leads to the Direct Iteration, and a local exponential model to the Natural Iteration.

It is useful to interpret the local model of first order in the above defined form as a transformation rule of the continuous growth rate Fx to the discontinuous growth rate.

With this interpretation the following conditions posed on f(D,u) in

$$g(D, u) = 1 + D f(D, u)$$

should be fulfilled

1 $sign(f(D, u)) = sign(u)$

2 $f(D, u)$ and $g(D, u)$ are monotonously increasing

   functions of u

3 $\lim_{D \to 0} f(D, u) = u$

Surely, instead of first order local models also higher order models can be applied depending then not only on the first derivative but on higher order derivatives of y(t). But also for the higher order local models we demand that they obey the conditions 1 to 3.

We have still to answer the question how we get a recursive simulation model following the local approach. Generally, for the purpose of system simulation we have to express the continuous derivatives contained in the transformation function g(...) by expressions taken from the model description with the aim to reduce the arguments in g(...) down to expressions only in state variables and input functions.

Some Examples

Local Models from Taylor-Series Expansions

We truncate the Taylor-series expansion with the second order term and get the following local model

$$x(t + D) = x(t) + D p x + \frac{D^2}{2} p^2 x$$

with the short-hand notation p = d/dt. In this form the given local model can be used immediately for simulation purposes. It turns out that the corresponding integrator is de facto the trapezoid rule for integration. But the derived mode can be easily reinterpreted with the help of Natural Iteration

$$x' = e^{DFxe^{\frac{D}{2}F(px)}}$$

which again gives us another integration formula.

Following this type of consideration, and admitting higher order Taylor-series expansions as local models we can derive the following pleasant formula

$$x' = xe^{DFxe^{\frac{D}{2}F(px)\cdots e^{\frac{D}{k}F\left(p^{k-1}x\right)}}}$$

More examples for local models are given in [4].

Control Concepts

With the above described modelling and simulation concept using local models it is possible to formulate and solve non-linear and linear control problems.

In the following we inform about some possibilities to use the Direct Iteration for the introduction of some simple non-linear control principles. We first start with some quasi-linear control principles.

For simplicity we consider a control loop with a control plant being a simple following plant with the identity operator the feedback of which is determined by a dynamical control factor U which yields the following behaviour of the closed control loop

$$x' = xU$$

or more general

$$x' = \phi(x)U$$

with

$$x_* = \phi(x_*)$$

again with x' = x(t+D) and x=x(t). In any case the control factor U must contain the prescribed demanded value x_* for the control variable x. We propose an aggregated discontinuous control concept

$$U = \frac{1 + K_{R1}\left(\frac{x_*}{x}\right)^{k_1}}{1 + K_{R2}\left(\frac{x}{x_*}\right)^{k_2}}.$$

Now we meet rather interesting phenomena in the corresponding most simple control loop

$$x' = xU.$$

We observe for increasing values of K_{R1} or K_{R2} period doubling cascades of bifurcations condensing into chaos as it is well-known from the famous Feigenbaum cascade.

It is quite natural to pass to corresponding control principles with the help of the Natural Iteration concept. For more detailed information see [5].

Network Concepts

Continuous network concepts need a prescription of the behaviour of the modules corresponding to the nodes of an oriented graph. Three generally applicable concepts are based on

Lotka-Volterra Description

$$F x = \sum g_i x_i$$

Ricatti Description

$$F x = K \prod x_r^{k_r}$$

S-System Description (M. Savageau)

$$F x = a \prod x_r^{g_r} - b \prod x_r^{k_r}$$

leading to the universal network description. Any network formerly described by a set of ordinary differential equations can be transformed in such a way that one of these unified descriptions will be achieved.

With our simulation techniques based on local models we gain especially the following discontinuous network concepts for example for the continuous Lotka-Volterra networks

$$x'_i = x_i \frac{\prod_{G_{ij}>0} \left(1 + D G_{ij} x_j\right)}{\prod_{G_{ij}<0} \left(1 + D |G_{ij}| x_j\right)}$$

and

$$x'_i = x_i e^{D \sum G_{ij} x_j}$$

Surprisingly, this discontinuous networks are met if we try to introduce new neural net concepts referring to natural laws for the relation between excitation and sensorial perception.

If we assume the Weber-Fechner-Law

$$E G = \ln\left(\frac{R}{R_s}\right)$$

with E being the perception, R the excitation, R_s the threshold of the excitation R, and G a proportional factor, and if we demand that within the neural network this information compression will be inverted (Nernst-Law), we get the following concept

$$x'_i = x_i e^{\sum G_{ij} x_j}.$$

If we in contrary assume Stevens-Power-Law

$$E G = \left(\frac{R}{R_s}\right)^k - 1$$

we get under the same conditions

$$x'_i = x_i \frac{\prod_{G_{ij}\geq 0} \left(1 + G_{ij} x_j\right)^{\frac{1}{k}}}{\prod_{G_{ij}<0} \left(1 + |G_{ij}| x_j\right)^{\frac{1}{k}}}.$$

Both concepts should be enhanced by introducing for the first factor x_i on the right-hand side of these expressions a chaotic mapping

$$\phi(x_i)$$

to propose a chaotic elementary behaviour of the single neuron.

Another rather interesting network concept we get if we try to model a system of coupled growth processes by a system of coupled EVOLON oscillators. A single sigmoid growth curve described by the hyper-logistic differential equation

$$\dot{x} = K x^{k} \left(B^{w} - x^{w} \right)^{l}$$

can be discontinuously simulated after some simple substitutions by

$$s' = s\cos(D \cdot CO) + c\sin(D \cdot CO)$$

$$c' = c\cos(D \cdot CO) - s\sin(D \cdot CO)$$

with

$$CO = K |s|^{k'} |c|^{l'}$$

$$s = \left(\frac{x}{B} \right)^{\frac{w}{2}} \quad , \quad c = \sqrt{1 - s^2}$$

We call this normalized description of an EVOLON an EVOLON-oscillator. An EVOLON network is then given by the following system of non-linearly coupled EVOLON oscillators

$$s'_{i} = s_{i}\cos(D \cdot CO_{i}) + c_{i}\sin(D \cdot CO_{i})$$

$$c'_{i} = c_{i}\cos(D \cdot CO_{i}) - s_{i}\sin(D \cdot CO_{i})$$

with the interaction forces

$$CO_{i} = K_{i} \prod_{r} |s_{r}|^{k_{ir}} \prod_{r} |c_{r}|^{l_{ir}} .$$

If in such a network all growth indicators can be measured synchronously, in principle all the parameters of an EVOLON network can be estimated by a linear regression method.

Because of the compensation tendency in the interactions by s_i and c_i variables we meet some difficulties in determining the unknown parameters under conditions where some random noise is present.

References

[1] **Peschel, M. & W. Mende** Predator-Prey-Models: Do We Live in a Volterra World ? Akademie-Verlag Berlin & Springer Verlag Wien 1986

[2] **Savageau, M.** Biochemical Systems Analysis. Addison-Wesley Publ. Co. 1976

[3] **Peschel, M., W. Mende & F. Breitenecker** Anwendung der diskreten Theorie ratengekoppelter Systeme in der Regelungstechnik. msr, Berlin, 30(1987)7, 322-324

[4] **Peschel, M., H.-M. Voigt, W. Mende & F. Breitenecker** Modellbildungsrahmen von Zeitreihen als Grundlage fuer die Simulation dynamischer Systeme. msr, Berlin (in preparation)

[5] **Peschel, M., H.-M. Voigt, W. Mende, F. Breitenecker & P. Kopacek** Nichtlineare Reglerkonzepte entsprechend der lokalen Modellierung der Theorie ratengekoppelter Systeme. msr, Berlin (in preparation)

[6] **Voigt, H.-M.** Evolution und Optimierung. Dissertation B (Habilitation). Academy of Sciences of the GDR 1986

[7] **Voigt, H.-M.** Optimization by Selection Pressure Controlled Replicator Networks. Systems Analysis, Modelling, and Simulation 5(5) 1988

Modelling of Large Processes
Containing Continuously and Binarily Controlled Parts

J. Alder [1] and K.J. Reinschke [2]

Summary - It is typical of most industrial plants that both continuous and binary control are applied. This paper concerns the coordination between both kinds of control by appropriate models. A "global model" of a controlled process may be described by a sequence of its different process stages. Each process stage may be characterized by a certain realization of a binary process vector which determines both the control structure and the set values of state variables. The structure of the process under consideration is reflected by a so-called "structure graph". An example of a chemical process is chosen to show the applicability of the suggested method.

Keywords - Large scale systems, multivariable control, binary control, structural properties, graph-theoretic approach

1. Introduction

Modelling of processes that are scheduled to be controlled automatically leads to an "interface" between the plant and the automation equipment. Seen from this interface the investigator is able to model the input output behaviour of the plant for control purposes. This is the starting point of this paper.

It is typical of large processes to consist of both continuously controlled partial processes and of binarily controlled ones, which are interconnected in a complicated way. The control engineer has to cope with a suited coordination by appropriate means and combinations between the two basic kinds of control. This problem is treated in this contribution in some detail.

2. A global model

By a "global model" of a controlled process we mean a representation of the sequence of its different process stages which are determined by process states influencing each other mutually. For the description of such a sequence, Petri nets may be used as an appropriate means /3/. Then, the places (balloons) are defined as process stages whereas the transitions (bars) are determined both by the quantities exciting the process and by the process states. Starting with a given process stage, the occurrence of a certain process state demands a well defined outer excitation in order to reach the next process stage. The global model (Fig. 1) does not take into account the properties of control devices

[1] Technische Hochschule Leipzig, Sektion Automatisierungsanlagen
7030 Leipzig, K.-Liebknecht-Straße, Deutsche Demokratische Republik

[2] Ingenieurhochschule Cottbus, Sektion Mathematik/Naturwissenschaften
7500 Cottbus, Karl-Marx-Straße, Deutsche Demokratische Republik

to be installed later on, comp. /6/.
Fortunately, the technological demands admit of distinguishing between
tasks for continuous and for binary control. So, process state quantities which must be stabilized within small tolerance intervals may be
recognized by their set values.

3. Models of selected partial processes

Fig. 2 shows the structure graph (according to /5/) for temperature
pressure stabilization as partial process to be controlled continuously. Fig. 3 illustrates the reactor control as partial process to be
controlled binarily. In the following, the coupling between these two
partial processes will be explained.

3.1. Binary reactor control

In /2/, the binary process modell has been discussed in detail. It may
be summarized as follows (comp. Fig. 3). The rectangles contain all the
input variables of the partial process under consideration. The contents
of each rectangle is denoted as an "operation". Those variables whose
binary evaluation has been changed (in comparison with the preceding
operation) are written down explicitly.
The "starting" operation is O1. It is characterized by variables that
ensure a stable situation of rest for the partial process. There is a
binary process variable p1 that immediately reflects this situation.
During the succeeding operation O2, a stirring device is switched on
up to the time when a required VC-volume has been reached (binary process variable p2). Then operation O3 is executed as long as the temperature remains less than 30° C (binary process variable p3) and the device for manual control of temperature is not switched on (binary process variable p4). As for the coupled continuous reactor control corresponding to this operation O3, automatic control of temperature with a
set value of 30° C is to apply. In Fig. 3, the binary variables r_1, r_2
and s_1 have been written down, each of which attains the binary value 1
in Fig. 2. After attaining a temperature value of 30° C further devices
must be switched on. When the pressure gets a minimal value (binary
process variable p5) the set value of temperature is changed to 50° C,
i.e., $s_1 = 0$, $s_2 = 0$.
BAsed on Fig. 2 and Fig. 3., the further discussion of the process may
be left to the reader.
The meaning of the scheme outlined is that it reflects not only a sequence of "switch-on" and "switch-off" for valves and pumps, but also
the main aspects for continuous reactor control. It should be realized
(see Fig. 3) that the time interval during which a binary process variable is fixed at 1 or at O is not indicated in most cases. However, p11
is a binary variable representing a time interval during which a "rela-
xation" takes place. Such binary quantities can be appropriately used

if a detailed state description is either impossible or too expensive.

3.2. Continuous temperature pressure stabilization

The graph-theoretic approach yields a promising way for analysis and synthesis of processes under continuous control (see /5/). This theory is a qualitative one and investigates primarily structural properties. Usually, it starts with the concept of state space description of a plant. All the variables - m input variables encompassed in an m-dimensional vector u, n state variables encompassed in an n-dimensional vector x, and l output variables encompassed in an l-dimensional vector y - are reflected by m+n+l vertices of a digraph. The existing direct coupling between input variables and state variables, between state variables, and between state variables and output variables are expressed by input edges, by state edges, and by output edges, respectively.
The control law may be reflected by state feedback (= edges from state vertices to input vertices), output feedback (= edges from output vertices to input vertices), and/or output injection (= edges from output vertices to state vertices).
Each process stage is determined by a certain subset of the state space. Provided the state vector x(t) running along a continuous orbit in the state space does not leave such a subset, then the process stage will be the same. Each process stage may be characterized by a certain realization of the binary process vector p (see Fig. 4). Both the control structure and the set values of the state variables depend on p. Assume the control structure and the set values are fixed. Then the state variables of the controlled process are continuous time functions up to the instant when a new process stage is reached. At this moment, the control law described by a set $\{r^1\}$ of the state feedback edges, the set $\{r^2\}$ of output feedback edges and the set $\{r^3\}$ of output injection edges as well as the set $\{s\}$ of set values must be newly fixed. Then, the controlled process runs again continuously with respect to time until another process stage is reached...

The structure graph shown in Fig. 2 may be regarded as an illustrative example of the generalized structure graph exhibited in Fig. 4.
For this example we have m = 1, n = 4, l = 3, and

$$u = V_{HK}, \quad x = \begin{pmatrix} T_m \\ T_R \\ C_{PVC} \\ P_R \end{pmatrix}, \quad y = \begin{pmatrix} T_M^{\cdot} \\ T_R^{\cdot} \\ P_R^{\cdot} \end{pmatrix}, \quad p = \begin{pmatrix} P_5 \\ P_8 \\ P_9 \end{pmatrix}, \quad w = \begin{pmatrix} W_{30} \\ W_{70} \\ W_{50} \\ W_P \end{pmatrix}, \quad r = \begin{pmatrix} r_1 \\ r_2 \\ r_3 \end{pmatrix}, \quad s = \begin{pmatrix} s_1 \\ s_2 \\ s_3 \\ s_4 \end{pmatrix}.$$

There are no state feedback edges, i.e., the set of $\{r^1\}$-edges is empty. The set of $\{r^3\}$-edges is formed by the edges r_2 and r_3, the set of $\{s\}$-edges by the edges s_1, s_2, s_3, and s_4.

4. Coupling between binarily and continously controlled processes

Consider two partial processes T_a and T_b coupled by a control relation (T_a, T_b).

Provided that both T_a and T_b are binarily controlled, then there are two possibilities of implementing (T_a, T_b).

(a) In T_a there is an operation variable o_{ka} which influences a transition from an operation to its succeeding operation in T_b, for short,
$$o_{ka} = p_{kb}.$$

(b) In T_a there is a binary process state variable p_{ia} which influences a transition from an operation to its succeeding operation in T_b, for short, $p_{ia} = p_{ib}$.

We shall say, the variable p_{ia} is an instruction sending variable with respect to T_b whereas the variable p_{ib} is an instruction receiving variable with respect to T_b.

Now, let T_b be a partial process controlled continuously. Then, comp. Fig. 4, the coupling variables $\{s\} \cup \{r\}$ are instruction receiving variables with respect to T_b, the variables p are instruction sending variables with respect to T_b.

In this way, two binary coupling vectors have been introduced as follows, the binary process state vector p containing the natural coupling variables (according to /2/) and characterizing the process stage, the binary coupling vector formed by $\{r\}$ and $\{s\}$ and characterizing the control structure and the set values, respectively.

Literature

/1/ Alder, J.: Über die Aufteilung von Steuerungsaufgaben auf eine
    dezentrale Automatisierungseinrichtung.
    Wiss.Z.d.TH Leipzig 10 (1986) 5, 293-305

/2/ Alder, J.: Aufgabenstellung und Entwurf von Binärsteuerungen.
    Band 222 der Reihe Automatisierungstechnik.
    VEB Verlag Technik Berlin 1986

/3/ Henry, M., et al.: Sequences in the control of processes.
    The Chemical Engineer 405 (1984), 9-12

/4/ Reinschke, K., et al.: Strukturmodell für komplexe Automatisierungs-
    anlagen und seine Anwendung in der Kraftwerksautomatisierung.
    Z.msr 27 (1984) 1, 21-23

/5/ Reinschke, K.J.: Multivariable Control - A Graph-theoretic Approach
    LNCIS, vol.108, Springer-Verlag
    Akademie-Verlag Berlin 1988

/6/ Hanisch, H.-M.: Mathematische Modellierung diskreter Steuerungs-
    aufgaben in diskontinuierlichen, verfahrenstechnischen Systemen.
    Diss.A, TH Leuna-Merseburg, Sektion Verfahrenstechnik, 1986

/7/ Wilkin, H.: Anpassen von Automatisierungseinheiten in einem Mikro-
    prozessorsystem. S.579-592 in Bd.5 der Fachberichte Messen,
    Steuern, Regeln. Springer-Verlag Berlin Heidelberg New York 1980

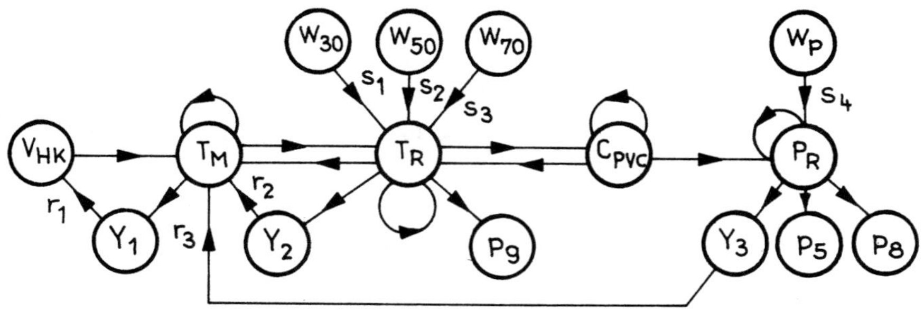

V_{HK} valve-opening ratio of heating and cooling feed

W_{30}, W_{50}, W_{70} set values of temperature

W_P set value of pressure

$s_1, s_2, s_3, s_4 \in \{0, 1\}$ set-point adjustment by binary control

T_M temperature in jacket

T_R temperature in reactor

Y_1, Y_2 measured values of temperature for controller

Y_3 measured value of pressure for controller

$r_1, r_2, r_3 \in \{0, 1\}$ control structure caused by binary control signals

P_5, P_8, P_9 values of pressure and temperature evaluated binarily
for the purpose of binary control

Fig. 2   Structure graph of reactor control

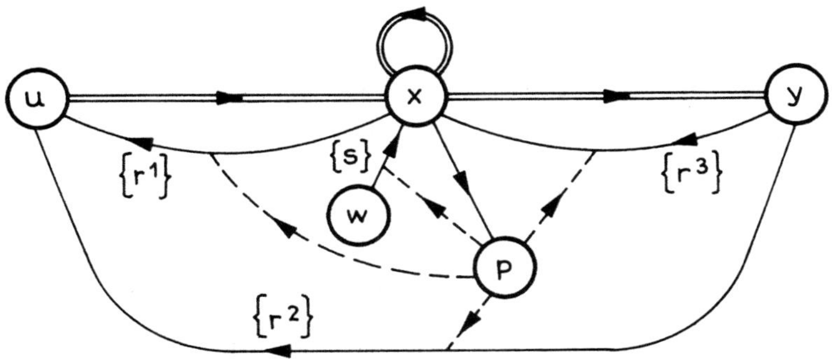

The directed dotted lines reflect the effect of possible binary
control signals.

Fig. 4   Generalized structure graph

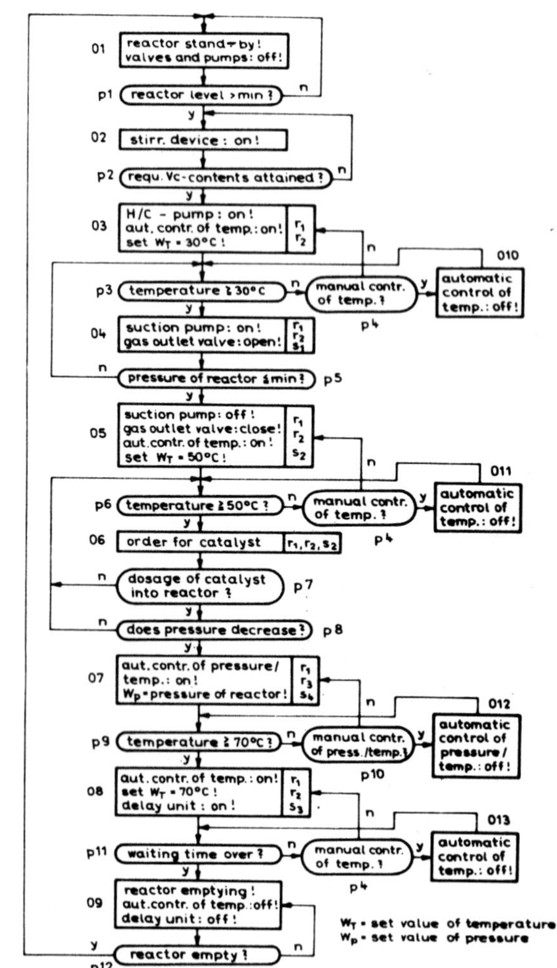

Fig. 1 Vinylchloride polymerization (Petri-net)

Fig. 3 Control of a reactor

Integral Riccati Equations for a Feedback Solution of LQCP with a Terminal Inequality Constraint

Zbigniew Emirsajlow [1]

Abstract

This paper considers the linear quadratic control problem (LQCP) for systems defined by evolution operators with a terminal state inequality constraint. It is shown that under a suitable assumption the optimal control exists and has a feedback structure. A synthesis of the feedback involves two integral Riccati equations.

1. Introduction

It is well known that a synthesis of a feedback in the linear quadratic control problem (LQCP) for systems defined by evolution operators without state constraints strongly involves infinite dimensional integral Riccati equations (see [1], [2], [4]). The present paper shows that this type of equations provides also an effective tool for LQCP with a terminal state inequality constraint. This optimal control problem has been originally investigated in [3], where the feedback synthesis required one to solve two integral equations: one linear and one Riccati's. Main advantage of the present approach is that under some additional assumptions (see [1], [2]) integral Riccati equations lead to differential Riccati equations what is not possible in the case of [3].

In order to state our results precisely we introduce the following notation and assumptions. Let U, H be real Hilbert spaces, $0 < T < \infty$, $T(t,s)$ be an evolution operator defined as a function from

$$\Delta = \left\{ (t,s) \; : \; 0 \leqslant s \leqslant t \leqslant T \right\}$$

to $L(H)$ such that :

(i) $T(t,r)T(r,s) = T(t,s)$, $0 \leqslant s \leqslant r \leqslant t \leqslant T$,

(ii) $T(t,t) = I$,

(iii) $T(t,.)$ is strongly continuous on $[0,t]$ (t is fixed) and $T(.,s)$ is strongly continuous on $[s,T]$ (s is fixed),

(iv) $\| T(t,s) \|_{L(H)} \leqslant M$ on Δ .

The control system we consider is described by

$$x(t) = T(t,0)x_0 + \int_0^t T(t,s)Bu(s)ds \; , \quad t \in [0,T] \; , \tag{1.1}$$

where $x_0 \in H$, $u(.) \in L^2(0,T;U)$, $B \in L(U,H)$.

For fixed $T \in (0,\infty)$, $\alpha \in (0,\infty)$ we define the set of admissible controls

$$\mathcal{U}_\alpha = \left\{ u(.) \in L^2(0,T;U) \; : \; \| x(T) \|_H^2 \leqslant \alpha \right\} \tag{1.2}$$

and consider the following optimal control problem.

(P) : Find a control $u_P(.) \in \mathcal{U}_\alpha$ minimizing the cost

$$J(u) = \int_0^T \left\{ (x(t),Wx(t))_H + (u(t),Ru(t))_U \right\}dt \; , \tag{1.3}$$

[1] Institute of Control Engineering, Technical University of Szczecin, Gen. Sikorskiego 37, 70-313 Szczecin, Poland

where $W \in L(H)$, $R \in L(U)$ are selfadjoint, W is nonnegative and R is coercive. $\square$

The main aim of this paper is to show that under the existence and uniqueness condition from [3] a feedback for the optimal control $u_p(t)$ can be derived using integral Riccati equations.

2. Auxiliary results

Some known results for LQCP without state constraints as well as for Problem (P) will be necessary in our consideration. This is why we include them in this section.

In order to synthetize a feedback in LQCP without state constraints ($\mathcal{U}_\alpha = L^2(0,T;U)$) one needs to solve the following integral equation

$$P(t)h = \int_t^T T^*(s,t)[W-P(s)BR^{-1}B^*P(s)]T(s,t)h \, ds \, , \quad h \in H \tag{2.1}$$

or one of its equivalent form ([1], [2], [4]). This equation has a unique solution $P(t)$ in the class of selfadjoint, nonnegative and strongly continuous functions on $[0,T]$ with values in $L(H)$.

Moreover, if we substitute in (1.1)

$$u(t) = v(t) - R^{-1}B^*P(t)x(t) \, ,$$

where $v(.) \in L^2(0,T;U)$, then we get

$$x(t) = S(t,0)x_0 + \int_0^t S(t,s)Bv(s)ds \, , \quad t \in [0,T] \, . \tag{2.2}$$

In this expression $S(.,.)$ is an evolution operator with the same properties as $T(.,.)$. It is called the first perturbed evolution operator and is characterized as a unique solution of the integral equation

$$S(t,s)h = T(t,s)h - \int_s^t T(t,r)BR^{-1}B^*S(r,s)h \, dr \, , \quad h \in H \, . \tag{2.3}$$

The system (2.2) is called the perturbed system.

Following [3] we introduce the functional defined on $L^2(0,T;U)$

$$G_\varepsilon(v) = \| \int_0^T S(T,s)Bv(s)ds + S(T,0)x_0 \|_H^2 + \varepsilon \int_0^T (v(s),Rv(s))_U ds \, , \tag{2.4}$$

where $\varepsilon \in (0,\infty)$, and recall the theorem.

Theorem 2.1 : Let $T(T,0)x_0 \in \overline{\text{range}}(\int_0^T T(T,s)B...ds)$.

(a) If $\alpha \in (0, \|S(T,0)x_0\|_H^2)$, then

$$u_p(t) = v_\varepsilon(t) - R^{-1}B^*P(t)x(t) \, , \tag{2.5}$$

where $v_\varepsilon(.)$ minimizes $G_\varepsilon(.)$ over $L^2(0,T;U)$ for $\varepsilon \in (0,\infty)$ such that

$$\varepsilon^2 \| \int_0^T S(T,s)Bv_\varepsilon(s)ds + S(T,0)x_0 \|_H^2 = \alpha \, , \tag{2.6}$$

and

$$J(u_p) = (x_o, P(0)x_o)_H + \int_0^T (v_\varepsilon(s), Rv_\varepsilon(s))_U \, ds \ . \tag{2.7}$$

(b) If $\alpha \in [\|S(T,0)x_o\|_H^2, \infty)$, then

$$u_p(t) = -R^{-1}B^*P(t)x(t)$$

and

$$J(u_p) = (x_o, P(0)x_o)_H \ . \ \square$$

3. Feedback solution of Problem (P)

It has been proved in [3] that $v_\varepsilon(t)$ in the formula (2.5) can be expressed as a linear function of the state $x(t)$. In this section we obtain this result but in a different way, making use of an integral Riccati equation.

<u>Theorem 3.1</u> : Let $T(T,0)x_o \in \overline{\text{range}}(\int_0^T T(T,s)B...ds)$. If $\alpha \in (0, \|S(T,0)x_o\|_H^2)$, then

$$u_p(t) = -R^{-1}B^*(N_\varepsilon(t) + P(t))x(t) \ , \tag{3.1}$$

where $N_\varepsilon(t) \in L(H)$ is a selfadjoint, nonnegative and strongly continuous solution of the integral Riccati equation

$$N_\varepsilon(t)h = S^*(T,t)S(T,t)h - \frac{1}{\varepsilon} \int_t^T S^*(r,t)N_\varepsilon(r)BR^{-1}B^*N_\varepsilon(r)S(r,t)h \, dr \ , \quad h \in H \tag{3.2}$$

for $\varepsilon \in (0,\infty)$ such that

$$\varepsilon^2 \|\emptyset_\varepsilon(T,0)x_o\|_H^2 = \alpha \ , \tag{3.3}$$

where $\emptyset_\varepsilon(.,.): \Delta \to L(H)$ is an evolution operator, called the second perturbed, defined as a unique solution of the integral equation

$$\emptyset_\varepsilon(t,s)h = S(t,s)h - \int_s^t S(t,r)BR^{-1}B^*N_\varepsilon(r)\emptyset_\varepsilon(r,s)h \, dr \ , \quad h \in H \ . \tag{3.4}$$

Moreover,

$$J(u_p) = (x_o, (N_\varepsilon(0)+P(0))x_o)_H \ . \tag{3.5}$$

<u>Proof</u> : By the assumption this theorem corresponds to part (a) of Theorem 2.1. According to [1], [2] the problem of minimization of $G_\varepsilon(.)$ over $L^2(0,T;U)$ can be solved using an integral Riccati equation. It is easy to check that in this case it has the form of (3.2). Since the system is now described by means of the evolution operator $S(.,.)$ (first perturbed) we define by the formula (3.4) the second perturbed evolution operator $\emptyset_\varepsilon(.,.)$. Making use of this operator and the expression (2.6) we get (3.3). And finally, the expression (3.5) follows easily from (2.7) and [1], [2]. $\square$

4. Final remarks

The main advantage of the approach presented here in comparison with the results of [3] is that under some additional assumptions (see [1], [2]) the integral Riccati equations (2.1) and (3.2) lead to differential Riccati equations and the latter are more convenient for numerical computations.

It is also easy to extend the presented results to the case of state constraints in the form

$$\|x(T)-x_1\|_H^2 \le \alpha \,,$$

where $x_1 \in H$ is fixed.

<u>Literature</u>

[1] Curtain R.F., Pritchard A.J.: The infinite dimensional Riccati equation for systems defined by evolution operators. SIAM J. Control and Optimization, 14(1976), 951-983.

[2] Curtain R.F., Pritchard A.J.: Infinite dimensional linear systems theory. Springer-Verlag, Berlin 1978.

[3] Emirsajlow Z.: A feedback in LQCP with a terminal inequality constraint. Control Theory Centre Report No. 140, University of Warwick, Coventry (to appear in Journal of Optimization Theory and Application).

[4] Gibson J.S.: The Riccati integral equations for optimal control problems on Hilbert spaces. SIAM J. Control and Optimization, 17(1979), 537-565.

Stable Variant of the Simplex Method
for Solving Supersparse Linear Programs[*]

Jacek Gondzio[**]

Abstract. An efficient method for solving supersparse large-scale linear programs is presented. The method develops Bisschop's idea of matrix augmentation and partitioning. A numerically stable procedure of updating LU factorization of a dense sensitivity matrix is described. The method gives substantial savings of storage requirements and opens the possibility of solving on IBM PC microcomputers real-life problems of dimensions up to 2000x3000.

Key words. Sparse large-scale linear programs, dense LU factorization

Introduction

The most efficient commercial LP packages use advanced implementations of the Bartels-Golub LU decomposition ([4],[6],[7]) for basis inverse representation. The method is fast and accurate and has relatively small storage requirements. The first basis inverse representation needs room for little more then the number of nonzero elements of the basis. The number of new nonzeros added at every iteration grows quickly, and this is the main disadvantage of the method (in the worst possible case a single iteration may add m new nonzeros, where m is the basis dimension).

We shall present a new method in which the storage requirements for basis inverse representation are drastically reduced. The method combines sparse and dense numerical techniques. Its storage needs depend only on the average number of nonzeros in a single column of the constraint matrix and the number of iterations between refactorizations.

Basis inverse representation

Let us consider a linear programming problem of the following form

$$\text{minimize} \quad c_B^T x_B + c_N^T x_N , \tag{1}$$

$$\text{subject to} \quad B_0 x_B + N_0 x_N = b , \quad \underline{x} \leq x \leq \bar{x} , \tag{2}$$

where $x_B \in \mathbb{R}^m$, $x = (x_B, x_N) \in \mathbb{R}^n$ and B_0 is a lower triangular starting basis (at the beginning it is a basis constructed totally from slacks).

A crucial numerical problem of the simplex method (see:[4]) is performing forward transformation (i.e. solving equation $B\beta = b$) and performing backward transformation (i.e. solving equation $B^T \pi = c$). For a

---

*   supported by Program CPBP.02.15
**   Systems Research Institute, Polish Academy of Sciences
      Newelska 6, 01-447 Warsaw, Poland

triangular basis B_0 FTRAN and BTRAN operations can be done in a fast and stable way and B_0 does not need any additional room. Since B_0 consists of columns of the constraint matrix only the pointers to the appropriate columns have to be stored (see:[5]).

Basis B after a few iterations differs from B_0 only in l columns. Let us denote by N an $m{\times}l$ submatrix of N_0 containing these new l basic columns and by V an $l{\times}m$ matrix which contains unit row vectors, where the ith row of V has a one in the position of the column of B_0 which is to be replaced by the ith column of N. The actual basis can be written as

$$B = B_0 + (N - B_0 V^T) V \ . \tag{3}$$

It is shown in [1] that FTRAN and BTRAN operations with that new basis can be performed by solving systems of equations with an augmented matrix of the form

$$A = \begin{bmatrix} B_0 & N \\ V & 0 \end{bmatrix} \ . \tag{4}$$

FTRAN operation is replaced by solving the equation

$$\begin{bmatrix} B_0 & N \\ V & 0 \end{bmatrix} \begin{bmatrix} x \\ y \end{bmatrix} = \begin{bmatrix} b \\ 0 \end{bmatrix} \ . \tag{5}$$

We have

$$\beta = x + V^T y \ , \quad \text{where} \tag{6}$$

$$y = (V B_0^{-1} N) V B_0^{-1} b \quad \text{and} \quad x = B_0^{-1}(b - Ny) \ . \tag{7}$$

Analogously BTRAN operation is replaced by solving the equation

$$\begin{bmatrix} B_0^T & V^T \\ N^T & 0 \end{bmatrix} \begin{bmatrix} x \\ y \end{bmatrix} = \begin{bmatrix} c \\ 0 \end{bmatrix} \ . \tag{8}$$

We have

$$\pi^T = x^T + c^T V^T (V B_0^{-1} N)^{-1} V B_0^{-1} \ , \quad \text{where} \tag{9}$$

$$y^T = c^T B_0^{-1} N (V B_0^{-1} N) \quad \text{and} \quad x^T = (c^T - y^T V) B_0^{-1} \ . \tag{10}$$

Let us observe that instead of inverting matrix B we only have to solve a sequence of equations with B_0 and an $l{\times}l$ sensitivity matrix

$$S = (V B_0^{-1} N) \ . \tag{11}$$

Our method needs storage only for a dense, in general, inverse representation of matrix S. Let us now evaluate its maximal dimension.

Any basic matrix of the problem (1)–(2) can be transformed by row and column permuting to a nearly triangular form. The number of bumps does not usually exceed the average length of columns of the constraint matrix. Replacing bumps by appropriate slack columns gives a new triangular basis B_0 . All bump-columns create new matrix N and new sensitivity matrix S (see (11)). Further growth of dimension of S is limited by the number of iterations between refactorizations.

The dimension of S does not exceed the sum of the average number of nonzeros in any column of the constraint matrix and the number of iterations between basis refactorizations. Storage requirements of the method are then equal to the square of the above sum and do not depend on the dimension of the original problem.

Inverting the sensitivity matrix

The accuracy of computing β and π^T from equations (6),(7) and (9), (10) depends strongly on the numerical implementation of solving equations with matrices B_0 and S. Since B_0 is triangular the corresponding equations can be solved in a stable way.

Matrix S is decomposed into the form

$$S = PLUQ , \tag{12}$$

where P and Q are orthogonal matrices of row and column permutations and L and U are lower and upper triangular factors, respectively.

Changes of matrix S depend on the type of variables that leave and enter the basis. There are four cases possible (see [1]):

Case 1) a column of N_0 enters the basis in place of a column of B_0 (causing row and column addition in S).

Case 2) a column of N_0 replaces another column of N_0 (causing column replacement in S).

Case 3) a column of B_0 replaces another column of B_0 (causing row replacement in S).

Case 4) a column of B_0 enters the basis in place of a column of N_0 (causing row and column deletion in S).

A numerically stable method of updating decomposition (12) in any of the four cases of changes S is discussed in [3] in detail. We shall here present only main ideas.

Modifications of (12) in case 1 are strightforward and shall not be discussed.

In case 2 we can define new matrices $\hat{V}$ and $\hat{N}$ as follows

$$\hat{V} = V , \quad \hat{N} = \left[n_1, n_2, \ldots, a_j, \ldots, n_l \right] , \tag{13}$$

where a_j is the column of N_0 replacing in basis column a_i, which was the pth column of N.

Modifications of N can be written as

$$\hat{N} = N + (a_j - Ne_p)e_p^T , \tag{14}$$

where e_p is the pth unit vector of $\mathbb{R}^l$.

Equations (11), (13) and (14) define the new sensitivity matrix

$$\hat{S} = S + VB_0^{-1}(a_j - Ne_p)e_p^T =: S + ke_p^T , \tag{15}$$

which involves adding column k to column p of S. Next we have

$$\hat{S} = PLUQ + ke_p^T = PL(U + L^{-1}P^{-1}ke_p^TQ^{-1})Q =: \tag{16}$$

$$=: PL(U+k_u e_{\tilde{p}}^T)Q = PL\tilde{\tilde{U}}\tilde{\tilde{Q}} \ . \tag{17}$$

where k_u denotes the column that has to be added to the $\tilde{p}$th column of U, forming a bump in it, and $\tilde{U}$ is an upper Hessenberg matrix (see [2]), which differs from U only by a column permutation. This new matrix $\tilde{U}$ is obtained from U when column p of U is put after the last column and all the columns $\tilde{p}+1, \tilde{p}+2, ..., l$ are moved one place to the left. The above permutation creates $l-\tilde{p}$ nonzero elements in $\tilde{U}$ positioned just below its main diagonal and causes changing Q to $\tilde{Q}$.

Elements of U that lie below the diagonal are zeroed by elementary transformations that do not destroy the triangular structure of matrix L. There are two possible transformations (see [3]). Their numerical stability properties can be easily compared by analysing their condition numbers. The updating procedure of [3] chooses better one.

Similarly in cases 3 and 4 the triangular decomposition (12) can be updated in a stable way. The procedure that performs it in case 3 is equivalent to the one described above. A stable updating procedure for case 4 is a combination of those of cases 2 and 3.

Rules for choosing the best elementary transformation developed in [3] result in the accuracy of the factorization (12).

Conclusions

We have presented here basic ideas of a new linear programming method in which the storage requirements for the basis inverse representation do not grow with the growth of the dimension of the problem to be solved.

Stability and low memory needs of the method make it especially attractive for implementing on a microcomputer. Numerical examples of applying the method to some multistage problems are presented in [3].

References

[1]   J.Bisschop, A.Meeraus, "Matrix augmentation and partitioning in the updating of the basis inverse" *Mathematical Programming* 13 (1977) 241-254.

[2]   G.H.Golub, C.F.Van Loan, "Matrix Computations", North Oxford Academic, 1983.

[3]   J.Gondzio, "Specialized methods for solving multistage linear programming problems", in preparation.

[4]   B.Murtagh, "Advanced Linear Programming", McGraw-Hill, 1981.

[5]   A.F.Perold and G.B.Dantzig,"A Basis Factorization Method for Block Triangular Linear Programs", in I.S.Duff and G.W.Stewart eds., *"Sparse Matrix Proceedings* 1978", SIAM Philadelphia 1979, 283-312.

[6]   J.K.Reid, "A sparsity-exploiting variant of the Bartels-Golub decomposition for linear programming bases", *Mathematical Programming* 24 (1982) 55-69.

[7]   M.A.Saunders,"A Fast, Stable Implementation of the Simplex Method Using Bartels-Golub Updating",in "Sparse Matrix Computations",ed., J.R.Bunch and D.J.Rose, Academic Press, New York (1976) 213-226.

Some Remarks on Optimizing Simulated Systems

J. Fischer*

1 Description of the problem

We consider the problem of optimizing a performance criterion of a computer simulation model exhibiting stochastic behavior. The main characteristics of these optimization problems are:

- The objective function is not known analytically. For a given set of parameters, we can only observe a stochastically perturbed function value by performing a simulation experiment.

- Every evaluation of the objective function requires a run of the computer simulation model and may be therefore very costly in terms of computing time.

Thus any optimization method for these problems should have the following properties:

- The optimization algorithm may be arbitrarily complicated, but it should make a very "economical" use of observed function values.

- The method should take into account the stochastic nature of the problem. As a consequence, every "solution" provided by an algorithm is only a statistical estimate with inherent random error.

This shows that the problem of optimization in simulation is an extremely difficult one, since it combines the problems of deterministic mathematical programming and of statistical estimation [5, p.381]. Therefore, instead of looking for a "universal" algorithm that may be equally well applied to all kinds of these problems, we will try — in a first step — to identify "good" algorithms that are applicable to a restricted class of "simple" problems. Following the approach in deterministic optimization, we make the following assumptions:

- We consider only problems with continuously varying parameters that are subject to lower and upper bounds.

- The (unperturbed) objective function, though its analytical form is unknown to the experimenter, is assumed to be at least twice continuously differentiable.

- We try to find only a local extremum.

The intention of this paper is to investigate how existing methods behave if they are applied to these "simple" problems and to obtain hints for developing methods that are possibly more efficient.

Fraunhofer-Institut für Arbeitswirtschaft und Organisation, Nobelstr. 12, 7000 Stuttgart 80, FRG

2 Properties of existing methods

There are essentially three different classes of methods that are applied to solve problems of the form described above. They are all iterative in the sense that they generate a sequence of points in the parameter space that is intended to "converge" to a solution of the optimization problem.

2.1 Direct search methods

These methods have originally been developed for the deterministic case when derivatives need not exist or need not to be smooth. Well known examples are the pattern search method of Hooke and Jeeves (1961) and the sequential simplex method of Nelder and Mead (1965). Since these methods use only function values and are rather simple to implement, they are frequently applied to the class of problems considered here.

The main drawback of these methods is that they are not designed for the case of stochastically perturbed functions. Therefore, they usually make rapid early progress toward an "optimum", but iterate laboriously as a solution is neared [1]. This is due to the fact that their behavior depends strongly on the comparison of single function values, which may become crucial if the stochastic "noise" is of the same order of magnitude as the difference of the function values.

2.2 Response surface methodology

This approach is described e.g. in [2] and based on fitting a linear or quadratic function to the observed values via regression analysis (method of least squares). Since for a general objective function such an approximation is valid only locally, at each stage of the procedure a new approximation has to be computed. Usually in a first phase a linear approximation is used in order to estimate the gradient direction, while in a second phase a quadratic approximation provides also information about the Hessian.

By using the statistical method of regression analysis, this approach takes into account the stochastic nature of the problem. On the other hand, at each stage of the process a complete first– or second–order design has to be evaluated, while the previously observed values are all thrown away. This indicates that the principle of "economy" may be violated here. Furthermore, the response surface methodology does not yield a completely determined algorithm, but rather a collection of procedures that have to be selected and used by a skilled human experimenter.

2.3 Stochastic approximation methods

The procedure of Kiefer and Wolfowitz (1952) estimates in each stage the gradient at the current point by differences of observed function values. Using this information the method tries to find a point where the gradient is zero.

This procedure may be interpreted as a special response surface method where the sequence of experiments is completely determined by the results of previous experimentation and by step and design sizes that are fixed in advance. The resulting algorithm seems to be the only optimization method for which convergence results have been proved. These proofs, however, are of a highly technical nature, and the results concern mainly the asymptotic behavior. Furthermore, this conceptual form of the algorithm is not well suited for practical applications, since it is too inflexible. To enable the algorithm to adapt to properties of the function under investigation, several modifications have been proposed:

- To improve the behavior during the initial stages a step size is introduced that is not fixed in advance but determined by the observations.

- To make the estimation of the gradient by differences of perturbed function values more stable, the size of the design is held fixed and does not tend to zero in the limit, since otherwise the variance of the estimate tends to infinity.

- To make it similar to Newton's method, one can try to incorporate second order information, as in [3] and [6].

The effects of these modifications can be investigated only by Monte Carlo experiments, and the proofs of convergence usually are no longer valid.

Furthermore, the best asymptotic rate of convergence of the Kiefer–Wolfowitz procedure is only $n^{-1/3}$, while in special cases a rate of $n^{-1/2}$ can be obtained [7, p.165].

3 Conclusions

The remarks made in the previous section seem to indicate that, at the present state, there does not exist an algorithm that clearly outperforms all other methods. This agrees with the various results of comparisons using Monte Carlo techniques, though these are difficult to evaluate due to the fundamental differences of the algorithms. One main problem seems to be the lack of a locally convergent method that uses observed function values in an "economic" way to generate information about the Hessian.

Imagine a problem with ten variables, where a quadratic approximation requires estimating 66 coefficients. If this is done by a full 2^k factorial design, 1024 simulation runs have to be performed. Let a single run take one minute of CPU time. Then it requires about 17 hours of CPU time to determine the quadratic approximation, and this aggregated information about the objective function is completely discarded in the next stage, where a new approximation has to be computed.

Since there does not exist a fully automated procedure that exhibits a uniformly good behavior, it is frequently recommended to let a human experimenter perform the optimization interactively. This is clearly necessary with the response surface methodology, but also suggested in [7, p.165] and [5, p.381]. But that of course requires, especially for problems with more than two variables, a sophisticated environment of supporting software (e.g. with graphical or knowledge–based components).

In the simulation context there are several possibilities to obtain more information than merely a perturbed function value:

- In steady–state simulations, the variance of the error can usually be controlled by the run length. Thus the experimenter may choose the accuracy of the observed function values according to specific features of the algorithm. Furthermore, information about the variance of the error may be incorporated into the algorithm.

- If the gradient is estimated by differences of perturbed function values, the use of common random numbers may considerably reduce the variance of the estimate [6].

- The technique of perturbation analysis [4] provides, based on only one simulation run, an estimate of the function value and also an estimate of the gradient. However, the application of this technique seems to be restricted — at least in the present state — to certain classes of simulation models.

This additional information may be used to improve the behavior of the existing algorithms or to design new optimization methods.

References

[1] Biles, W. E., and J. J. Swain: Mathematical Programming and the Optimization of Computer Simulations. Mathematical Programming Study 11 (1979), 189–207.

[2] Box, G. E. P., and N. R. Draper: Empirical Model–Building and Response Surfaces. Wiley, New York 1987.

[3] Fabian, V.: Stochastic Approximation. In: Optimizing Methods in Statistics (J. S. Rustagi ed.), 439–470. Academic Press, New York 1971.

[4] Ho, Y. C., and C. Cassandras: A New Approach to the Analysis of Discrete Event Dynamic Systems. Automatica 19 (1983), 149–167.

[5] Law, A. M., and W. D. Kelton: Simulation Modeling and Analysis. McGraw–Hill, New York 1982.

[6] Ruppert, D., R. L. Reish, R. B. Deriso and R. J. Carroll: Optimization Using Stochastic Approximation and Monte Carlo Simulation (with Application to Harvesting of Atlantic Menhaden). Biometrics 40 (1984), 535–545.

[7] Wetherill, G.: Sequential Methods in Statistics. Chapman & Hall, London 1975.

D-Controllability and Strong D-Controllability and Control of Multiparameter and Multiple Time-Scale Singularly Perturbed Systems*

Xu Kekang[1] and Wang Zhenquan[1]

Abstract. Concepts of D-controllability and strong D-controllability are introduced, in terms of which controllability of multiparameter and multiple time-scale singularly perturbed systems is investigated, even regardless of the relative magnitudes of the singular perturbation parameters.

1. Introduction

Singular perturbation problems including several small parameters have in the past attracted considerable attention from engineering and applied mathematics. This is due in part to the difficulty of these problems and in part to their wide applications in engineering. In practice the small singular perturbation parameters are represneted as unmodeled "parasitic" capacitances, inductances, time constants, inertia constants, inverses of high feedback gains, etc ([1]).

In the last decade the stability of multiparameter singularly perturbed systems and stabilizability of two time scale singularly perturbed systems (with single parameter) were discussed in a lot of literatures, such as [2,3]. Decentrallization control of some special systems was investigated([4]). This paper presents concepts of (block) D-controllability and strong (block) D-controllability of singularly perturbed systems and investigates the controllability of them.

In remains of this section, preliminary of singularly perturbed systems with several singular perturbation parameters is presented. In section 2, concepts of D-controllability and block D-controllability are introduced. Concepts of strong D-controllability and strong block D-controllability are introduced in section 3.

Consider LTI singularly perturbed system

$$\dot{x} = A_{oo}x + \sum_{j=1}^{N} A_{oj}\, y_j + B_1 u \; , \tag{1a}$$

$$\varepsilon_i \dot{y}_i = A_{io}x + \sum_{j=1}^{N} A_{ij}y_j + B_{i2}u \; , \qquad (i=1,2,\cdots,N), \tag{1bi}$$

where $x \in R^{n_o}$, $y_i \in R^{n_i}$ ($i\, 1,2,\cdots,N$), $u \in R^m$; while $\varepsilon_i > 0$ ($i=1,2,\cdots,N$) are small parameters and A_{oi}, A_{io}, A_{oo}, A_{ij}, B_{i2} and B_1 are time invariant matrices with appropriate dimensions, and $n_1 + n_2 + \cdots + n_N = n$.

Recall that Sys.(1) possesses two time scales if the ratios $\varepsilon_i/\varepsilon_j$ are bounded and possesses multiple time scales otherwise([2,5]). Stability of Sys.(1) with $u=o$ was probed in both cases above. In [2,6] concept of D-stability was introduced and some results of stability of the overall system were obtained in the case that the mutual ratios of parameters ε_i's are bounded. Control of Sys.(1) was considered merely in a special case([4]).

* Project supported by the National Natural Science Foundation of China.
1) Institute of Systems Science, Academia Sinica, Beijing, 100080, China.

It is useful to rewrite Sys.(1) in the following compact form

$$\dot{x} = A_{oo}x + A_{of}\, y + B_1\, u \,, \tag{2a}$$

$$E(\varepsilon)\, \dot{y} = A_{fo}\, x + A_f\, y + B_2\, u \,, \tag{2b}$$

where $y = (\, y_1^T, y_2^T, \cdots, y_N^T\,)^T$, $A_{oo} = (\, A_{01}, A_{02}, \cdots A_{ON}\,)$, $E(\varepsilon) = \text{block diag}(\varepsilon_1 I_{n_1}, \varepsilon_2 I_{n_2}, \cdots, \varepsilon_N I_{n_N})$,

$$A_{fo} = \begin{pmatrix} A_{10} \\ A_{20} \\ \vdots \\ A_{NO} \end{pmatrix} \,, \qquad A_f = \begin{pmatrix} A_{11} & A_{21} & \cdots & A_{N1} \\ A_{21} & A_{22} & \cdots & A_{N2} \\ \cdots & \cdots & \cdots & \cdots \\ A_{N1} & A_{N2} & \cdots & A_{NN} \end{pmatrix} \,, \qquad B_2 = \begin{pmatrix} B_{12} \\ B_{22} \\ \vdots \\ B_{N2} \end{pmatrix} \,, \qquad \varepsilon = \begin{pmatrix} \varepsilon_1 \\ \varepsilon_2 \\ \vdots \\ \varepsilon_N \end{pmatrix}$$

here I_k denotes $k \times k$ identify matrix.

By use of transformation ([4])

$$\begin{pmatrix} \eta \\ \xi \end{pmatrix} = \begin{pmatrix} I - M E b & - M E \\ L & I \end{pmatrix} \begin{pmatrix} x \\ y \end{pmatrix} \,, \tag{3}$$

Sys.(1) can be decoupled into

$$\dot{\eta} = (\, A_{oo} - A_{of}\, L\,)\eta + [\, B_1 - M(\, E L B_1 + B_2)]\, u \,, \tag{4a}$$

$$E(\varepsilon)\, \dot{\xi} = (\, A_f + E L A_{of}\,)\xi + (\, B_2 - E L B_1\,)\, u \,. \tag{4b}$$

It is shown in [4] and [7] that

$$L = A_f^{-1}A_{fo} - \Gamma \,, \qquad M = A_{of}A_f^{-1} + \mathcal{P} \,, \tag{5}$$

and $\|\Gamma\| = O(\|\varepsilon\|)$, $\quad \|\mathcal{P}\| = O(\|\varepsilon\|)$.

The assumption $\det A_f \neq 0$ is prior throughout this paper. In [2,4] control of Sys.(1) was probed without considering controllability of the systems. As defined in [8] for single parameter perturbation and in [9] for several parameter perturbations, the system

$$\dot{x}_s = A_0\, x_s + B_0\, u_s \,, \tag{6}$$

$$E(\varepsilon)\, \dot{x}_f = A_f\, x_f + B_f\, u_f \,, \tag{7}$$

is defined as the reduced order subsystem and boundary layer subsystem of Sys.(1), respectively, where $B_f = B_2$, $A_0 = A_{oo} - A_{of}\, A_f^{-1}\, A_{fo}$.

2. D-controllability and Multiparameter Systems

In terms of (5), Sys.(4) can be expressed as

$$\dot{\eta} = (\, A_0 + A_{of}\Gamma\,)\eta + (\, B_0 - M E(\varepsilon)\, L B_1 + \mathcal{P} B_2\,)\, u, \tag{8a}$$

$$E(\varepsilon)\, \dot{\xi} = (\, A_f + E(\varepsilon)\, L A_{of}) + (\, B_f - E(\varepsilon)\, L B_1)\, u \,, \tag{8b}$$

where $B_0 = B_1 - A_{of}\, A_f^{-1}B_2$, and A_0, A_f, B_f, M, L are defined as above.

For single parameter singular perturbation, controllability of the reduced order subsystem and boundary layer subsystem implies the controllability of the overall system for $\varepsilon > 0$ sufficiently small ([8]). But for multiple parameter singular perturbations , this consequence is not true as shown in the example below.

It can be shown that (8b) is controllable for all $\varepsilon = \mu \varepsilon^*$ with $\mu > 0$ sufficiently small if Sys.(7) is controllable for a given $\varepsilon^* \in R_+^N$. It is natural for us to investigate the controllability of sys.(7).

$\underline{Defintion\ 1}$. For matrices $A \in R^{n \times n}$, $B \in R^{n \times m}$ and $D = \text{diag}\,(\alpha_1, \alpha_2, \ldots, \alpha_n)$ (or $D =$ block diag$(\alpha_1 I_{n_1},$ $\alpha_2 I_{n_2}, \ldots, \alpha_N I_{n_N})$), system (A,B) is said to be (block) D-controllable (relative to the index n_1, n_2, $\ldots$, n_N) if (DA, DB) is controllable for any $\varepsilon \in R_+^n$ (or $\varepsilon \in R_+^N$).

It is well known that controllability of LTI system (A,B) is robust to the perturbation of matrices A and B. Unfortunately, (block) D-controllability has not this property.

Example. Suppose that $A_\mu = A + \Delta A$, $\quad B_\mu = B + \Delta B$ $\quad$ and

$$A = \begin{pmatrix} 1 & 1 & 0 \\ 0 & 1 & 1 \\ 0 & 0 & 1 \end{pmatrix}, \quad B = \begin{pmatrix} 0 \\ 0 \\ 1 \end{pmatrix}, \quad \Delta A = \begin{pmatrix} 0 & 0 & -\mu \\ 0 & 0 & 0 \\ 0 & 0 & 0 \end{pmatrix}, \quad \Delta B = \begin{pmatrix} 0 \\ 0 \\ 0 \end{pmatrix}, \quad D = \text{diag}(\alpha_1, \alpha_2, \alpha_3).$$

It is easy to verify that (A,B) is D-controllable, and

$$C(DA_\mu, DB_\mu) = D \begin{pmatrix} 0 & -\mu\alpha_3 & (-\mu\alpha_1 + \alpha_2 - \mu\alpha_3)\alpha_3 \\ 0 & \alpha_3 & \alpha_3(\alpha_2 + \alpha_3) \\ 1 & \alpha_3 & \alpha_3^2 \end{pmatrix}.$$

It clear that (DA_μ, DB_μ) is uncontrollable for any $\alpha_3 > 0, \mu > 0$ when

$$\alpha_2 = \frac{\mu}{1 + \mu}\,\alpha,_2.$$

Sufficient condition for the controllability of Sys.(8) is due to the controllability of $(\widehat{A}_o, \widetilde{B}_o)$ and $(\widehat{A}_\mu, \widetilde{B}_\mu)$, in which $\widehat{A}_o = A + A_{of}\Gamma$, $\widetilde{B}_o = B_o - M E L B_1 + \mathscr{P} B_2$, $\widehat{A}_\mu = E^{-1}(A_f + E L A_{of})$, and $\widetilde{B}_\mu = E^{-1}(B_f - E L B_1)$. From above analysis one can know that $(\widehat{A}_\mu, \widetilde{B}_\mu)$ may be uncontrollable for some $\varepsilon \in R_+^N$, no matter how small $\|\varepsilon\|$ is, even if (A_f, B_f) is D-controllable.

Denote by H and G that

$$H = \left\{ \varepsilon \in R_+^N : \; 0 < c_{ij} \leq \varepsilon_i / \varepsilon_j \leq C_{ij}, \; i,j = 1,2,\cdots, N \right\}, \quad G = \left\{ \varepsilon \in R_+^N : \|\varepsilon\| = 1 \right\}. \tag{9}$$

where c_{ij} and C_{ij} are positive finite constants. With these notation, we can obtain some results to D-controllability and Sys.(8).

$\underline{Propositin 1}$. Suppose $(E^{-1}A_f, E^{-1}B_f)$ is controllable for all $\varepsilon \in G$, then (A_f, B_f) is D-controllable, where $E = E(\varepsilon) = $ block diag $(\varepsilon_1, \varepsilon_2, \cdots, \varepsilon_N)$.

$\underline{Theorem\ 1}$. Sys.(1) is controllable for all $\varepsilon \in H$ with $\|\varepsilon\|$ sufficiently small if

$\qquad$ (i). (A_o, B_o) is controllable,

and

$\qquad$ (ii). (A_f, B_f) is block D-controllable relative to the index $n_1, n_2, \cdots, n_N$.

Note that the set $G_H = G \cap H$ is compact in R^N and $E(\varepsilon)$ is bounded for all $\varepsilon \in G$, theorem 1 can be proved by use of proposition 1 to 3.

$\underline{Proposition\ 2}$. Suppose condition (ii) in theorem 1 holds then for any $\varepsilon \in G$ there is a small constant $\mu_\varepsilon > 0$ such that Sys.(8b) is controllable for all $\alpha = \mu\varepsilon$ with $0 < \mu \leq \mu_\varepsilon$.

$\underline{Proposition\ 3}$. suppose condition (ii) in theorem 1 holds, then there is a constant $\mu_0 > 0$ such that Sys.(8b) is controllable for all $\varepsilon \in H$ with $0 < \|\varepsilon\| \leq \mu_0$.

3. Strong D-controllability

As be shown in section 2, D-controllability is not robust to the perturbation of matrix. This may result in that Sys.(1) is uncontrollable for some $\varepsilon \in R_+^N$, no matter how small $\|\varepsilon\| > 0$ is, even though (A_f, B_f) were D-controllable. In order to overcome this shortage, we ought to consider a kind of systems for which the **D-control**lability is robust to the perturbations of system matrices.

Definition 2. System (A,B) is said to be strongly (block) D- controllable (relative to the index $n_1, n_2, \ldots, n_N$), if there is a constant $\mu^* > 0$ such that the perturbed system $(A + \Delta A, B + \Delta B)$ is D-controllable for any matrix perturbations ΔA and ΔB of matrices A and B, respectively, with $\|\Delta A\| + \|\Delta B\| < \mu^*$.

With definition 2, we can easily get the following result.

Theorem 2. Sys.(8) is controllable for all $\varepsilon \in R_+^N$ with $\|\varepsilon\|$ sufficiently small if

(I). (A_o, B_o) is controllable,

and

(II). (A_f, B_f) is strongly block D-controllable relative to the index $n_1, n_2, \ldots, n_N$.

By use of condition (II) the proof of this theorem can be completed as did in the proof of theorem 1.

4. conclusion

Concepts of (block) D-controllability and strong (block) D-controllability are introduced in this paper, and controllability of singularly perturbed systems with several perturbation parameters is investigated.

Results given in this paper are qualitative, namely, the design of feedback control or feedback matrix is not considered. Because of the complexity of testing the D-controllability of tha "parasitic" subsystems, difficulty we are faced with is to establish the criteria of D-controllability and strong D-controllability for the given systems.

5. References

[1]. Saksena, V.R., J. O'Reilly and P.V. Kokotovic: Singular perturbations and time-scale methods in control theory: survey 1976—1983, Aut. 20,(1984),273—297.

[2]. Khalil, H.K and P.V. Kokotovic: Control of linear systems with multiparameter singular perturbations, Atu. 15. (1979),197—207.

[3]. Abed, E.H. and A.L. Tits: On the stability of multiple time-scale systems, Int. J. Control, 44, 1 (1986), 211—218.

[4]. Khalil, H.K. and P.V. Kokotovic: Control strategies for decision makers using different models of the same system, IEEE Trans. AC-23:2, (1978).

[5]. Ladde, G.S. and D.D. Siljak: Multiparameter singular perturbations of linear systems with multiple time scales, Aut. 19,(1983), 385—394.

[6]. Khalil, H.K. and P.V. Kokotovic: D-stability and multiparameter singular perturbation, SIAM. J. Contr. Opti. 17, 1, (1979),

[7]. Abed, E.H.: Strong D-stability, Systems & Control letters, 7, (1986), 207—212.

[8]. Xu, K.K.: Singular perturbations in control systems, Academia Press, Beijing,1986.

[9]. Abed, E.H.: Decomposition and stability of multiparameter singular perturbation problems, IEEE Trans. AC-31:10, (1986), 925—934.

On-Line Optimal Control of Nonlinear Systems by Singular Perturbation Techniques

J. Doležal[1]

Introduction

Application of necessary optimality conditions for determination of an optimal control law for nonlinear systems results in a solution of nenlinear two-point boundary-value problems (TPBVP). In general, numerical algorithms of considerable complexity are to be invoked to obtain at most open-loop solution (program) to the respective problem. Also the computer time needed prevents to apply such an approach in many practical situations. Moreover, the optimal feedback is usually essential in most real-world engineering applications.

In certain cases, when the problem in question exhibits an appropriate time scale separation, the methodology of singular perturbations can be successfully applied. The order of such approximation depends on the respective problem and is, as a rule, acceptable only during the limited interval of time. However, the considerable less numerical effort required makes it possible to repeat such procedure in a successive way taking into the account always the actual values of all variables. Then an approximate feedback is constructed with accuracy given by the reachable frequency of the actualization cycle. For example, if only zero-order expansions of outer and inner solution are sufficient, then in each actualization cycle only a solution of the set of nonlinear algebraic equations can be needed. Thus fairly accurate, on-line implementable nonlinear feedback can be expected also for considerably fast dynamical systems.

Technique of Singular Perturbations

Singularly perturbed dynamic systems are characterized by a small parameter ε multiplying the derivatives of some state variables. These variables behave as "fast" ones compared to the other "slow" variables of the system exhibiting the phenomenon of time scale separation. If this parameter becomes zero, the order of the dynamic system is reduced. The solution of the reduced-order system is called the zero-order outer solution. The reduced-order solution is unable to meet the end conditions imposed in the original problem on the "fast" variables. This discrepancy is corrected by inner solutions allowing rapid changes of these variables using a streched time scale. The inner solutions have to satisfy the violated end conditions and match the outer solution. An additive composite of outer and inner solutions represents a uniformly valid

[1] Institute of Information Theory and Automation, Czechoslovak Academy of Sciences, 182 08 Prague, Czechoslovakia.

zero-order approximation (for $\varepsilon=0$) of the original problem. If all variables of the problem can be expanded in a form of asymptotic power series of ε for the outer and the inner solution as well, the accuracy of the zero-order solution can be improved by matched higher order terms.

The technique of singular perturbations (SP) has been applied with considerable success in several nonlinear optimal control problems to replace the solution of the respective TPBVP. In many cases the zero-order approximation of the optimal control function has been expressed in a feedback form, see [8], where it is also shown that a "forced" singular perturbation model (obtained by artificial insertion of the perturbation parameter ε) results in the same zero-order composite feedback solution as a classical singularly perturbed model (where a small parameter ε appears as a consequence of a scaling transformation). The accuracy of the zero-order feedback approximation depends in both cases on the actual time scale separation of the variables.

The existing possibility to treat also fast nonlinear systems in this way resulted in a number of applications of in flight mechanics, mainly when development of nonlinear feedback control law was necessary [2], [3], [9]. Examples of optimal pursuit manouvers [8], [9] suggest further perspective applications also in differentiable game setting and to overcome difficulties when e.g. method of neighbouring extremals is used to determine an approximate optimal feedback strategy [4].

Nonlinear Pursuit-Evasion Problem

Let us consider somewhat realistic persuer P dynamics related to the plane of rotation of the evader E, as used also in [9].

$$dx/dt = v_P \cos\gamma \cos\phi_P - v_E \cos\phi_E \tag{1}$$

$$dy/dt = v_P \cos\gamma \sin\phi_P - v_E \sin\phi_E \tag{2}$$

$$dz/dt = v_P \sin\gamma \tag{3}$$

$$dv_P/dt = (T - D)/m \tag{4}$$

$$d\phi_E/dt = a_E/v_E \tag{5}$$

$$d\phi_P/dt = a_P \sin\sigma /(v_P \cos\gamma) \tag{6}$$

$$d\gamma/dt = a_P \cos\sigma / v_P \tag{7}$$

where x, y, z define the relative position of the pursuer from the evader, v_P and v_E denote the respective speed, a_P and a_E acceleration magnitudes, T is thrust, D is drag, m is mass of the pursuer, ϕ_P and ϕ_E the respective headings, γ is pursuer's flight path angle, σ is the orientation of the pursuer's acceleration. For simulation parabolic drag was used [5], [9]. The thrust T is nonzero till certain burn-out time t_T. The control in this problem is lift $L \leq L_{max}$, which imposes a constraint on $a_P(t)$. The

situation is simplified by the assumption that the other control variable
σ is governed by the autopilot (bank-to-turn navigation). As such manouver
takes some time, the full acceleration in the pitch plane is available
with some delay after rolling is finished. This is reflected by
constraints on the rate of change on a_P during the early stage of pursuit.
Aim is to bring the point (x,y,z) to the origin in minimum time. The
analysis of the pertinent TPBVP shows that its solution cannot be easily
obtained by classical method. However, the use of "forced" SP technique
leads to the solution of a system of four nonlinear algebraic equations in
each time instant to produce reasonable feedback control strategy – see
[5] and [9] for details of construction which assumes the persuer's
knowledge of v_E and a_E.

The results of simulation are presented in Fig. 1-4. Trajectories are
parametrized by 0.1 sec. The acceptable miss distance was assumed to be
not over 1 m. The initial values for the critical head-on approach in
plane were selected as folows: $x(0)=0$, $y(0)=600$, $a_{max}= 30$ g (gravitational
constant), $a_E= 8$ g, $v_E=300$ m/sec, $v_P(0)=300$ m/sec. In Fig. 1 a comparison
with a proportional navigation (PN) scheme is performed in favour of SP
approach. This solution reflects the theoretical (see Fig. 2) optimal
behaviour, i.e. full initial turn followed by long coasting period [4], as
the SP a priori assumes optimal strategy (full evasive turn) of the
evader. Needless to say that the difference in optimal capture time is
negligible, as singular optimal control problems are little sensitive in
functional value to the fist-order changes of control variable. Observe
considerably lower requirement on a_P in SP approach being thus technically
more acceptable. The final increase in a_P is caused by the fact that
approximations become poorer as time-to-go is very small.

Fig.3 illustrates the efficiency of SP technique for spatial intercept
with $z(0)=50$ m, $y(0)=0$. The vertical distance of persuer's trajectory and
the dotted line represents one fifth of the corresponding $z(t)$, i.e. the
height above the plane of the evader's rotation. Finally, in Fig. 4 one
can evaluate the sensitivity of the algorithm to a_E estimate, which is not

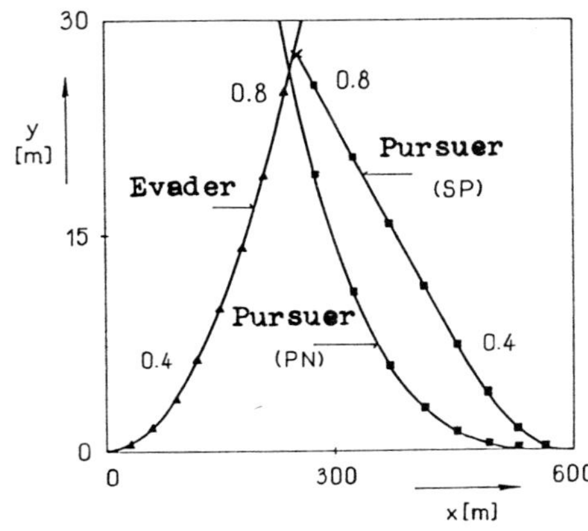

Fig. 1

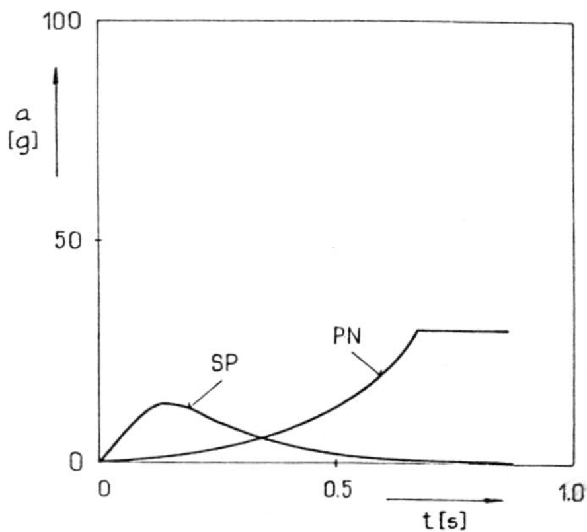

Fig. 2

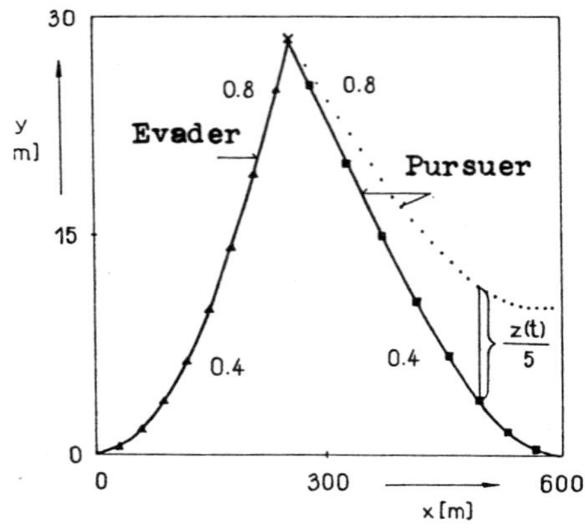

Fig. 3

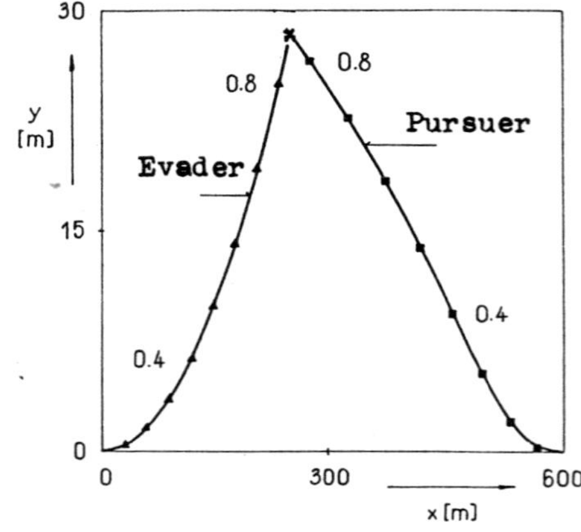

Fig. 4

exact value now, but equals 12 g, i.e. 150 % of the real value. In fact, this estimate is not crutial and can be set to zero if there is no other information available. On the other hand, estimate of v_E should be more precise ($\pm$30 % of its value still gives satisfactory results).

Conclusions

The performed analysis and extensive simulation runs confirmed reported applicability of SP techniques for fast determination of nonlinear optimal control law. In principle, such an approximate feedback solution can be implemented in on-line environment and it can substitute classical techniques. Crutial step is the existence of time scale separation. Sometimes higher-order expansions of both, outer and inner solution may be necessary to produce higher corrections to the optimal solution.

References

[1] Ardema, M.D. (Ed.): Singular Perturbations in Systems and Control. Springer-Verlag, Berlin 1983.
[2] Calise, A.J.: Singular perturbation analysis of optimal aerodynamic and thrust magnitude control. IEEE Trans. Autom. Control AC-24 (1979), 5, 720-730.
[3] Calise, A.J.: Singular perturbation techniques for on-line optimal flight path control. J. Guidance & Control 4 (1981), 4, 398-405.
[4] Doležal J.: A Numerical Approach to the Solution of Differential games. Research Report No. 704, ÚTIA ČSAV, Prague 1976.
[5] Doležal J., Al-Faiz M.Z.: Application of Singular Perturbation Method in Guidance. Research Report No. 1448, ÚTIA ČSAV, Prague 1987.
[6] Kokotovic, P.V., O'Malley, R.E., Sannuti, P.: Singular perturbations and order reduction in control theory - An overview. Automatica 12 (1976), 123-132.
[7] Saksena, V.R., O'Reilly, J., Kokotovic, P.V.: Singular perturbations and time-scale methods in control theory: Survey 1976-1983. Automatica 20 (1984), 273-293.
[8] Shinar, J.: On applications of singular perturbation techniques in nonlinear optimal control. Automatica 19 (1983), 203-211.
[9] Sridhar, B., Gupta, N.K.: Missile guidance laws based on singular perturbation methodology. J. Guidance & Control 3 (1980), 158-165.

Design of Optimal Feedback Controllers for Some Classes of Nonlinear Systems

M. Schwaar [1]

Abstract

A design method for optimal controllers for both linear and nonlinear systems is presented. Quadratic optimization criterions are used. The fundamental principle is based on the optimal state transition calculation of the system where the controller itself is left out of consideration. In a second phase, feedback controllers are designed by means of the signals gained in the the above mentioned way so that their loops will contain the calculated optimal signals if a state transition occurs. The structures of these controllers are also determined during the design process. First of all the design process is demonstrated by example of a linear system where a closed solution is gained. The same principle can also be used to design nonlinear system controllers. Especially, systems are outlined being describable by hyperlogistic differential equations.

Assuming that the dynamic model of a system is given, an optimal feedback controller shall be designed. The objective functional is based on tests using deterministic signals, especially step functions.

In the simplest case a single-input single-output follow-up control is considered. To define the objective functional we assume that some controller is given. The control circuit is defined by the equations

$$X(p) = G(p) * U(p)$$
$$U(p) = - G_R(p) * X(p) + W(p) \tag{1}$$

Here $G(p)$ is the transfer function of the system. $X(p)$, $U(p)$, and $W(p)$ are the Laplace transforms of the output signal, the input signal, and the set point, respectively. Beginning with a steady state of the whole control circuit a step function at the setpoint $w(t)$ produces a step response function $x(t)$. In a first step the control task is now considered as an open loop optimal control problem, whereby the following objective functional Q is used.

[1] Akademie der Wissenschaften der DDR, Institut f. Mechanik
Pf 408, Karl - Marx - Stadt, 9010

$$Q = 1/2 \int_0^\infty ((x-x)^2 + e^2(u-u)^2) \, dt \tag{2}$$

The first term of Q is a measure of the follow-up behaviour of the control. The second term is a measure of the change of the actuating variable u, if the control circuit is excited at the set point. The weighting factor e^2 determines the balance between these two terms. Reducing the value of e^2 implies a stronger controller and the resulting step response approaches to the step function, but the actuating variable increases.

If the optimal open loop policy u(t) is calculated, the optimal controller transfer function G_R is defined by

$$G_R * (1/p - X(p)) = U(p) \tag{3}$$

If the transfer function G(p) may be represented by a quotient of polynoms G(p) = Z(p)/N(p), the transfer function of the optimal controller is $N(p)/(N^*(p)-Z(p))$. The polynom $N^*(p)$ is calculated by

$$N^*(p)*N^*(-p) = N(p)*N(-p) + Z(p)*Z(-p)/e^2 \tag{4}$$

For instance, if the system transfer function is given by
G(p) = 1/ (1 + 2DTp + T^2p^2), the optimal follow-up controller is a PID controller with

$$G_R(p) = \sqrt{(1+1/e^2)} \ \frac{1/Tp +2D + Tp}{Tp + (4D^2 - 2 + 2 \ (1+1/e^2))}$$

This method can be generalized for multivariable control and both following-up and disturbance suppression control. We consider systems G(p) with n input and n output signals. Assuming some multivariable controller is given, n tests are carried out in parallel. These tests differ each from other by the disturbance vectors, which are connected with the steady state control circuit at time t_o. If X,V,U, and Z are matrices, the columns of which are the Laplace transforms of the corresponding signal vectors during the different tests, we get

$$X = G * V + G_z * Z \tag{5}$$
$$V = - G_r * X \tag{6}$$

Here G_z is the disturbance transfer function. Now we carry out a second series of tests with a modified disturbance connection

$$X^* = G * U + W$$
$$U = - G_R * X^* \tag{7}$$

The here used disturbances are the columns of the matrice W. Now the following relation holds:

$$V^T(p) = Z^T * G_z(p) * (W^{-1})^T * U^T(p) \tag{8}$$

Thus we get an optimization problem which can be solved without any knowledge about the controller transfer function. We use an objective functional

$$Q = \int_0^\infty tr\ (X^T * F^2 * X + U^T * E^2 * U)\ dt \tag{9}$$

Here E and F are diagonal matrices. The first term of the objective functional is a measure of the output caused by the disturbances during the first series of tests. The second term describes the output of the controller if the step disturbances of the second series of tests work immediately at the input of the controller. With the weighting matrices E and F these two aspects of optimal control are balanced. The system itself is represented by the equations (8) and (5). Equation (6) may be used for defining the optimal controller transfer function.

The principle described above is now transferred to nonlinear systems. At first an optimal follow-up feedback controller for logistic systems shall be designed. A logistic system is described by

$$x = x * (u - x) \tag{10}$$

The input signal is u and we use the objective functional stated in equation (2). The Hamilton function is

$$H = 1/2\ (x-x_\infty)^2 + 1/2\ e^2 * (u-u_\infty)^2 + \lambda * x * (u-x) \tag{11}$$

With $\partial H/\partial x = -\dot\lambda$, $\partial H/\partial u = 0$, $x_\infty = u_\infty = w$, and $\mu = \lambda * x$ we get the following two differential equations

$$\dot\mu = -x\ (\ (x-w) - \mu)$$
$$x = x\ (\ w - \mu/e^2 - x\ )$$
$$u = w - \mu/e^2 \tag{12}$$

With $\dot\mu/\dot x = \mu/(x-w) = e^2 * (-1 + \sqrt{1+1/e^2}) * (x-w)$ these differential equations are reduced to

$$\dot x = \sqrt{1+1/e^2} * x * (w-x) \tag{13}$$

We see that the transition from one state to another will be accelerated
by using the controller. The controller itself is described by

$$u = w + (w-x)*(-1 + \sqrt{(1+1/e^2)}\ )$$
$$= (w-x) * \sqrt{(1+1/e^2)} + x$$

$$= \sqrt{(1+1/e^2)} * (\ (w-x) + \int x*(w-x)\ dt\ ) \tag{14}$$

The controller is similar to a PI controller, but there is a nonlinear
integral part.

A general nonlinear first order system $x=f(x,u)$ is treated in a similar
way. With $x_\infty = u_\infty =w$ (linear characteristic) we get a Hamilton function

$$H = 1/2 * (x-w)^2 + 1/2 * e^2 * (u-w)^2 + \lambda *f(x,u) \tag{15}$$

Now the substitution $\mu=\lambda*\partial f/\partial u$ results in

$$\dot{\mu} = (w-x)*\partial f/\partial u + \mu(\partial/\partial t\ [\ln(\partial f/\partial u)] - \partial f/\partial x)$$
$$\dot{x} = f(x,u)$$
$$u = w - \mu/e^2 \tag{16}$$

This system may be reduced by finding a stationary function $\mu=\mu(x,w)$
so that all points of $\mu(x,w)$ are placed on a path to the steady state
$x=w$. This stationary function is calculated by solving the differential
equation

$$\frac{d\mu}{dx} = \frac{\dot{\mu}}{\dot{x}} = -\frac{2}{w} * \frac{(w-\mu/e^2)((w-x)(w-\mu/e^2) + x*\mu)}{(w-\mu/e^2)^2 - x^2} \tag{17}$$

The integration procedure starts at $x=w$. This is a singular point and
the right hand side of the equation is calculated as a limit. The
obtained controller is

$$u = w-x - \mu(x,w) + \int f(x,u(x,w))\ dt \tag{17}$$

References

[1] Weihrich,G.: Optimale Regelung linearer dynamischer Prozesse,
    Oldenbourg Verlag, Munchen, 1973
[2] Korn, U., H.H. Wilfert: Mehrgrossenregelungen, Verlag Technik,
    Berlin 1982
[3] Ray, W.H.: Advanced Process Control, McGraw Hill, New York 1981
[4] Peschel, M., W.Mende:Leben wir in einer Volterra-Welt?
    Akademie-Verlag, Berlin 1982

An Algorithm for Optimal Control of Nonlinear Systems

Petr Javorský[1]

1. Introduction

The optimal control of nonlinear systems is not often possible with classic methods from the theory of control. The Bellman dynamic programming method requires for multidimensional systems a large memory capacity because it is necessary to perform a discretization of state space and permissible region of the control space. A new algorithm, which is described in this paper, reduces shortcomes mentioned above. This algorithm for optimal control of nonlinear systems is involved as a tool in the ACM method (Automated construction of system dynamics models [1]).

2. Problem formulation

Let us have a nonlinear system desribed by a difference state equation:

$$\underline{x}(T+1) = \underline{f}(\underline{x}(T), \underline{u}(T)) \qquad /1/$$

where $\underline{x}$ is n-dimensional state vector $[x_1, x_2, \ldots x_n]^T$

          $\underline{u}$ is r-dimensional control vector $[u_1, u_2, \ldots u_r]^T$

          $\underline{f}$ is nonlinear function vector $[f_1, f_2, \ldots f_n]^T$

          $\underline{T}$ is $0, 1, 2, \ldots$

We have a restricted control $\underline{u} \in U$, where U is the given permissible region in the control space. So that

$$\forall \underline{u} \in U : u_i \in \langle u_{i\,min}, u_{i\,max}\rangle \qquad i = 1, 2, \ldots r$$

We are to find a control sequence $\underline{u}(i)$ $(i=0,1,2, \ldots K-1)$, which transforms the initial state of the system $\underline{x}(0)=\underline{x}^o$ to the desired final state $\underline{x}(K)=\underline{x}^d$ and on condition that a criterion function

$$J = \sum_{i=0}^{K-1} \underline{u}^T(i) \underline{R} \underline{u}(i)$$

would be minimized. $\underline{R}$ is a positive semi-definite weighing matrix ($r_x r$).

3. The algorithm

The proposed algorithm converts problem mentioned above into two subproblems:

---

[1] Petr Javorský, Institute for Application of Computing Technique in Control, Revolucni 24, Praha 1, Czechoslovakia

- A finding of the control sequence $\underline{u}(i)$ (in the permissible region of the control space), which transforms system /1/ from the initial state $\underline{x}(0)=\underline{x}^o$ to the final state $\underline{x}(K)=\underline{x}^d$; that is determination of the "basic" trajectory.
- The control sequence $u(i)$ is modified step by step by means of the adaptive random searching in surroundings of "basic" trajectory so that the criterion J would be minimized.

Now we describe both subproblems in detail.

a) Determination of the "basic" trajectory

In this part we want to reach the desired final state $\underline{x}^d$. The "reaching criterion" the final state may be different, e.g.

$$\| \underline{x}(K) - \underline{x}^d \| < \delta$$

$$\text{or} \qquad \frac{\| \underline{x}(K) - \underline{x}^d \|}{\| \underline{x}^d \|} < \delta$$

$$\text{or} \qquad \sum_{i=1}^{n} \frac{|x_i - x_i^d|}{|x_i^d|} < \delta$$

where $\| \cdot \|$ means the euclidean norm.

First the algorithm chooses the time invariant control sequence $\underline{u} \in U$ (i.e. $\underline{u}(0)=\underline{u}(1) \ldots \underline{u}(K-1)$) and then it attempts to reach the desired final state $\underline{x}^d$ by performing adaptive random search. The success of this process is dependent on the given value of δ (i.e. required precision of "reaching" final state) and on what algorithm of the adaptive random search is used. If the process of finding the "basic" trajectory is not successful, the best trajectory (in sense of "reaching criterion") is used and the process of reaching the final state x^d is repeated from the discrete time points of the best trajectory.

b) Minimizing of the criterial function J

At this moment we have the "basic" trajectory and control sequence $\underline{u}(i) \in U$, $i=0,1, \ldots K-1$, which transforms the system from the initial state $\underline{x}^o$ to the desired final state $\underline{x}^d$. A value of the criteron J for this trajectory is as follows:

$$J = \sum_{i=0}^{K-1} \underline{u}^T(i) \, \underline{R} \, \underline{u}(i) = J_0 + J_1 + \ldots + J_{K-1}$$

On Fig.1 is depicted further action in modifying the "basic" trajectory. The computation proceeds backward in time from $i=K-1$ to $i=1$. Let we are in the state $\underline{x}(i-1)$, we search permissible control $\underline{u}'(i-1)$ and $\underline{u}'(i)$ so that we reach state $\underline{x}(i+1)$ and a sum of criterial addition $J'_{i-1} + J'_i$ was minimized and at the same time

$$J'_{i-1} + J'_i < J_{i-1} + J_i$$

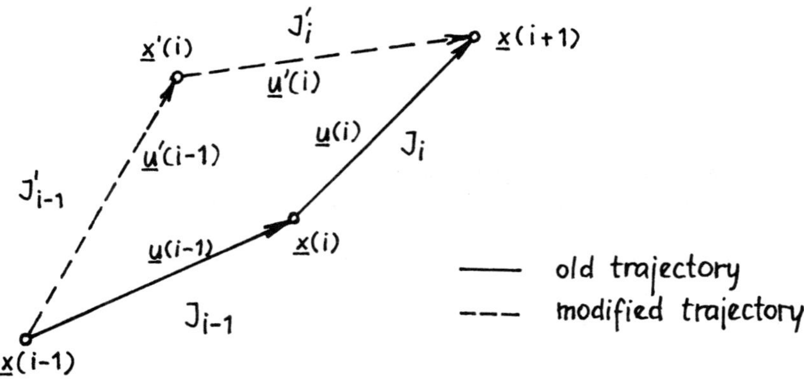

Fig. 1

This one run of modifying process finishes at initial state $\underline{x}^\circ$ and the whole process is repeated provided the value of the criterion function J changes. In other words, we finish if

$$J' - J < \delta$$

where δ is required precision, J' is value of criterion of the modified trajectory and J is the same of the old trajectory.

4. Conclusion

The algorithm for optimal control of nonlinear systems, which is described in this paper, has two main advantages. Firstly it does not carry out a discretization of the state space and secondly it uses only the simulation equations of the system, initial state vector, permissible range control and the given criterion function. This algorithm can be widely used for many nonlinear systems. On the other hand we must mention two disadvantages of this algorithm. It find an optimal control and the optimal trajectory as a local minimum in the surroundimgs of the "basic" trajectory and it also requires more computing solution time.

References

[1] Kejak,Javorský: Method of automated construction of simulation
                    system dynamics models (ACM). In: Proceedings of
                    European Congress on Simulation, Prague, 1987,
                    Academia.

Design of a Combustion Controller

J. ČRETNIK, S.STRMČNIK [1]

ABSTRACT: The aim of this paper is to show some possibilities of improving the efficiency of the combustion process by means of an advanced controller. For control design purposes a mathematical model of combustion process was developed first. Its response was then compared with measurements on an industrial plant. The verified model was used to design a predictive PI state controller and feedforward compensator.

1. INTRODUCTION

Depending on the specific constituents of a fuel, a certain rate of air supply is necessary for complete oxidation according to the rules of stoichiometry. Too small air supply leads to incomplete combustion with emission of soot and carbon monoxide. This not only causes energy losses but also increases the maintenance costs. On the other side, excess of oxygen supply effects unnecessary heating of surplus air and by that way consumes energy (Gilbert, 1976). Moreover, the production of nitric oxide and sulphur trioxide, which are harmful for the natural environment and also for the furnace itself, may be facilitated with air excess.

In the real combustion process, a complete oxidation at the theoretically needed air supply cannot be achieved. Rather, there is still some residual carbon monoxide and soot in the exhaust gas, together with some part of unconsumed oxygen. Supplying more air excess reduces the remaning concentration of carbon monoxide, while at the same time increases the surplus concentration of oxigen in the exhaust gas. Obviously, there exists an optimal value of air rate, at which combustion is optimal. However, the optimal operating point depends on a variety of operational parameters like composition of fuel, load level, construction, condition of burners, and others. This establishes a feedback control problem with the objective to adjust the air supply automatically whenever a change in the operational state occurs. Several authors propose control concepts which rely on excess oxygen measurement. Since the conditions of combustion are different at different load levels, the optimal efficiency requires load dependent set points for the surplus oxygen in the flue gas. This control concept is used in the presented research work.

2. COMBUSTION PROCESS MODEL

The composition of fuel can be expressed by corresponding concentrations of carbon, hydrogen, oxygen, nitrogen, sulphur, ash and water. Air composition is expressed by volume fractions of oxygen and nitrogen. Only the most important components of the flue gas are concidered, those are oxygen, carbon dioxide, carbon monoxide, sulphur dioxide, nitrogen and water. Fundamental equations of the mathematical model are based on stoi-

[1] "Jožef Stefan" Institute, Yugoslavia, 61000 Ljubljana, Jamova 39

chiometric chemical reactions of combustion. The latter are supplemented with factors m and b, which are functions of air rate and denote the ratio of CO_2/CO and soot, respectively. Let the simbol ϕ_f^x represent the equivalent of gas flow coming from fuel. Then the static balance equation for gases taking part in the combustion process can be written as follows:

$$\phi_f^x + \phi_a = \phi_g \ , \tag{1}$$

where simbols are norm fuel flow in kg/s, norm air flow in m^3/s and norm flue gas flow in m^3/s respectively. The corresponding dynamic balance equation is given in the form:

$$\frac{dV}{dt} = \phi_f^x + \phi_a - \phi_g \ , \tag{2}$$

where V is the overall volume of gases in the combustion chamber. The equivalent of gas flow coming from fuel can be expressed by:

$$\phi_f^x = \phi_f \ (V_g - V_{O_2}) \ , \tag{3}$$

where V_g is the volume of flue gas being developed by 1kg of fuel and V_{O_2} volume of oxygen required for complete combustion of 1kg of fuel. Both parameters are functions of the fuel composition and the parameters m and b. Because the volume of the combustion chamber is usually large, it can be supposed that changes of V (due to rapid fluctuation in the flows ϕ_f, ϕ_a and ϕ_g) are much faster than the change of concentrations of components of flue gas. Therefore dV/dt=0 and V is set to be a constant, depending on combustion chamber dimensions. The corresponding volume balances for the individual components of the flue gas can be seen together in the block scheme of combustion process model in Fig.1 (Čretnik, 1985). Balance of gas flow through the combustion chamber is taken into account in stationary state:

$$\phi_g = \phi_a + \phi_f \ (V_g - V_{O_2}) \tag{4}$$

There exists a certain time delay of flue gas before it comes to the stack. This time delay is variable and depends on volume flow of flue gas ($T_d = f \ (1/\phi_g)$). It includes also the dead time of oxygen analyzer.

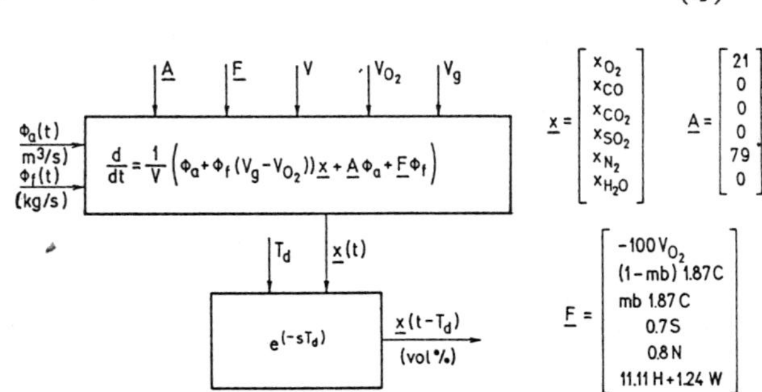

Fig.1. Block scheme of combustion process model.

3. MODEL VALIDATION

The theoretically developed model was verified by series measurements in a distric heating plant "Toplarna Ljubljana", on its hot-

water boiler. Because the
air flow in the boiler
cannot be measured, the
mathematical model was
augmented by the block
for estimation of air
flow, as it can be seen
in Fig.2. The inputs of
the model were fuel oil
flow and air-pressure
drop between the fan and
the combustion chamber.
Parameters of the model
are supposed to be con-
stant. The measurements
of oxygen concentration
in the flue gas, the fuel
oil flow, the air-pressu-
re drop and simulated
oxygen concentration are
shown in Fig.3. The simu-
lation was made on a PDP
11/34 computer.

4. DESIGN OF A COMBUST-
   ION CONTROLLER

The verified model
augmented by the identi-
fied model between the
control input and air-
pressure drop was then
linearised (Čretnik, 1986):

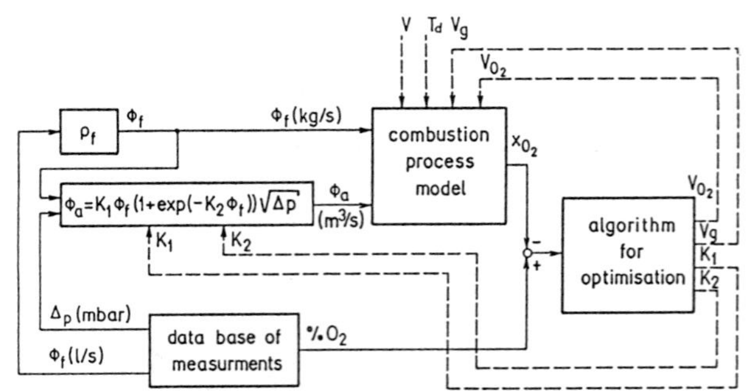

Fig.2. Block scheme for verification of
the combustion process model.

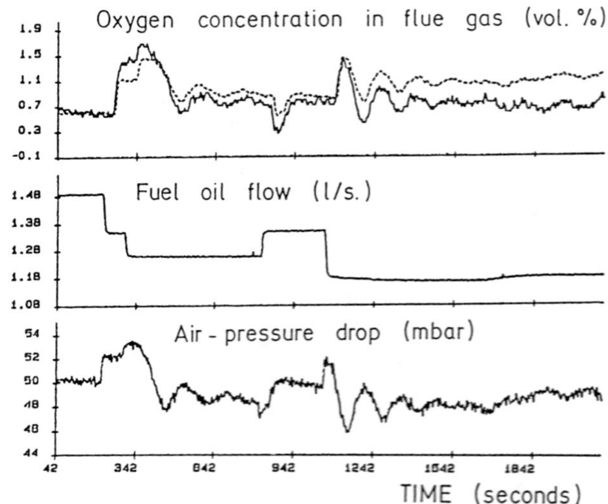

Fig.3. Oxygen concentration in flue gas
(—— measurements, --- simulation).

$$\underline{x}(k+1) = \underline{\underline{A}}\ \underline{x}(k) + \underline{\underline{B}}\ \underline{u}(k-d)$$
$$y(k) = x_1(k) \tag{5}$$

where x_1 is vol.% of oxygen in flue gas, x_2 is air-pressure drop in mbar, u_1 control input in %, u_2 fuel oil flow in l/s and d time delay. First, a PI state controller was designed for the system as it was without delay. The control input for the system (5) without delay is:

$$u_1(k) = \underline{K}\ [\underline{x}(k)-\underline{x}^*(k)] + K_I\ \sum_{n=0}^{k} [x_1(k) - x_1^*(k)]\ T_0 , \tag{6}$$

where $\underline{x}^*$ are state references and T_0 is sample time (3s). But system (5) has a time delay and its control input must be different:

$$u_1(k) = \underline{K}\ [\underline{x}(k+d)-\underline{x}^*(k)] + K_I\ \sum_{n=0}^{k} [\ x_1(k) - x_1^*(k)\ ]\ T_0 , \tag{7}$$

where $\underline{x}(k+d)$ is state prediction and is calculated from (5) as follows:

$$\underline{x}(k+d) = \underline{\underline{A}}^d \, \underline{x}(k) + \sum_{i=1}^{d} \underline{\underline{A}}^{i-1} \, [\, \underline{\underline{B}}^- \, \underline{u}(k-i)\,] \quad . \tag{8}$$

In this way, a robust predictive PI state controller was designed. The controller was designed for a constant time delay (42s), but it is robust for large changes in time delay, too.

Much better results were achieved when the predictive compensator was added to the predictive PI state controller. Its control signal is:

$$u_1^+(k) = [\, \underline{\underline{B}}_1^T \underline{\underline{B}}_1\,]^{-1} \, \underline{\underline{B}}_1^T \left[\, [\underline{\underline{I}} - \underline{\underline{A}}]\, \underline{x}^*(k) - \underline{\underline{B}}_2 \, u_2(k)\,\right] \quad , \tag{9}$$

where $u_2(k)$ is the measurable disturbance, in this case fuel oil flow. The response of oxygen concentrations in flue gas, controlled by the predictive PI state controler with and without the predictive compensator is seen in Fig.4.

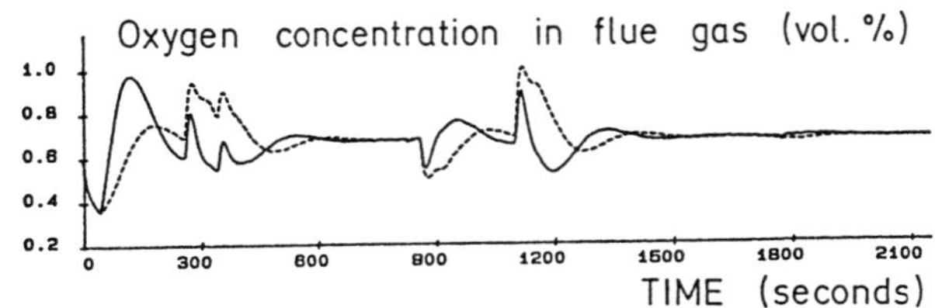

Fig.4. Oxygen concentration in flue gas, controlled by the predictive PI state controller without (---) and with (——) the predictive compensator. Fuel oil flow is shown on Fig.3.

5. CONCLUSION

The aim of this paper was to show some possibilities of improving the efficiency of the combustion process by means of an advanced controller. It is shown that for the processes with a large time delay and in cases when disturbances are measurable, the best results can be achieved by feedforward compensators, specially when the process is not completely deterministic to take the advantage of the model prediction. The predictive controller alone is too slow to fulfill the technical requirements of the combustion process.

6. LITERATURE

[1] Čretnik, J., S. Strmčnik, and B. Zupančič: A model for combustion of fuel in the boiler. 3rd Proceedings of the Symposium Simulationstechnik (1985), Bad Münster am St.-Ebernburg, 469-473.
[2] Čretnik, J. and S. Strmčnik: Control ofprocesses with a large time delay. Proceedings of XXXI.JUREMA, 1 (1986), Plitvička jezera, 71-74 (in slovene).
[3] Gilbert, L.F.: Precise combustion control saves fuel and power. Chemical Engineering (1976) June 21, 145-150.

Solving Assignment Problems by Selection Pressure Controlled Replicator Networks

Hans-Michael Voigt
Ivan Santibanez-Koref

Academy of Sciences of the GDR
Centre for Scientific Instrumentation (ZWG)
Rudower Chaussee 6, Berlin-Adlershof, 1199

Keywords

Neural Networks, Replicator Networks, Biological Evolution, Natural Selection, Combinatorial Optimization, Travelling Salesman Problem, Graph Partitioning Problem, Module Placement Problem, Diversified Replication, Selection Pressure Control

Introduction

Following **Darwin's** famous work "On the Origin of Species", evolution processes of living nature can be characterized by **replication, mutation** and **selection**. Natural selection means the the evolution of species to ever higher stages with respect to function and structure. This evolution implies the application of natural selection mechanisms for optimization purposes.

For the description of evolutionary processes a large variety of models has been established. One of the most fundamental is the **Lotka-Volterra-Equation**. The transformation of this equation to the unit simplex yields the so called **Replicator Equation**

$$\dot{x}_i = x_i \big(f_i(x) - \Phi(x) \big) \quad , \quad x'_i = x_i \frac{f_i(x)}{\Phi(x)}$$

for **overlapping** and **non-overlapping population generations** which is defined on the probability simplex

$$S_n = \left\{ (x_1, \ldots, x_n) : x_i \geq 0, \sum_i x_i = 1 \right\}.$$

This equation reflects replication and selection. Mutation terms may be added. It is worth to note that this equation plays an essential role in the description of the evolution of **macromolecules**, the evolution of **alleles** in a **gene pool**, the evolution of species in **animal populations**, and evolution processes in **large scale systems**.

In terms of genetics, x_i means the proportion of allele i in the gene pool, $f_i(x)$ the mean fitness of alleles i, and $\Phi(x)$ the mean fitness of the total gene pool. For linear fitness functions $f_i(x)$

$$f_i(x) = \sum_j G_{ij} x_j = (Gx)_i$$

with real, symmetric, non-negative interaction matrices G this is the well-known **Fisher-Wright-Haldane-Model** of population genetics.

Stable structures of a population or gene pool are given by stable attractors, in the most simple case by stable fixed points.

Underlying a homogeneous potential function U(x) of degree s and using as fitness functions $f_i(x)$ the Euclidean gradients

$$f_i(x) = \frac{\partial U(x)}{\partial x_i}$$

we have for the mean total fitness $\Phi(x)$ the relation

$$\Phi(x) = s U(x).$$

In this case a **maximum principle** in the sense of Shahshahani holds, and
the increase rate of the mean fitness of the total gene pool

$$\Phi(x) \geq 0 \quad , \quad \Phi' \geq \Phi,$$

respectively, is non-negative. This is the fundamental theorem of population genetics.

Conversely, we have with the above given settings for the **Inverse Replicator Equation**

$$\dot{x}_i = x_i\left(\frac{1}{f_i(x)} - \Phi(x)\right) \quad , \quad x'_i = x_i \frac{\left(\frac{1}{f_i(x)}\right)}{\Phi(x)}$$

a **minimum principle**.

Diversified Replication

If we assume that evolution proceeds not only for a single gene
locus, as in the above case, but for m different connected loci, then a
Diversified Replicator Equation

$$\dot{x}_{ik} = x_{ik}\left(f_{ik}(x) - \Phi_i(x)\right) \quad , \quad x'_{ik} = x_{ik}\frac{f_{ik}(x)}{\Phi_i(x)}$$

which is defined on the simplices

$$S_m^i = \left\{(x_{i1},\ldots,x_{im}) : x_{ik} \geq 0, \sum_k x_{ik} = 1\right\}$$

may be used as an appropriate model. Here x_{ik} is the proportion of
alleles i at gene locus k, $f_{ik}(x)$ the mean fitness of alleles i at locus
k, $\Phi_i(x)$ the mean fitness of alleles i, and

$$\Phi(x) = \sum_i \Phi_i(x)$$

the mean fitness of the total gene pool with i = 1,...,n and k =
1,...,m. As in the single locus model for homogeneous potential functions U(x) and Euclidean gradients

$$f_{ik} = \frac{\partial U(x)}{\partial x_{ik}}$$

as fitness functions the relation

$$\Phi(x) = sU(x)$$

holds, and a **maximum principle** in the sense of Shahshahani can be shown.
The increase rate of the mean fitness of the total gene pool is also
non-negative. That's the fundamental theorem for diversified replication.

In much the same way as in the one locus model an **Inverse Diversified Replicator Equation** can be derived for which a **minimum principle**
holds.

The application of diversified replication to combinatorial optimization problems will be demonstrated by the module placement problem.
The approach was equally used for the graph partitioning problem, the
module placement problem, the mapping problem of processes to processors
for multi-processor systems etc.

The Module Placement Problem

In case of the module placement problem the potential function can
be derived in the following way. Let us assume that there are given
modules i, i = 1,...,n. The modules are connected by wires where G_{ij} is
the number of wires between module i and module j. The modules should be
placed on a regular grid with k rows and l columns. The total length of
the wires for a placement of the modules to grid points has to be
minimized according to a Manhattan distance.

This can be formalized as a maximization problem with

$$U = \sum_{k=1}^{row} \sum_{l=1}^{col} \left(C_{kl,kl-1} + C_{kl,kl+1} + C_{kl,k-1l} + C_{kl,k+1l}\right) \rightarrow \max$$

with

$$C_{kl,rs} = \sum_i x_{ikl} \sum_j G_{ij} x_{jrs}.$$

The Euclidean gradients are given by

$$\frac{\partial U}{\partial x_{ikl}} = \left(G x_{.kl-1}\right)_i + \left(G x_{.kl+1}\right)_i + \left(G x_{.k-1l}\right)_i + \left(G x_{.k+1l}\right)_i.$$

Furthermore we have to consider the constraints

$$X_{kl} = \sum_i x_{ikl} = 1.$$

Using the **Diversified Replicator Equation** and introducing a **Selection Pressure** control u we have as fitness functions $f_{ikl}(x)$

$$f_{ikl}(x) = \frac{\sum_j G_{ij}\left(x_{jkl-1} + x_{jkl+1} + x_{jk-1l} + x_{jk+1l}\right) + u x_{ikl}}{X_{kl}} \qquad for \quad u \geq 0$$

$$f_{ikl}(x) = \frac{\sum_j G_{ij}\left(x_{jkl-1} + x_{jkl+1} + x_{jk-1l} + x_{jk+1l}\right) - u\left(1 - x_{ikl}\right)}{X_{kl}} \qquad for \quad u < 0$$

x_{ikl} is the probability with which module i is at the k-th row and the l-th column of the grid.

Replication and Selection are given by the **recurrence equation**

$$x'_{ikl} = x_{ikl} \frac{f_{ikl}(x)}{\Phi_i}.$$

Mutations are introduced by a slight random perturbation for each new generation

$$x_{ikl} := \frac{x_{ikl} + \mu y_{ikl}}{1 + \mu \sum_r \sum_s y_{irs}}$$

where y_{ikl} are uniformly distributed random numbers in $(0,1)$ and μ is the **mutation rate**.

The **evolution process** works as follows:

```
1  Start with a uniform distribution of the x_ikl
2  Set the selection pressure u sufficiently low
3  Disturb the x_ikl slightly ( mutation )
4  Iterate the recurrence equation ( replication and selection )
5  If the probabilities x_ikl are stabilized go to 6, else go to 3
6  If the assignments x_ikl are sufficiently crisp then stop
7  Increase the selection pressure u slightly and go to 3
```

Some Results

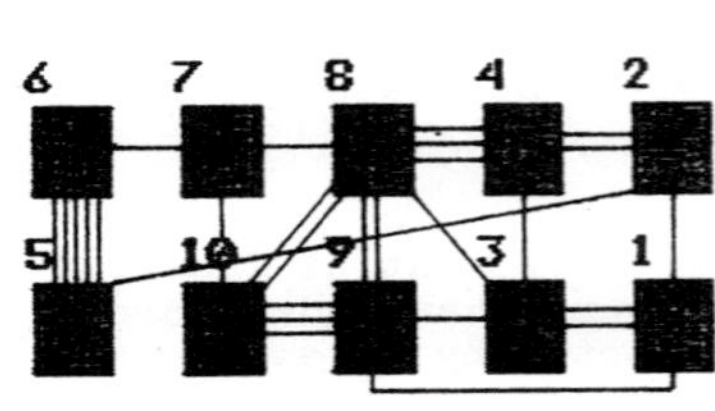

FIGURE 1.

PLACEMENT OF MODULES TO
GRID POSITIONS

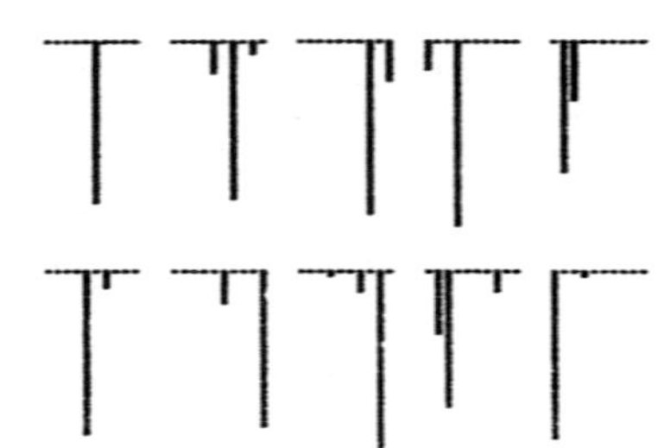

FIGURE 2.

ASSIGNMENTS X(I,K,L) OF MODULES
TO GRID POSITIONS , ITERATION 73

In fig. 1 there is shown a 10-module problem which was computed within 45 ... 85 iterations for 30 runs. For every run a minimal length placement was obtained at u = -1.0. Assignments of modules to grid positions are shown in fig. 2.

Conclusions

In case of the **module placement problem** good results were obtained by the **replicator network approach**. At present the approach will be compared with other methods. In case of the **travelling salesman problem** the Replicator network approach is superior to the Hopfield & Tank neural networks and to the Boltzmann machine approach.

Acknowledgment

We constantly benefited from discussions with M. Peschel from the Academy of Sciences of the GDR. Discussions with H. Muehlenbein from the Gesellschaft fuer Mathematik und Datenverarbeitung about his experiences with replicator network algorithms were very stimulating.

References

Aarts E.H.L. & J.H.M. Korst Solving Travelling Salesman Problems with Massively Parallel Networks Based on the Boltzmann Machine. European Journal of Operations Research

Brady R.M. Nature 317, 804-806(1985)

Darwin Ch. On the Origin of Species (1859) (Reprint Harvard University Press 1964, Knoener 1906)

Durbin R. & D. Willshaw Nature 326, 689-691(1987)

Ewens W.J. Mathematical Population Genetics (Springer 1979)

Ebeling W. & R. Feistel Physik der Selbstorganisation und Evolution (Akademie-Verlag 1982)

Eigen M. & P. Schuster The Hypercycle (Springer 1979)

Fisher R.A. The Genetical Theory of Natural Selection (Clarendon Press 1930)

Garey M.R. & D.S. Johnson Computers and Intractability (Freeman 1979)

Hadeler K.P. Mathematik fuer Biologen (Springer 1974)

Hofbauer J. & K. Sigmund Evolutionstheorie und dynamische Systeme (Parey 1984)

Hopfield J.J. & D.W. Tank Biol. Cybern. 52,141-152 (1985)

Kirkpatrick S., C.D. Gelatt & M.P. Vecchi Science 220,671-680 (1983)

Lotka A.J. Proc. N.A.S. 6,410-415(1920)

Maynard Smith J. Evolutionary Game Theory (Cambridge University Press 1982)

Muehlenbein H., M. Gorges-Schleuter & O. Kraemer Parallel Computing 4,269-279(1987)

Papadimetriou C.H. & K. Steiglitz Combinatorial Optimization (Prentice-Hall 1982)

Peschel M. & W. Mende Predator-Prey-Models (Akademie-Verlag & Springer 1986)

Shahshahani S. Memoirs AMS 211 (1979)

Sigmund K. Mathematical Research 23, 63-71 (Akademie-Verlag 1985)

Svireshev Yu.M. & D.O. Logofet Stability of Biological Communities (Mir Publishers 1983)

Volterra V. Lecons sur la theorie mathematique de la lutte pour la vie (Gauthiers-Villars 1931)

Voigt H.-M. Evolution und Optimierung. Diss. B (Habilitation) (Acad. Sci. GDR 1986)

Voigt H.-M. Lect. Notes in Economics and Mathematical Systems 273, 117-125 (Springer 1986)

Voigt H.-M. (Ed.) Structure Formation Based on Self-Organisation Principles with Applications to Electronics Design (Acad. Sci. GDR, Inst. for Informatics, Report 12, 1985)

Voigt H.-M. Population Genetics and Combinatorial Optimization. Proc. Int. Tagung Opt. und Anwendungen,53-56 (Eisenach, Nov.1986)

Voigt H.-M. Wiss. Z. TH Ilmenau 2,117-132(1985)

Voigt H.-M. Report R-MATH-01/83,242-249 (Acad. Sci. GDR, Inst. for Mathematics 1982)

Voigt H.-M. Mathematical Research 23,228-231 (Akademie-Verlag 1985)

Voigt H.-M. Optimization by Selection Pressure Controlled Replicator Networks. Systems Anal. Mod. Sim. 5 (5) 1988 (to appear)

Wilson G.V. & G.S. Pawley On the Stability of the Travelling Salesman Problem Algorithm of Hopfield and Tank. Univ. of Edinburgh, Dep. of Physics

Zeeman E.C. Lect. Notes in Mathematics 819, 249-270 (Springer 1980)

Local Area Networks with Different Topological Structures, Analysis of Qualitative and Quantitative Behaviour

Adam Grzech[1]

ABSTRACT: The aim of this paper is to investigate and examine interdependence between the quality of services which are delivered by local area networks and local area network architectures. The relation is examined for networks with different topological structures. The problem of proper topological structure selection has been formulated as a multitask problem. Solution, obtained for the whole set of users (globally optimal solution) is compared with interconnected set of solutions obtained for distinctive subsets of users (locally optimal solution). The obtained solutions are compared using well-known performance measure.
An example, which illustrates the investigated problem is presented.

KEYWORDS: local area networks, topological structure, performance measure

1. INTRODUCTION

A local area network, being a distributed computer system, usually contains some number of general- and special-purpose autonomous entities (stations) which are interconnected by a communication subsystem [5,7]. Besides allowing communication among aforenamed entities, the network facilitates resource sharing that allows a local tasks to be transferred and processed at a remote location of the network. Tasks are migrated because the local processor does not have the required computational capabilities or data or has failed. A task may be also processed remotly if the expected turnaround time is better [3]. It is evident that different communication subsystems offer different service quality. It depends on users requirements/communication subsystem matching. The users requirement are usually divided into qualitative and quantitative requirements. The former express the functional description of automated data processing system (ADP) and allow to determine the area of feasible solutions within which the proper solution, after taking into account quantitative requirements, should be found [1,2].

In this paper, assuming that the users set and users characteristics are the same, the dependence between quality of services (measured by average customer delay) and network topological structures is investigated.

2. PROBLEM FORMULATION

Let us assume that the quantitative requirements of users community are defined by a square matrix $A = [a_{ij}]$ (i,j = 1,2, ... ,n) where a_{ij} is equal to an average number of access requires generated at i-th station and destinated to j-th station (in specified time period) an n is a number of stations in which the users are clustered. It is also assumed that the users requirements can be satisfied by local area networks having different architectures (defined by: topological structure, access-control discipline, channel configuration and signalling technique [7]) which offer different grades of delivered services.

The problem we are interested in is presented below as a multitask problem:

1. For known matrix A, find a solution R_i $(R_i \in \{R_1, R_2, ... , R_N\} = \mathcal{R}_f)$ - $\mathcal{R}_f$ is a set of all feasible solutions (all feasible network architectures) - which minimizes assumed performance measure. This solution is called globally optimal solution.

---

[1] Institute of Control and Systems Engineering,
Technical University of Wrocław, Wrocław, Poland

2. Divide the entire set of stations (X) into two distinctive subsets X_1 and X_2 $(X_1 \neq \emptyset, X_2 \neq \emptyset, X_1 \cup X_2 = X$ and $X_1 \cap X_2 = \emptyset)$ such that the division minimizes the following sum:

$$\sum_{v \in X_1} \sum_{u \in X_2} \left(a_{vu} + a_{uv} \right) = a(X_1, X_2) + a(X_2, X_1)$$

   i.e., that it minimizes the sum of average number of customers originated in the first subset and destinated to the second and vice versa.

3. For each subset of stations (X_1 and X_2) and for known traffic attached to the subsets, find solution R_{ij} ($R_{ij} \in \left\{ R_{i1}, R_{i2}, \ \dots \ , R_{iN(i)} \right\} = \mathcal{R}_{if}$, $i = \left\{ 1,2 \right\}$ and where $\mathcal{R}_{if}$ is a set of all feasible solutions) which, for the X_i subset of stations, which minimize assumed performance measure.

   The traffic attached to each i-th subset of stations is given by matrices A_i ($i = 1$ and 2) elements of which are given below:

$$a_{1k}^{i} = \begin{cases} a_{1k} & \text{if} \quad 1,k = 1,2, \ \dots \ , n_i \\[2ex] \displaystyle\sum_{j \in X\backslash X_i} a_{1j} & \text{if} \quad 1 = 1,2, \ \dots \ , n_i \quad \text{and} \quad k = n_i + 1 \\[2ex] \displaystyle\sum_{j \in X\backslash X_i} a_{j1} & \text{if} \quad 1 = n_i + 1 \quad \text{and} \quad k = 1,2, \ \dots \ , n_i \end{cases}$$

   where $n_1 + n_2 = n$.
   The obtained solutions are called locally optimal solutions.

4. Find the values of coordinates between locally optimal solutions (between two subnetworks which have been found as the best for known s ets of stations and for known amount of traffic) which allow to compare the two obtained solutions and select the best one.

3. PERFORMANCE MEASURE

Let us assume that the average delay of customer encountered in the local area network communication subsystem is a measure of a solution quality [4].
The average customer delay in globally optimal solution, being a function of R_i and A is denoted by $d_g(R_i, A)$.
In the second case, i.e., when the users set is served by interconnected subnetworks, it is evident that the average customer delay within i-th subnetwork ($i = 1$ and 2) depends on selected solution (R_{ij}) and traffic matrix (A_i). To obtain the average customer delay in the two connected subnetworks, the traffic between subsets of stations (X_1 and X_2) and its service discipline should be taken into account. To evaluate the average customer delay the following delays are considered:

- $d_1(R_{1j}, A_1)$ and $d_2(R_{2j}, A_2)$ - average customers delays encountered within the first and the second subnetworks, respectively, and

- $d_{12}(S_{12}, a(X_1, X_2))$ and $d_{21}(D_{21}, a(X_2, X_1))$ - customers delays encountered on the connection between the two subnetworks on their way from the first subnetwork to the second to the first, respectively.

By S_{12} and S_{21} we denote the connection fashions between the considered subnetworks and $a(X_1,X_2)$ and $a(X_2,X_1)$ denote an average traffics between the two separeted subsets of stations. Taking into account the aforenamed delays, the average customers delay in the second network composition of two distinctive subnetworks), denoted by d_c, is given by the following expression:

$$d_c = \frac{1}{a}\left\{ a(X_1,X_1)\,d_1(R_{1j},A_1) + a(X_2,X_2)\,d_2(R_{2j}A_2) + \right.$$
$$+ a(X_1,X_2)\left[d_1(R_{1j},A_1) + d_{12}(S_{12},a(X_1,X_2)) + d_2(R_{2j},A_2)\right] +$$
$$\left. + a(X_2,X_1)\left[d_1(R_{1j},A_1) + d_{21}(S_{21},a(X_2,X_1)) + d_2(R_{2j},A_2)\right]\right\}$$

where $a = a(X_1,X_1) + a(X_1,X_2) + a(X_2,X_1) + a(X_2,X_2)$.

For known: matrix A (describes users requirements), all feasible sets of solutions $\mathcal{R}_f$, $\mathcal{R}_{if}$ and $\mathcal{R}_{2f}$ (specify all possible local area network architectures which can be applied as global and local solutions) and subnetworks interconnection manners (describe how the customers are served when flow through the interconnection device or devices) and for known, assumed, performance measure, values of d_g and d_c can be obtained. When compared, allow to select the best solution.

4. EXAMPLE

Let us discuss two networks which serve the same users community (Fig. 1). In the first case (Fig. 1a), the users set (X) is served by one bus-type network (globally optimal solution). In the second case, the users set is divided into two distinctive subsets (X_1 and X_2) (Fig. 1b). Each users subset is served by one bus-type subnetwork (locally optimal solutions). It is assumed that in both considered cases CSMA/CD (Carrier Sense

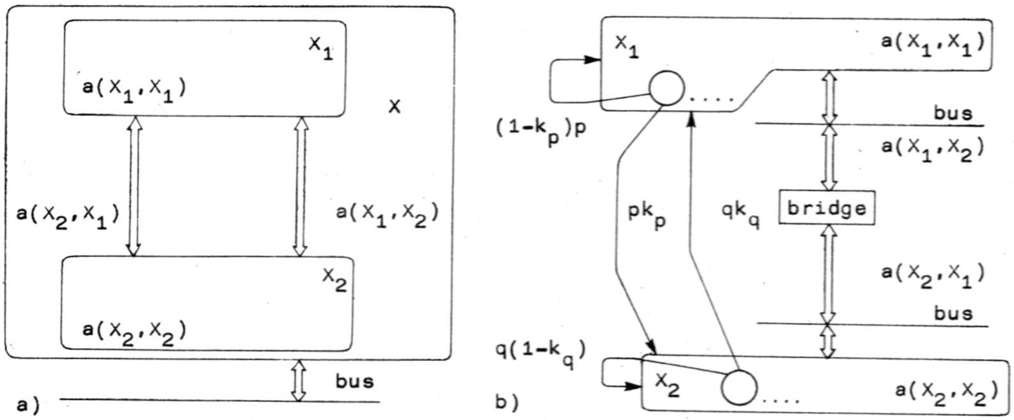

Fig. 1. Two networks compared in the discussed example

Multiple Access with collision detection) [6] access-control discipline is applied. The two subnetworks are interconnected via birdge.

It is also assumed that the users quantitative requirements are given by probabilities that the i-th user require an access to the channel (p and q for stations which belong to X_1 and X_2 subsets, respectively) (bus) in a slot (slotted non-persistent version of CSMA/CD is considered $[5,6]$). To describe the traffic between considered subsets two ratios: k_p and k_q are introduced. It means that with probability pk_p (qk_q) the access-require, originating in i-th station which belong to subset X_1 (X_2), is directed to the subset X_2 (X_1) and with probability $(1-k_p)p$ $((1-k_q)q)$ is directed to station which belong to the X_1 (X_2) subset.

The two networks, presented above, were examined and compared assuming that the average customer delay is a performance measure. To obtain analytical form of expressions for average delay, the interconnected bridge has been modelled as a tandem $M|M|1$ queueing system.

To illustrate relations between performance measure and parameters which describe considered networks and set of users, some numerical experiments have been done. They allow to compare the two discussed, in the example, networks and determine conditions in which the first, or the second, solution gives better quality of delivered communication services.

5. CONCLUSIONS

In this paper the problem of users requirements/local area network topological structure matching has been discussed. It has been shown that changes in local network architecture (globally and locally optimal solutions) may produce changes in quality of delivered services.

Results, obtained from numerical experiments, shown that the services' quality are strongly influenced by structure and values of customers streams which are exchanged among network stations as well as by assumed solution and its topological structures. In the future work some complex structures and general design tasks will be investigated in order to obtain general relations between requirements and networks architecture.

Obtained results will be presented at the conference.

6. REFERENCES

[1] L. Borzemski, A. Grzech, A. Kasprzak, L. Koszałka, "Development of distributed applications for experiment and education support based on the NETEX local area network", COMNET'85, Budapest 1985.

[2] A. Grzech, "Matching the medium access-control discipline to applications in local area network", Wissenschaftliche Beitage zur Informatik, TU Dresden, 2/1987.

[3] L. Kleinrock, "Distributed systems", Communication of the ACM, vol. 28, No. 11, November 1985.

[4] M. Reiser, "Performance evaluation of data communication system", Proceedings of the IEEE, vol. 70, No. 2, February 1982.

[5] W. Stallings, "Local network performance", IEEE Communications Magazine, vol. 22, No. 2, February 1984.

[6] F.A. Tobagi, W.B. Hunt, "Performance analysis of Carrier Sense Multiple Access with Collision Detection", Computer Networks 4, 1980.

[7] C.D. Tsao, "A local area network architecture overview", IEEE Communication Magazine, vol. 22, No. 8, August 1984.

Bicriterial Optimization of Structure of Complex Network[1]

Ewa Szlachcic [2]

1. Introduction

The rapid complexity increase of the technical systems applying the complex networks caused the necessity of the increase of the networks reliability parameters. In the case the reliability of the network's elements were insufficient the amelioration of global reliability by structural method can be realized. When the new, basic topology did not assure the reliability parameters or new structure could not be used because of structural restrictions the redundant structure for the complex network had to be applied.

In the paper some network's parameters and optimization problems will be presented in section 2. In section 3 the bicriterial, nonlinear, integer optimization problem and its properties are shown. It allows to formulate special two- level algorithm. In section 4 the results are presented.

2. Model description

Let us consider complex network with fixed, basic topology with transmission lines as arcs u_i. These lines interconnect all w_i perfect nodes using partial structure. There exists x_i spare lines for one arc u_i /group G_i/, where n – number of arcs in the basic structure [3].

In the paper the fixed, basic topology of complex network had to be assumed. The mentioned problem can be described as maximization of availability function $A(x)$ formulated as follows:

$$A(x) \longrightarrow \max \tag{2.1}$$

$$\sum_{i=1}^{n} c_i x_i \leqslant d \tag{2.2}$$

$$T(x) \geqslant T_g \tag{2.3}$$

$$\left(\forall\, i \in N_1^n\right) \left(0 \leqslant x_i \leqslant s_i\right) \quad \text{and} \quad \left(x_i\text{-integer}\right) \tag{2.4}$$

where: c_i – cost enlargement of arc u_i,

$T(x)$ – mean time among two following failure in network,

T_g – minimal permissible $T(x)$,

s_i – maximal number of redundant elements for u_i,

or minimization of cost enlargement function $C(x)$:

$$C(x) = c\,x \longrightarrow \min \tag{2.5}$$

$$A(x) \geqslant A^x \tag{2.6}$$

where: A^x – sufficient value of $A(x)$ with restrictions /2.3/ - /2.4/

1/ This research was supported by the Scientific Program PR.02 "Theory of Control and Optimization of Continuous Systems and Discrete Processes " .

2/ Institute of Technical Cybernetics Technical University, Poland 50-370 Wrocław, Wybrzeże Wyspiańskiego 27.

For the complex network with fixed or not fixed terminal nodes $A(x)$ takes the same, polynomial form once described on a set of efficient paths or on a set of efficient trees in the form [3]:

$$A(x) = \sum_{j \in N_z} r_j \prod_{k \in I_j^z} A_k(x) \qquad (2.7)$$

where: A_k – availability of group G_i with $x_i + 1$ arcs [5],
r_j – integer coefficient depending on basic structure,
N_z, J_j^z – sets of arcs index.

The terminal form of /2.7/ depends on the types of the network structure. A_k is an increasing function of x_i. It can be shown that an increase of x_i gives an increase of $A_k(x)$ and $A(x)$. Simultaneously the cost enlargement function increases. The above mentioned functions are the conflicting goals. It means it is impossible to find the reliability network's structure optimal simultaneously for two considered goals.

In many cases the solution of one of the above problems /2.1/-/2.4/ or /2.3/-/2.6/ is insufficient. As the solutions of /2.1/-/2.4/ one can obtain the optimal availability value for some different reliability structures. Frequently the final selection of structures would be accidental. It depends on values of restrictions and on the used method of solution. The solution of /2.3/-/2.6/ problem is very difficult to obtain and to fix the sufficient value A^x

3. Bicriterial optimization problem

Finally the bicriterial approach have to be used in the following form:

$$v_{max}\left\{ f(x) = \left[f_1(x), -f_2(x) \right] \,\middle|\, x \in X \right\} \qquad (2.8)$$

$$X = \left\{ x \,\middle|\, T(x) \geqslant T_g, \; x_i - \text{integer}, \; 0 \leqslant x_i \leqslant s_i \; \text{for} \; \forall i \in N_1^n \right\} \qquad (2.9)$$

where: $f_1(x) = A(x)$; $f_2(x) = C(x)$.

As the result of /2.8/-/2.9/ problem one could obtain many effective solutions taking a set of optimal solutions on Pareto definition.
For the set of effective solutions marked X_p one can described the set of Pareto goals Y_p, defined the set of values of vectorial function calculated on the set X_p as follows:

$$Y_p = \left\{ y = \left[f_1(x), -f_2(x) \right] \,\middle|\, x \in X_p \right\} . \qquad (2.10)$$

Problem /2.8/-/2.9/ has bicriterial, nonlinear, integer form. In this case it is impossible to formulate the utility function, which uses the linear order in R^k.
In the paper the whole Pareto sets X_p and Y_p are calculated.
For /2.8/-/2.9/ problem the special, two-level method using the changing restrictions and the branch and bound technique is constructed. That method uses an extention of sensibility function.

On higher level the monooptimal problem taking one goal function to the set of restrictions is realized. On the local level applying the idea of branch and bound method an algorithm for integer nonlinear programming problem is constructed [4],[5].

On higher level two problems can be obtained:

Problem I

$$\max_{x \in X} \ f_1(x) = f_1(x^o) \tag{2.11}$$

$$-f_2(x) \leqslant \triangle \tag{2.12}$$

Problem II

$$\max_{x \in X} \ -f_2(x) = -f_2(x^o) \tag{2.13}$$

$$f_1(x) \geqslant f_1(x^o) \tag{2.14}$$

Solving one of the above tasks the noneffective solutions can be received, characterized with weak bicriterial optimality. Elimination of noneffective solutions can be realized taking the condition of uniform domination in a set [2].

The set of effective solutions X_p can be received by solving the problem I or problem II on the set of restrictions /2.3/-/2.4/ with changing restriction /2.12/ or /2.14/ on the whole range of $f_2(x) \in \left[\underline{f_2(x)}, \overline{f_2(x)} \right]$ or $f_1(x) \in \left[\underline{f_1(x)}, \overline{f_1(x)} \right]$. Such order allowed to solve only one of the two considered problems.

The property of gradual saturation to one of availability function values allowed to prove some properties, which accelerate the calculating algorithm.

Property 1

If for some $x^x \in X \ f_1(x) = 1$ and it exists $\tilde{x} \in X$ such that for all $i \in N_1^n$ it exists one $j \neq i$ and $\tilde{x}_i \geqslant x_i^x$ and $\tilde{x}_j > x_j^x$ then

$$f_1(\tilde{x}) = 1 \ \text{and} \ f_2(\tilde{x}) \geqslant f_2(x^x).$$

All vectors $\tilde{x}$ with the above property could be rejected from the set of solutions.

Let $x' \in X$ and $x'' \in X$ and $P^x(x^o) = \left(x^o, f_1(x^o), f_2(x^o)\right)$ the solution of /2.1/-/2.4/ then the property 2 can be formulated as follows:

Property 2

If $P^x(x') = (x', f_1(x'), f_2(x'))$ and $P^x(x'') = (x'', f_1(x''), f_2(x''))$ there exists $x \in X$ such that for $d' \leqslant d''$ then $f_1(x') \leqslant f_1(x'')$ but only for vectors $x \in X$ like: for all $i \in N_1^n$ there exists $j \neq i$ such that $x_i' = x_i''$ and $x_j' < x_j''$. The proofs of all above properties are shown in [5].

The above properties allowed to solve one optimization problem /2.11/-/2.12/ or /2.13/-/2.14/. The simple linear form of cost enlargement function leaded to take the availability function to the set of restrictions. Complicated, polynomial form of $f_1(x)$ with expressions like C_a^b made using the nonlinear, continuous programming methods impossible. $f_1 x$ is not sensitive for change of vector x and it is very difficult to create the structure of network sufficient for restriction /2.14/. All difficulties leaded to solve the problem I.

At last the bicriterial optimization problem /2.8/-/2.9/ was changed
to monoptimal problem /2.1/-/2.4/. Changing restrictions method can be
used solving some problems /2.1/-/2.4/ on the whole range of cost enlar-
gement values $f_2(x) \in \left[\, f_2(x),\ \overline{f_2(x)}\, \right]$. In each great iteration the effective solution is obtained using the special algorithm for the lower le-
vel, applying the idea of branch and bound metodology. In the local
problem the recurrent procedure of estimating the least upper bound of
availability function for integer variables is proposed [4].

4. Results

The computer program calculating the set of effective solutions and
the set of Pareto goals was written in FORTRAN - 1900 [4],[5]. The number of effective solutions strongly depends on restriction /2.3/. It is
why in the numerical experiences the low value T_g was taken to receive
the numerous set of effective solutions, Having regards to the received
effective solutions structures it can be shown that it is not necessary
to apply one redundant element for each basic arc in the complex network
The same or better value of availability function and the lower cost value can be obtained for the calculated structures using the lower number of arcs.

The proposed bicriterial optimization problem is specially actual for
the discussed type of availability function. That function is not very
sensitive for the change of vector x, while the cost enlargement function change considerably. Having regards to the properties of the bicriterial problem the changing restrictions method can be used to received the set of Pareto solutions. The properties of discussed vectorial problem gave the possibility of construction two-level algorithm.

Literature

[1] Banerjee, S.K.: Optimal redundancy allocation for non series-paral-
lel network. IEEE Trans. on Reliability. R-25 /1976/ 6, 149-157.

[2] Bowman, V.I.: On the relationship of the Tchebycheff norm and the
efficient frontier of multiple-criteria objectives. Proc. of conf.
Multiple Objective Decision Problems. Jonay-en-Josas. France /1974/.

[3] Szlachcic, E.: The reliability structures optimization of the computer networks. Proc. of conf. RELCOMEX. Książ Castle, Poland
/1984/, 337-384.

[4] Szlachcic, E.: Metoda poszukiwania optymalnej struktury niezawodnoś-
ciowej dla sieci złożonych. Arch. Autom. i Telem/1988/ [printed].

[5] Szlachcic, E.: Dwukryterialna optymalizacja struktury niezawodnoś-
ciowej na przykładzie sieci komputerowej metodą zmiennych ograni-
czeń. Preprint 83/81. Politechnika Wrocławska /1981/.

OPTPACK – An Interactive Optimization Software Package for Personal Computers

V. Sima [1]

1. Introduction

OPTPACK is a comprehensive series of high-quality integrated software programs for linear and nonlinear optimization available for professional/personal computers. The package is user-friendly and operates in an interactive-conversational mode, including help options.

OPTPACK provides many procedures for solving various optimization problems. The computational facilities offered by the package include :

- unconstrained optimization (via cvasi-Newton methods, with or without analytical derivatives, or via conjugate gradient algorithm with restart) ;

- nonlinear least squares (via a derivative-free Levenberg-Marquardt algorithm) ;

- linear programming (via the revised simplex algorithm) ;

- strictly convex quadratic programming (via numerically stable dual algorithm proposed recently [2]) ;

- nonlinearly constrained optimization (via Powell's constrained variable metric algorithm [4], incorporating the watchdog technique [1]).

Efficient and well-tested computational subroutines from the IMSL library [3] are used. The codes for quadratic programming and for non-linearly constrained optimization are new or substantially improved, respectively, and include features to enhance the numerical proprieties (e.g., updating matrix factorizations by orthogonal transformations), the efficiency and flexibility [5].

An important design objective of the package has been to make the powerful and sophisticated optimization software available to even an unexperienced user. OPTPACK operates in a user-friendly menu-driven environment. Many facilities are included to assist the user in various stages of the optimization problems solving process: entering/editing data and options, programming the problem functions (objective function, constraints functions, and their derivatives, if necessary), testing the consistency of user's defined derivatives, printing the (intermediary) results, assessment of final results and performances. A specialized matrix-oriented "data-base" is attached to the package and appropriate programs to enter, edit, list, or delete matrices are provided. Facilities for optimization problem defining, editing and automatic solving are included.

Previous computer experience is almost unnecesary. The user has only to edit eventually some fragments of Fortran code which define the

[1] Research Institute for Computers and Informatics, Bd.Miciurin, No 8-10, 71316 Bucharest, Romania.

concrete problem functions. The mechanics of OPTPACK can be learn
easily. A self-documenting menu-driven environment guides the user
through the solution process.

OPTPACK is written almost entirely in a portable Fortran IV
language and operates in a CP/M environment. The first version of the
package is implemented in double precision arithmetic on the Romanian
family of 8-bits microcomputers. An IBM-PC compatible version is under
development.

2. OPTPACK structure and operation

OPTPACK system includes four main components: Multilevel Exploitation Subsystem (MES), Database Manager Subsystem (DMS), Problems
Manager Subsystem (PMS), and Optimization Solvers Subsystem (OSS), which
provide together a full range of capabilities for handling optimization
problems and data, solving such problems and presenting the results.

OPTPACK MES creates the environment under which data or problem
management and optimization programs may be loaded and executed. MES is
an extension of the CP/M operating system which ensures an easy work
with menu trees and offers some additional facilities over the standard
CP/M functions. These facilities determine in fact the main operational
attributes of the OPTPACK package :
- select the desired function using menus ;
- guide the user through the menu tree by help functions ;
- solve completely a problem, from its specification up to
  displaying the results.

When a menu is displayed on the terminal screen, by pressing an
appropriate figure or letter, the user has the possibility to execute
a program, to obtain information about the system and programs, and to
advance or to return in any point of the menu tree. A position in a
menu corresponds to either another menu or to an application program.
In the last case, the control is given to the selected program and this
becomes responsible for the subsequent conversation.   When the program is completed, the control is returned to the position in the menu
tree on which this program was called.

To load OPTPACK nucleus, the user must enter the command
SET OPTPACK
and press the "RETURN" key. The system will respond with the following
menu

OPTPACK Master Menu

```
- - - - - - - - - - - - - - - - - - - - - - - - - - - - - - - - - -

  0    Data Base Manager              1    Problems Manager
  2    Optimization Solvers
- - - - - - - - - - - - - - - - - - - - - - - - - - - - - - - - - -

  H - Help                           E - Exit
  Press "Return" for previous menu

  Choice  --> [ ]
```

and then it will be waiting for the user respond to specify the option.
By pressing 0, 1, or 2, the corresponding menu will be activated. The
"Exit" option permits to leave OPTPACK and to return immediately to the
operating system.

OPTPACK DMS is responsible for all "external" matrix data
handling. Matrix is the essential data structure for linear and quadratic programming solvers. DMS creates all matrix data files used by these
solvers. Thus, the data may be listed, verified and, if necessary,
edited prior to use. The main available options are listed in the following menu.

OPTPACK Database Manager Menu [1]

- -

0	Enter Matrix	1	List Matrix
2	Edit Matrix	3	Edit Matrix File Header
4	Delete Matrices	5	Precision Conversion
6	Delete Rows	7	Vertical Augment

- -

Four types of matrix storage modes are provided: general, symmetric, band and band symmetric [3]. In addition, a conversion facility,
from single to double precision and conversely, is included.

Any matrix has an associated header file, containing information about its dimensions, storage mode, column names and matrix label
etc. Part of these information, as well as all the numerical values of
the matrix entries can be easily modified.

OPTPACK PMS handles problem information needed for nonlinear
optimization (problem functions, problem dimensions, parameters and
options) and generates the corresponding executable program. The main
available functions are listed in the following menu.

OPTPACK Problems Manager Menu

- -

0	Enter/Edit Problem Data	1	Enter/Edit Problem Functions
2	Compile User's Subroutine	3	Task Building

- -

Any nonlinear optimization problem has an associated problem
header file, containing general information and options for solving it,
depending on problem category. The Enter/Edit Problem Data option creates or updates the problem header file in a conversational mode.
Default values are provided whenever possible. By selecting this option,
another menu will be displayed containing distinct entries for each
problem category.

The problem functions are defined in a subroutine edited by
the user. OPTPACK offers also the source codes of some "model" subroutines for each category. The segments of codes which must be modified

---

[1] The H, E and "Return" options have not been displayed here and in
the following menus.

are clearly marked. The Enter/Edit Problem Functions option creates or updates the user's subroutine.

Compile User's Subroutine option generates the object code of the subroutine defining problem functions, and Task Building option links this to the associate main program and to the library modules, creating the executable program for each concrete nonlinear problem.

OPTPACK OSS contains a comprehensive collection of optimization procedures. The main available options are listed in the following menu.

OPTPACK Optimization Solvers Menu

0	Unconstrained Quasi-Newton	1	Unconstrained Conj. Gradients
2	Nonlinear Least Squares	3	Linear Programming
4	Quadratic Programming	5	Constrained Quasi-Newton
6	Test Analytical Derivatives		

The Unconstrained Quasi-Newton option solves an unconstrained minimization problem using the algorithms implemented in subroutines ZXMIN [3], or QNDER [5]. The Unconstrained Conjugate Gradients option is based on ZXCGR [3]. The Nonlinear Least Squares option uses an algorithm implemented in ZXSSQ [3]. The Linear Programming option is based on ZX3LP [3]. The Quadratic Programming option solves a strictly convex problem by an extension of the algorithm in [2], implemented in PODPP [5]. The Constrained Quasi-Newton option solves a general nonlinearly constrained problem using VMCWDM and the related subroutines from [5]. The Test Analytical Derivatives option offers a quick check if the derivatives programmed by the user are consistent to the programmed functions.

Due to its capabilities, OPTPACK is a valuable tool in scientific research, engineering design, optimization techniques teaching and evaluation.

3. Literature

[1] Chamberlain, R.M., C. Lemaréchal, H.C. Pedersen, and M.J.D. Powell: The watchdog technique for forcing convergence in algorithms for constrained optimization. Math. Prog. Study, 16 (1982) 1-17.

[2] Goldfarb, D., and A. Idnani: A numerically stable dual method for solving strictly convex quadratic programs. Math. Prog. 27 (1983) 1-33.

[3] *** IMSL Library Reference Manual, IMSL Inc., Houston 1980.

[4] Powell, M.J.D.: VMCWD: A Fortran subroutine for constrained optimization. Report DAMTP 1982/NA4, University of Cambridge 1982.

[5] Sima, V., and A. Varga: Computer Aided Optimization Practice (in Romanian). Editura Technică, Bucureşti, 1986.

Approximate Performance and Sensitivity Analysis of Closed Queueing Networks

M.Aicardi, F.Davoli, R.Minciardi *

Abstract. A new heuristic algorithm called QNA (Queueing Network Analyzer), has been proposed by the authors for the evaluation of the average performance measures in a queueing network. This algorithm iteratively computes mean queue lengths, throughputs and machine utilizations by means of heuristically derived formulas. The same approach is extended to the computation of the partial derivatives of the same quantities with respect to the parameters of the routing policies. This can be accomplished by considering aside to the basic algorithm an additional set of equations which give, at convergence, the desired sensitivity coefficients.

1. Introduction

The field of approximate performance analysis of closed queueing networks is recognized as a promising research area, from an applicative point of view, in connection with several types of physical systems (e,g, transportation systems, communication networks and flexible manufacturing systems). The most commonly considered among approximate analysis methods is the so-called Mean-Value Analysis [1], which is a heuristic version of an exact algorithm originally developed by Reiser and Lavemberg [2]. The main drawback of this approach is related with the exponential services assumption, on the basis of which the method is derived. To overcome this fact, some other heuristic approaches have been proposed, which attempt at facing the problem of analyzing queueing networks with general or even deterministic processing times [3],[4].

In this connection, in [5] the authors have proposed an original algorithm whose development has been accomplished with the case of deterministic times in mind. The iterative algorithm has the scope of determining, in the convergence point, the mean values representing the performance measures.

Experimental tests carried out on the algorithm, have brought into evidence that the proposed method, which has been called QNA (Queueing Network Analyzer), is characterized by a good accuracy (its results have been compared with those provided by a conventional simulator [5]). This fact encourages to go further by enriching the basic algorithm with the capability of iteratively computing the partial derivatives (sensitivity coefficients) of the performance measures with respect to some parameters of the model. This is just the scope of this paper, with particular reference to the parameters characterizing the routing strategies, which we suppose to be of stochastic state-independent type.

The organization of the paper is the following. In the next section we shall detail the considered system model and recall the basic equations of the QNA. In the third section, we present the original part of the paper, related with the computation of the above sensitivity coefficients.

2. Network model and basic algorithm

We refer to a closed queueing network where:
a) every service center (machine) has infinite storage capacity;
b) the customers (pieces) transportation system requires zero transportation time for any couple of machines;
c) local scheduling is of FIFO type;
d) customers are stochastically routed according to a-priori fixed routing frequencies;
e) execution times are deterministic and depend only on the type of the piece, the operation to be performed and the machine where the operation is executed.

---

* The authors are with the Department of Communications, Computer and Systems Science, University of Genoa, Via Opera Pia 11/A, 16145, Genova, Italy.

Let us define:

M=number of service centers;

S(m)= set of servers of machine m with cardinality $\sigma(m)$;

P=number of part-types to be processed;

N(p)=number of pieces belonging to class p, p=1,...,P;

O(p)= ordered sequence of operations that have to be performed on class p pieces (without loss of generality we suppose that each operation in encountered only once in each operation sequence);

T(p,o,m)= deterministic processing time needed by machine m to execute operation o on class p workpieces, m=1,...,M, $o \in O(p)$, p=1,...,P (without loss of generality, we suppose T(p,o,m) > 1 $\forall$p, $\forall$o, $\forall$m);

λ(p,o,m)= probability of assigning to machine m the execution of operation o on a piece belonging to class p, m=1,...M, $o \in O(p)$, p=1,...P.

The scope of the algorithm is that of providing extimates of the following classical performance measures:

Q(p,o,m)=mean queue length relevant to workpieces of class p, on machine m, and operation o;

X(p,o,m)= mean execution rate (throughput) for machine m performing operation o on workpieces of class p.

The basic algorithm can be structured as follows:

<u>Step 1 (initialization)</u>: guess some initial values for Q(p,o,m), $\forall$p, $\forall$o, $\forall$m, such that

$$\sum_{m=1}^{M} \sum_{o \in O(p)} Q(p,o,m)=N(p) \qquad\qquad \forall p$$

<u>Step 2:</u> compute U(m), $\forall$m such that $\sigma(m) \neq \infty$, via the following equation

$$U(m)= \min \left\{1, \tilde{U}(m)\right\} \overset{\Delta}{=} \min\left\{1, \frac{\displaystyle\sum_{p=1}^{P} \sum_{o \in O(p)} Q(p,o,m)\, T(p,o,m)}{\sigma(m) \displaystyle\max_{p,o} T(p,o,m)}\right\} \qquad (1)$$

<u>Step 3:</u> compute X(p,o,m) $\forall$p, $\forall$o, $\forall$m, via equations

$$X(p,o,m)= \frac{Q(p,o,m)\, \sigma(m)\, U(m)}{\displaystyle\sum_{p=1}^{P} \sum_{o \in O(p)} Q(p,o,m)\, T(p,o,m)} \overset{\Delta}{=} \frac{Q(p,o,m)\, \sigma(m)\, U(m)}{\tilde{T}(m)} \qquad (2)$$

$$\text{if } \sigma(m) \neq \infty$$

$$X(p,o,m)= \frac{Q(p,o,m)}{T(p,o,m)} \qquad\qquad \text{if } \sigma(m) = \infty \qquad (3)$$

<u>Step 4:</u> compute the new set of values of Q(p,o,m) $\forall$p, $\forall$o, $\forall$m, via

$$Q^{new}(p,o,m)=Q^{old}(p,o,m) + \lambda(p,o,m) \sum_{k=1}^{M} X(p,1=\mathcal{O}_p(o),k) - X(p,o,m) \qquad (4)$$

where $1 = \mathcal{P}_p(o)$ is the operation preceding o in the sequence $O(p)$.

<u>Step 5:</u> if $\displaystyle\max_{p,o,m} \left| Q^{new}(p,o,m) - Q^{old}(p,o,m) \right| \leq \varepsilon$ then stop; else go to 1.   ◁

As pointed out in [5], equations (1),(2),(4) can be intuitively justified. Namely, equation (4) is a sort of mass balance, whereas (2) can be derived by considering the ratio between the mean queue content (seen as an actual queue length) and the corresponding emptying time. Finally, (1) comes out if one claims that the system eventually reaches a steady-state behaviour where the utilization of each machine is the maximum compatible with the determined queue lengths.

3. Sensitivity analysis

Whenever the above equations (1)-(4) are accepted as true, one can think to use them even to derive relations involving sensitivity coefficients (with respect to the routing parameters), of the quantities representing the system mean behaviour. To this end let us define the following set:

$$\Gamma(p,o) = \text{the set of machines with } T(p,o,m) \neq 0 \quad \text{(with cardinality } \gamma(p,o))$$

where we have implicitly admitted that $T(p,o,m)=0$ represents the incompatibility of machine m with operation o on part-type p. Let us consider, for each (o,p), the following set of independent variables

$$\lambda(p,o,m), \ m \in \widetilde{\Gamma}(p,o) \tag{6}$$

being $s(p,o)$ the largest machine index in $\Gamma(p,o)$, and $\widetilde{\Gamma}(p,o) = \Gamma(p,o) - \{ s(p,o)\}$. Obviously, $\lambda(p,o,s(p,o))$ is not an independent variable since it is given by

$$\lambda(p,o,s(p,o)) = 1 - \sum_{m \in \widetilde{\Gamma}(p,o)} \lambda(p,o,m) \tag{7}$$

Then, considering any $\lambda(p,o,j)$, $j \in \widetilde{\Gamma}(p,o)$, we derive the following equations from (1),(2),(3),(4), respectively.

$$\frac{\partial U(m)}{\partial \lambda(p,o,j)} = \left[\sum_{r=1}^{P} \sum_{u \in 0(r)} \frac{\partial Q(r,u,m)}{\partial \lambda(p,o,j)} T(r,u,m) \right] \frac{1}{\sigma(m) \ \max\limits_{p,o} \ T(r,u,m)} \qquad \text{if } \widetilde{U}(m)<1$$

$$= 0 \qquad\qquad\qquad \forall m \qquad\qquad \text{if } \widetilde{U}(m)>1 \tag{8}$$

if $\widetilde{U}(m)=1$, the partial derivatives does not exist and only left and right partial derivatives can be evaluated.

$$\frac{\partial X(r,u,m)}{\partial \lambda(p,o,j)} = \frac{\sigma(m)}{\widetilde{T}(m)} \left[\frac{\partial Q(r,u,m)}{\partial \lambda(p,o,j)} U(m) + Q(r,u,m) \frac{\partial U(m)}{\partial \lambda(p,o,j)} - \frac{Q(r,u,m) \ U(m)}{\widetilde{T}(m)} \right.$$

$$\left. \sum_{r=1}^{P} \sum_{u \in 0(r)} \frac{\partial Q(r,u,m)}{\partial \lambda(p,o,j)} T(r,u,m) \right] \qquad \text{if } \sigma(m) \neq \infty \tag{9}$$

$$\frac{\partial X(r,u,m)}{\partial \lambda(p,o,j)} = \frac{\partial Q(r,u,m)}{\partial \lambda(p,o,j)} \frac{1}{T(r,u,m)} \qquad \text{if } \sigma(m) = \infty \tag{10}$$

$$\forall r, \ \forall u, \ \forall m$$

$$\frac{\partial Q^{new}(r,u,m)}{\partial \lambda(p,o,j)} = \frac{\partial Q^{old}(r,u,m)}{\partial \lambda(p,o,j)} - \frac{\partial X(r,u,m)}{\partial \lambda(p,o,j)} + \lambda(r,u,m) \sum_{k=1}^{M} \frac{\partial X(r,1=\mathcal{P}_r(u),k)}{\partial \lambda(p,o,j)} +$$

$$+ \delta(r-p,u-o,m-j) \sum_{k \doteq 1}^{M} X(r,1=\mathcal{P}_r(u),k) \qquad \begin{array}{l} \text{if } m \neq s(p,o), \forall r, \forall u \\ \text{if } m=s(p,o), \forall(r,u) \neq (p,o) \end{array} \qquad (11)$$

(with $\delta(h,k,1)=1$ if $h=k=1=0$, $\delta(h,k,1)=0$ otherwise)

$$\frac{\partial Q^{new}(p,o,s(p,o))}{\partial \lambda(p,o,j)} = \frac{\partial Q^{old}(p,o,s(p,o))}{\partial \lambda(p,o,j)} - \frac{\partial X(p,o,s(p,o))}{\partial \lambda(p,o,j)} +$$

$$+ \left[1 - \sum_{v \in \tilde{\Gamma}(p,o)} \lambda(p,o,v) \right] \sum_{k=1}^{M} \frac{\partial X(p,1=\mathcal{P}_p(o),k)}{\partial \lambda(p,o,j)} - \sum_{k=1}^{M} X(p,1=\mathcal{P}_p(o),k) \qquad (12)$$

Equations (8),(9) and (10), even if they appear complex, do not require any particular consideration about their meaning. On the other hand, some remarks are necessary for equations (11) and (12). Actually, these equations naturally suggest the definition of an iterative algorithm which, in convergence, yields the partial derivatives

$$\frac{\partial Q(r,u,m)}{\partial \lambda(p,o,j)} \qquad \forall p, \forall o, \forall m, \forall r, \forall u, \forall j$$

These quantities, together with the equilibrium point of the basic QNA, allow us to compute, at the convergence point, the sensitivity coefficients

$$\frac{\partial X(r,u,m)}{\partial \lambda(p,o,j)} \quad ; \quad \frac{\partial U(m)}{\partial \lambda(p,o,j)} \qquad \forall p, \forall o, \forall m, \forall r, \forall u, \forall j$$

related with the considered configuration of the system. Without formally defining the structure of the sensitivity analysis algorithm (which of course resembles most of the features of the basic one), it can be easily understood that, due to the structure of the involved equations, the execution of the algorithm can be started either together with the basic QNA, or once QNA has reached its equilibrium point.

Moreover, note that, as a consequence of the constraint appearing in the Step 1 of the basic QNA, we must initialize the queue sensitivity coefficients in such a way that

$$\sum_{m=1}^{M} \sum_{u \in O(r)} \frac{\partial Q(r,u,m)}{\partial \lambda(p,o,j)} = 0 \qquad \forall p, \forall o, \forall j, \forall r$$

4.Conclusions

A sensitivity analysis algorithm has been proposed, taking the moves from a previously introduced one regarding the approximate analysis of closed queueing networks. The sensitivity analysis is based on an iterative procedute to compute the partial derivatives of the quantities representing the mean system performance, with respect to parameters defining the stochastic routing strategies. Basing on the same approach, sensitivity analyses with respect to variations of other system parameters, like stochastic routing parameters or machine speeds, are matter of present research.

References

[1] P.Schweitzer,"Approximate analysis of multiclass closed network of queues", Proc. Int. Conf. on Stochastic Control and Optimization, Amsterdam, The Netherlands,1979

[2] M. Reiser and S.S. Lavemberg,"Mean Value Analysis of closed multichain queueing networks", Journal of the A.C.M. , vol. 27 (1980),313–322.

[3] D.D. Yao and J.A. Buzacott,"The exponentialization approach to Flexible Manufacturing Systems models with general processing times", European Journal of Operation Research, vol. 24 (1986), 410–416

[4] J.B. Cavaille and D. Dubois," Heuristic methods based on Mean-Value Analysis for Flexible Manufacturing Systems performance evaluation", Proc. 21st IEEE Conf. on Dec. and Control, Orlando, Florida, 1982.

[5] M.Aicardi and R.Minciardi,"A new approach to performance analysis of closed queueing networks", IEEE Int. Conf. on C.I.M., Troy, N.Y.,,May 1988.

Computer-Aided Asynchronous Synthesis Procedure

DR. SANDOR VINCZE

Microelectronics Company /MEV/

1325 Budapest P.O.B. 21,Hungary

KEY WORDS:asynchronous logical design procedure,flow diagram,hazards, independent from speed,multiple input changes,programmable logical structure.

The applied asynchronous logical design procedure /1/ is based on a flow diagram.The operational sequence is divided into so-called phases.The phase transitions are triggered by the changes of the input combinations important to the output sequence.The transition 0 - 1 in the value of the triggering function F indicates the phase transition if the enabling function G enables the phase transition.

The hardware structure realised uses fixed building blocks.The important block is the asyncronous phase register in which the phases are coded in " 1 out of N ". The combination network block can be realised with PLA.The network designed by the procedure is free from essential hazards and critical race conditions, and can be considered independent from speed. The procedure takes the multiple input changes into account.

The PLS$^+$ synthesis program

The program consists of the following parts :

- Entering the flow diagram into the computer on a task description language /TDL/;
- Syntactical analysis of the TDL;
- Verifying the formal properties of the flow diagram;
- Procedure specific fast and simple logical simulation;
- Programming the hardware structure on the metallization mask of a PLS .

On the next page there is an example of the flow diagram.

The applied triggering and enabling functions are the followings :

FOO = XO,FO1 = XO X X1, GO1 = X2,F10 = X1 X X2,F20 = X1 + X2,G20 = XO X X1.

The syntax and the structure of the TDL can be found in the reference /6/.

PLS$^+$: Programmable logical structure

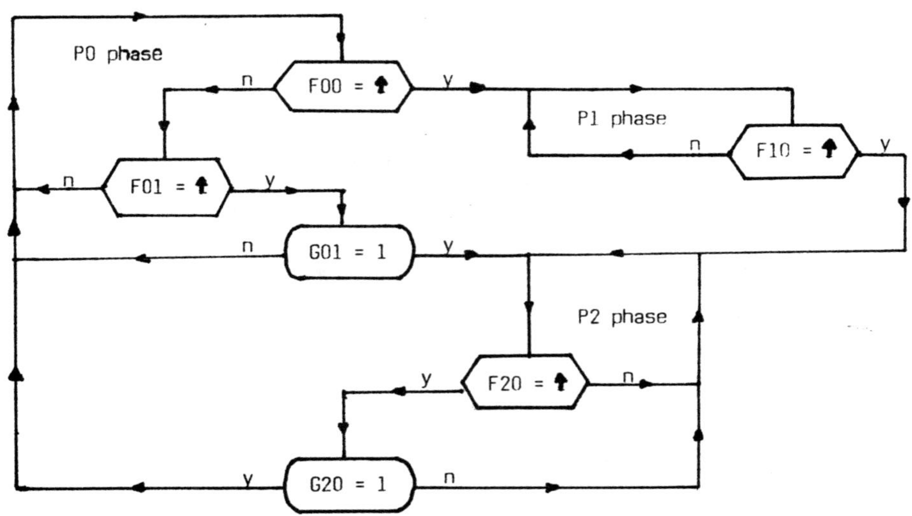

The respective TDL is:

```
TDL   MAIN        INS    XO,X1,X2.             SNI
                  PHA    PO,P1,P2.             AHP    NIAM
      PH    PO    FD     FOO,NEG XO.
                         FO1,XO AND NEG X1.    DF
                  GD     GO1,X2.               DG
                  FGNPH  FGOO,P1.
                         FGO1,P2.              HPNGF  HP
      PH    P1    FD     F10,X1 AND  X2.       DF
                  FGNPH  FG10,P2.              HPNGF  HP
      PH    P2    FD     F20,X1 OR X2.         DF
                  GD     G20,XO AND NEG X1.    DG
                  FGNPH  FG20,PO.              HPNGF  HP   LDT
```

Verifying the formal properties

The program for verifying the formal properties of the flow diagram
examines if certain conditions are fulfilled by the functions F and G.

a./ In the just existing phase the output of any function G has to remain
unchanged, if a transition 0 - 1 occurs in the output of the respective
function F.

b./ It is true for any actual phase that out of those functions F
participating in the determination of the phase whose pair functions G
are just enabling some phase transition always only on the output of one
single function F can a transition 0 - 1 take place upon the effect of
the given change of input combination.

c./ If in any actual phase, a transition 0 - 1 indicating a phase transition
arises on the output of any of the functions F determining the phase, no
transition 0 - 1 may arise on the output of any of those functions F
taking part in the determination of the thus determined next phase whose
function G pairs are just enabling some phase transition.

The cases which do not correspond to the properties mentioned above
will be taken into a so-called exclamation mark list/!-LIST/. Afterwards it
is examined if the states in the "!-LIST" can be reached starting from a
base state applying only adjacent input combinations /xk - xl/.Supposing
that they are accessible the formal condition derived from the properties
a.,b.,c., for the flow diagram can not be fulfilled.In this case the program
shows which part of the flow diagram is not correct formally.After
correcting the error, the examination of the formal properties have to be
repeated for the whole flow diagram.

The procedure specific logical simulation

Networks designed by procedure /1/ are free from hazards.Thus a fast
and simple simulation model can be created /6/ :
- Algorythmic /3/ level simulation;
- State representation with a two-level logic;
- The zero transitional delay /4/ used for timing of microevents;
- Both synchronous and asynchronous simulation.

According to the original interpretation the systematic procedure
/1/ operates with triggering instructions scheduled asynchronously.Of course
synchronous networks can also be created in the special case when the
triggering instructions are controlled by a synchronous clock signal.
Description of the logical simulation procedure:

The designer should assign unambiguously a simulation output
combination sequence to the input combination sequence concerned.The network
designed by the procedure can be evaluated by comparing the predefined
output combination sequence with the computed one.

Remarks:

1., The operation can only be examined starting from a real initial state
i.e. by switching on the power supply.This state is considered as the
initial state and phase for TDL.

2A.,The logical simulation algorythm for the case of changes in adjacent
input combination can be seen in reference /6/.

2B.,In asynchronous cases /1/ the multiple input changes should be divided
into adjacent input combination changes.The simulation should be done
for every possible combination obtained.

3., The networks designed by the procedure are free from any hazards.

Realization of hardware structure on PLS

A design of array type ICs seems to be the most economical, the realisation efficiency of the hardware structure /chip-size,degree of automation/ appeares to be the best in the case of a new design.Because of this a metal programmable master PLS chip design has been developed on an N-channel polysilicon gate HMOS technology at MEV .

The design of a / 18 X 24 X 8 / master PLS chip will be presented during the lecture. We are going to draw attention to the design of base elements like the PLA base cell /5/ or the phase register element as well as to the formation of the PLS structure.This research work has been carried out on the VLSI IC design system /2/ of MEV.

Conclusion:

So far the program for verifying the formal properties of the flow diagram has been implemented in "C" programming language /5/.This program runs on the PDP 1140 minicomputer under the UNIX operating system and is under laboratory testing.For realising the hardware structure a master PLS chip has been developed.The computer implementation of the PLS synthesis procedure results in fast and reliable design.The other programs of the program system are under development.

References:

/1/ S.Terplan : A Design Method of Asyncronous Sequential Circuits Based on
Flow Diagram
MTA SZTAKI Reports 137/1982 , Budapest.

/2/ A.Asztalos,L.Márkus : Some Elements of the IC Design System in the
Microelectronics Company /in Hungarian/
MP'82 Seminar,Pécs 1982,Hungary

/3/ M.R.Barbacci : Sintax and Semantics of CHDLs
North-Holland Publishing Company 1981.

/4/ M.A.Breuer : Diagnosis and Reliable Design of Digital Systems
Computer Science Press,INC 1976.

/5/ S.Vincze : Some Aspects of Computer Implementation of the PLA Synthesis
procedure.
Custom Circuit Conference 87 Gyöngyös,12-15 May 1987,Hungary

/6/ S.Vincze : Some Aspects of the Computer Implementation of the Logical
Simulation for the PLA Synthesis Procedure
European Congress on Simulation,Prague,20-24 September 1987
Czechoslovakia.

Performance Evaluation of Communication Services in CIM Environments

Klaus Irmscher [1]

Abstract

Communication services provide a supposition for manufacturing automation. They follow in general on base local area networks using
standard protocol proposals in CIM environments (MAP or TOP e.g.).
For performance evaluation modelling investigations are performed.
The foundation forms a three-layer-model. By using Norton's theorem
decomposition methods are applied (equivalence networks, isolation
techniques etc.). For support the analyzing tool LANEX is developed.

1. Communication processes in manufacturing automation

The progressive development of microelectronics and computer technique in connection with communication techniques enables a general
automation in industrial processes (manufacturing, design, planning,
management etc.). By using the increased computerization the aim
consists in a far extending perfect automation of all processes. The
realization follows by a high automatized factory, comprehended in
the definition of CIM (Computer Integrated Manufacturing).
An important role takes in the communication system in an automatized manufacturing system. It realizes the communication between man,
computers, computerized machines and between CAD/CAM-, CAP-, CAQ-
systems and management information systems. Communication techniques
provide the connection of decentralized computers and computerized
machines as well as the access about local or remote terminals or
work stations on base existing or new communication establishments
(e.g. telephone line, coaxial cable, fibre optic cable) /5,7,9/.
Following demands are placed to communication systems in CIM-envi-
ronments:
- data transfer between CAD and CAM stations, in the CAQ system
  and in the management information system,
- program transfer, e.g. remote program loading from a program data
  base into a metal-cutting machine.
- office communication, for instance mailbox-service,
- exchange of technical documents and designs,
- operator communication,

---

[1] Technical University at Dresden, Informatic Centre, Department
of Computer Systems. Mommsenstr. 13, Dresden, 8027, GDR

- exchange of control informations between decentral manufacturing
  processes,
- distributed processing.

2. Local area networks

The realization of communication services in manufacturing automation favourable follows by local area networks (LAN). They posses
generally a ring or bus structure. As fundamental architectural concept for LAN serves the OSI-reference model of ISO (ISO7498). For
CIM-typical applications two communication model proposals are developed international, both founded on the ISO standard communication
protocols:
1. MAP (Manufacturing Automation Protocol):
    - service for manufacturing control,
    - token-passing-bus medium access control (deterministic),
    - broadband transmission (coaxial cable);
2. TOP (Technical Office Protocol):
    - service for exchange technical documents and for management
      information system,
    - CSMA/CD medium access control (stochastic),
    - baseband transmission.

The quantity of data to be transmitted and the response times (in
range of manufacturing cell with real-time condition) give the industrial communication requirements. The size of messages or files
differs from 1 to 500 Kbyte, of graphical datas (CAD) from 1 to
10 Mbyte. The values of response time lies between 0.01 and 10 sec.
From this follow throughput requirements in the range from 0.1 to
10 Mbit/sec (in detail see /5,9/).
For manufacturing-oriented transmission processes with real-time conditions token-passing bus or ring nets are used whereas CSMA/CD-LAN's
(e.g. Ethernet) are applied for office communication transmission
processes. International is oriented on 10 Mbit/sec-LAN's.

3. Performance evaluation of communication services

By reason of high load and meaning of the communication services
methods are required to model and analyze the communication system
in order to determine performance quantities (e.g. throughput, mean
response time) or structure state.
Behavior models (analytical, simulative, empirical or hybrid) are

used to determine the performance quantities of existing or to dimensioned systems. The in computers and communication systems happened
processes are often stochastically, so that beside simulative or statistical techniques methods of probability theory frequently are applied. For determination of performance quantities as throughput, utilization, mean waiting time etc. the queueing theory is used in performance evaluation. Single queueing systems with different distribution
laws of probability and queueing disciplines and queueing networks
(opened, closed or mixed) with one or multiple classes of customers
are put in. This fact is supported by presence of corresponding analyzing tools (e.g. BNETD, PANACEA, QNET4, RESQ, PAWCS, SATURN) /3.4/.
Foundation of modelling of distributed systems on LAN-base forms a
three-layer-model (see fig. 1):

- data processing layer: it represents the distributed processing using
  the application services in a local area network (in model replaced
  by a closed queueing network);
- data transport layer: it characterizes the aggregation of the upper
  layers in the network reference model esspecially protocol and
  transport functions (modelled in good approximation by queueing
  networks);
- data transmission layer: it describes the transmission-typical processes in a LAN: connectionless or connection-oriented data transmission, different medium access control techniques (e.g. CSMA/CD,
  token-passing bus medium access control) and data transmission
  channels (e.g. coaxial or fibre optic cable); in this layer different
  queueing systems are applied.

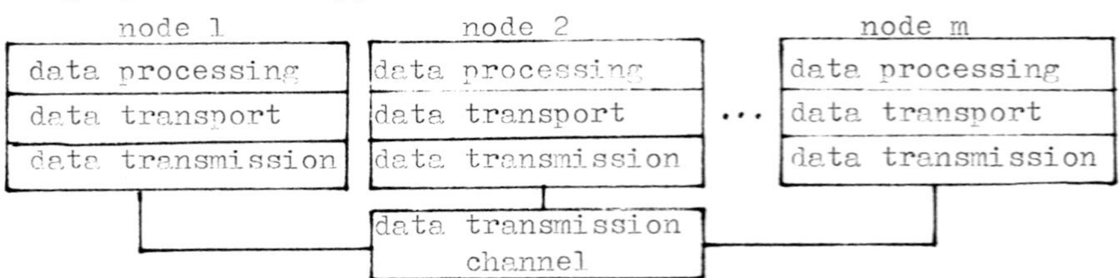

fig.1: Three-layer-model of communication-oriented systems

This model allows by application of decomposition and corresponding
interfaces to treat the layers isolated at which the evaluation
follows bottom-up beginning with the data transmission layer. In all
layers some modelling principles are applied: decomposition, stepwise
refinement, equivalence networks and isolation techniques by using
Norton's theorem. With this concept large systems are possible to
investigate /5/.

For regard of different transmission characteristics (e.g. connectionless or connection-oriented transmission, token-passing bus or CSMA/CD
medium access control) the layers 1 and 2 of the ISO/OSI reference
model are divided into two submodels:
- MAC (Medium Access Control; layers 1 and 2a)
- LLC (Logical Link Control; layer 2b).
Solutions on base simulation as well as analytic methods, e.g. nonex-
haustive cyclic polling, stochastic CSMA/CD medium access control,
token-passing are applied in MAC-submodel /1,8/. The model of LLC-
sublayer leads to a complicated queueing network. Analytic solution on
base decomposition method using Norton's theorem as well as simulation
experiments are performed /1,5,6/.
For support of performance evaluation of distributed systems on LAN-
basis the modelling tool LANEX is developed (written in Pascal,
running under CP/M-like operating system in preparation on MS-DOS).
LANEX uses computational algorithms from BNETD /3/ as well as further
algorithms and simulation technique. All included models are valided
by simulation experiments. The parametrization of models follows in
menue-technique by user with measured or estimated means.

References

/1/ Bux, W.: Local-Area Subnetworks: A Performance Comparsion. IEEE
    Transactions on Communications, VOL.COM-29,No.10,1981,pp.1465-1473
/2/ Gihr, O.; Kuehn, P.J.: Comparsion of communication services with
    connection-oriented and connectionless data transmission. Computer
    Networking and Performance Evaluation, Elsevier Science Publishers
    B.V. (North-Holland), IFIP, 1986,pp.173-186
/3/ Irmscher, K.: Programmsystem BNETD zur Analyse geschlossener Be-
    dienungsnetze. Wiss.Ztschr. TU Dresden, 34(1985)4,pp.87-94
/4/ Irmscher, K.: Rechnergestuetzte Modellierung und Bewertung. edv-
    aspekte, No.4/1986,pp.9-11
/5/ Irmscher, K.: Modelling of communication processes in manufactu-
    ring automation. Journal Systems-Analysis-Modelling-Simulation,
    Akademie-Verlag, Berlin, 1988
/6/ Kuehn, P.J.: Modelling of new services in computer and data tech-
    nics. Computer Networking and Performance Evaluation, Elsevier
    Science Publishers B.V. (North-Holland), IFIP,1986,pp.283-303
/7/ Loeffler, H.: Lokale Netze.
    Akademie-Verlag, Berlin, 1987
/8/ Reiser, M.: Communication-system models embedded in the OSI-Re-
    ference Model, A survey. Computer Networking and Performance
    Evaluation, Elsevier Science Publishers B.V. (North-Holland),
    IFIP, 1986,pp.85-111
/9/ Zschiesche. M.; Loeffler, H.: LAN-Transportstations in CIM-
    Environments . Report 12th. Seminar "Rechnergestuetzte Fertigungs-
    technik" - DAUS-Seminar 87. Gaussig, 16.-20.11.1987

New Simulation Approaches to Ill-Defined Systems

G.C. Vansteenkiste[1]

ABSTRACT

Ill-defined systems like in environmental, biomedical and biotechnological studies pose several challenging problems in system simulation. The impact they have on simulators are touched in this paper. Methodology is stretched as well as computer hard- and software needs come into play. It is seen how techniques from artificial intelligence might be promising. Finally a proposed simulator environment currently under investigation at the University of Ghent is described.

INTRODUCTION

Designing and using simulation for complex industrial plant studies is a demanding and time-consuming task, requiring extensive knowledge of both the plant and the simulation system. While the basic principles and tools for process flowsheeting (calculation of governing laws like heat and material balances) are well understood, difficulties arise when sufficient fundamental knowledge is lacking. In such cases, the validity of management and control studies through simulation comes into play. Processes as in biotechnological and medical systems are important examples. There is a definite need in these so-called ill-defined systems for improved man-machine communication and support in order to complete formalized process knowledge with human expertise. It is the purpose of this paper to provide a perspective over the manner in which simulation has evolved in recent years taking into account demands introduced from ill-defined system studies.

NEED FOR USER FRIENDLY SYSTEM OPERATION

Over the past 25 years several solid techniques have been developed for the analysis of systems using simulation. Mathematical formalisms constituted the widest used computer models. The extension of simulation from engineering applications towards the soft sciences in environmental, biomedical applications however, created a need for a less mechanistic deduction.

Every.abstract model represents an attempt to logically rationalize the behavior and perhaps even the mechanism of one or other system under study. The successes of representation showed to be very different according to the nature of the system, mostly because our mental constructs are too meager.

Although almost unlimited raw computing power became available, eventually shaped in specialized tools, one must be able to construct models using entirely the language and thought patterns of the specialized field under study. Enormous benefits both qualitatively and quantitatively result when interactive tools are made available. Computer-based graphics and statistics as well as data base management and networking should be specifically designed supporting the simulationist. The interactive environment fundamentally changes the way the data are perceived and the way the modeller behaves during problem analysis and model design.

---

[1] University of Ghent, Ghent, Belgium

Human behavior is strongly influenced by an interactive system's response time.
According to Doherty [1] it has been observed that each second of computer response degradation leads to a similar degradation added to the user's time for the next request.
This phenomenon is related to an individual's attention span. People seem to have a sequence of actions in mind, contained in a short-term memory buffer. Increases in computer
response time seem to disrupt their thought processes and this may result in a time consuming rethinking of what is to be done next. In addition they become emotionally upset
and make more mistakes.

As will be seen furtheron the complexity of the computer model is dynamic in the
sense that its content is continuously updated the more knowledge and data about the
system under study are detected. In order not to degrade computer response time by increasing the model complexity, the processor architecture for advanced simulators should
be capable to perform parallel information processing.

NEED FOR INCREASED SYSTEM INFORMATION REPRESENTATION POWER

Often the degree of isomorphism between model and system found in engineering applications is lacking in ill-defined systems ; the terms of the equation have a profoundly
different, mystical relationship to the object modeled. At less resolution the differential equation may simulate behavior perfectly but tell very little about the underlying
causality. Deland [2] illustrates how for instance in biomedical system models primitives
like diffusion, mass-balances, dilution, etc. are building blocks for models. Just as the
equations relating to mass, force and velocity deal with the primitives in engineering,
a mathematics that deals with the fundamental primitives of the system under study is
needed. Each of the primitives could itself be a model hidden for the user except for its
input-output and calibration parameters.

The concept of a model as a data base has not yet been exploited to its full potential. A model is a comprehensive representation of captured intelligence. Conceptual
models involve at their very essense the fundamental problem of pattern recognition.
Still today all the important pattern recognition is being performed by the human operator. Man is especially apt in reasoning and in recognizing patterns. His computational
power however is limited. The knowledge-based structures at the heart of the logical
schema are constructed almost exclusively of interactive rules to be matched against
factural question- and answer material. The future is to incorporate valid models into
the structure of the reasoning. It should be noted at this point that the concept of captured intelligence is extremely useful for raising the previously discussed man-machine
interface to a useful level for most users.

In the early 1960's already the computing industry speculated on the notion of artificial intelligence, whereby the computing system would heuristically determine the
solution to some problems. This computer support is now implemented in domain specific
"expert systems". The intelligent guidance in simulation through for instance model synthesis aid, is an example. A simulator becomes essentially a knowledge-based expert system utilizing a combination of symbolic reasoning and data-processing. They require, for
expert performance on problems, heuristic knowledge to be combined with the facts of the
discipline. Through "expertise modelling" human expert knowledge is transferred to

computer programs. "Expert systems" used in the context of simulation should contain in addition to à priori information, a modelbase, an experimental frame base and a database with experimental results : the so-called extended knowledge base.

A promising strategy in simulation of ill-defined systems is to build up modelbases of "primitives". Aspects concerning the relations between the various primitive submodels should also be taken into account : how subsystems interfere with their environment (inputs, outputs), whether they operate in parallel with other subsystems or not, etc... . This kind of knowledge can also be captured in rules. Note that "primitives" may appear in numerous different forms. For example, a simple mechanical primitive "mass/spring oscillator" may have different mathematical descriptions, dependent upon matters as : "Can the spring be assumed to be massless", "Is Hooke's law valid", "Does there exist a friction", "Is the frictional force assumed to be proportional to velocity", "Does there exist a driving force", etc... . To choose the right primitive requires expert knowledge. An "expert system" can support in this respect.

An "expert system" could furthermore support modelling and simulation in the following tasks :
- make the retrieval of an existant model from primitive submodels easier and more reliable ;
- relate diverse models of the same real-world system to each other ;
- impose the correct experimental conditions on each model ;
- build new models out of existing models ;
- simplify large-scale systems and help in the construction of lumped models for them ;
- assist in the structure characterization and the parameter identification ;
- assist in the validation of tentative models by suggesting useful experiments.

Note that attention must not only be paid to the dialogue with the user, but also techniques must be developed to extract the expertise from the human expert and to incorporate this knowledge in the system. The learning behaviour is one of the challenges of today's research in expert systems. The ultimate "intellectual task" of such an expert system should be the integration of knowledge from the model-level to the à priori level. For this purpose existing models must be generalized. Experiments have to be carried out to validate these general models and last but not least these general models have to be interpreted in laws and theories.

The functional tasks developed so far are summarized in figure 1. The importance is shown of the interactions with the process-expert as well as the process itself.

NEED FOR ENHANCED COMPUTER MODELLING METHODOLOGY

Modelling, the activity by which computer models are constructed based on the content of the extended knowledge base, has been shown to be system dependent. Biomedical systems require other modelling procedures than electromechanical systems. Parameter optimization techniques used in engineering systems will be blurred by system structure uncertainties : error minimization will indeed lead to a parameter updating which compensates structural error [3]. In addition the model formalism has also an impact on the methodology. Determining a discrete event model will require other procedures than specifying a distributed parameter model. Fortunately the dependence has its limits.

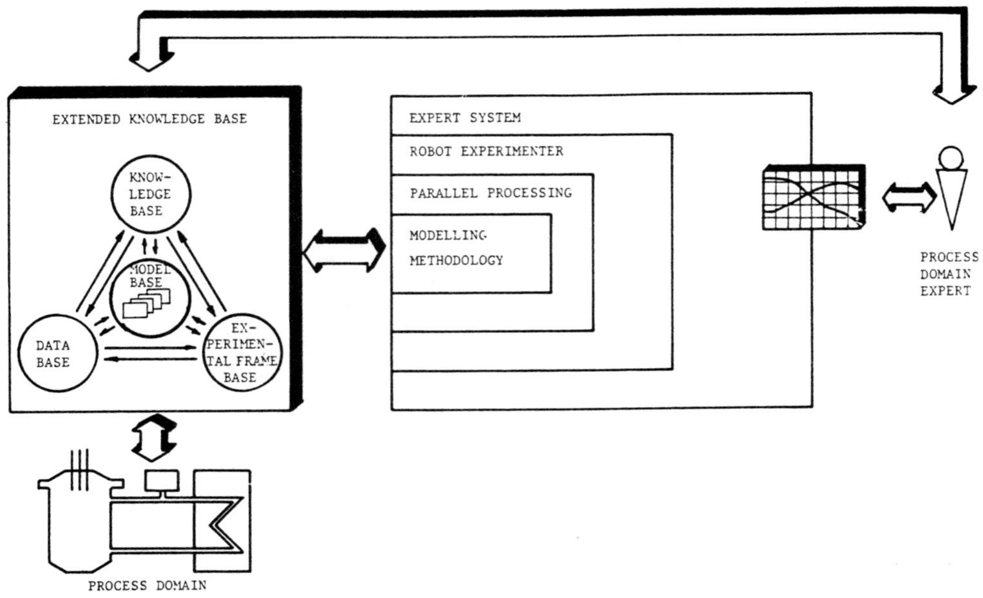

Figure 1. Simulation methodology for ill-defined system study

Considerable efforts have been put in systems theory and in model building theory to develop concepts and procedures that are to a certain extent broadly applicable. The aim is to exploit this generality. In the last decade, the amount of study, analysis and investigation of techniques for the modelling process has been large but very unequally spread over the different stages. Most attention has been paid to the parameter estimation problems while other issues like structure characterization, experiment design, model framework definition, model validation and goal incorporation have been neglected. Simulators for ill-defined systems should be equipped with the necessary algorithmics to make mathematical modelling transparent to the discipline expert (robot experimenter as indicated in figure 1).

As a rather simple but illuminating example consider a mathematical model building problem in microbiology. Consider microbial growth on a single substrate as food source within a fermentor vessel. Suppose further, that for the problem at hand the following model frame is sufficient : two-state variables X, S, respectively the biomass and the substrate concentration ; no inputs but only initial conditions ; the biomass is considered as the only output variable.

First a deterministic mathematical model is aimed for. The structure characterization process, discussed in detail in [4], results in a hierarchical decision process as shown in the following steps :

$$\dot{X} = f'(X,S).X$$
$$\dot{S} = f''(X,S)$$
$$Y = X \qquad\qquad X \geqslant 0 \quad \text{and} \quad S \geqslant 0$$

- model class 1 : f', f'' : two real continuous functions in two variables

- model class 2 : f' : one real continuous function in two variables

                 f" : $-\dot{X}/Y$

- model class 3 : f' : f(S) one real continuous monotonic never decreasing function in one variable

                 f" : $-\dot{X}/Y$

- model class 4 : f' : $f^*(S)$ with $f^* \in \{f_1, f_2, \ldots, f_c\}$ and f_i known

                 f" : $-\dot{X}/Y$

In model class 4 for instance is decided to restrict oneself to functional relationships already proposed by other researchers. During the coarse of the fermentation reaction it could very well be that the structure is shifted between different relationships. Parameter identification, the next step in modelling the growth, is computationally intensive due to the iterations needed and has to be performed for each f_i in class 4. Either a sophisticated algorithm with less internal parallelism, or a low order search routine suited to the processor architecture has to be selected.

In ill-defined system modelling, deterministic models may be too rigid and not adequate in the sense that their presumed validity is much higher than one is willing to accept in the context considered and for the purpose at hand. Stochastic models may be a way out here. Adding a stochastic factor which takes modelling errors and imperfections into account, may weaken the model rigidity by making it less valid. As the representation is now closer to our knowledge about the system, it may turn out that the goal can be reached in a more realistic manner. The degree of abstraction in case of a stochastic model is more pronounced. The Monod model, being one of the functional relationships f_i in model class 4 above, is heuristically transformed to a stochastic process. In [5] it is shown how the probability function associated with the random variable X at each instant in time t, can be found as the solution of the Fokker-Planck partial differential equation. The properties of the model and the relationship with the deterministic case (mean value of the probability function) are obtained correspondingly. Using a computer architecture with internal parallelism on-line display drawing can be obtained, providing direct statistical information to the biotechnological process expert.

It is not necessary and probably dangerous to try to automate completely in the robot experimenter all different steps involved in mathematical modelling. In the present state of development the optimal course seems to be to have the complete configuration done under the guidance of the human investigator, the expert of the problem under study.

As an example of methodology-based AI supported modelling, we shall assume that the modelbase consists of candidate models and that the problem is to find an appropriate model in a given experimental frame [6]. In the case of non-engineering systems, the model building activity relies almost fully on inductive methods (result from experiments). No general methods exist and the model building activity relies heavily on trial and error. Model construction could however be made easier by an intensive dialogue between the modeller and an expert system. In the first stage a tentative model formalism is chosen. The experimental frame further restricts the set of possible candidate descriptions. The next step will be the choice of a structure. This is done in a decision network, in which each decision step restricts the set of remaining candidate models. The procedure stops when a parametrized model is obtained. Two kinds of knowledge are required : mathematical properties of the model set that results from the decision made

(realization theory) and how to choose the correct decision steps, and how they are resolved [7] . Realization theory can be formalized into a number of rules and principles.
The decision process depends much on the field studied and the problem considered. It is
based on a lot of experience. A much harder task is the formalization of this knowledge
from the modelling expert in a set of rules. Heuristics will also be necessary to guide
the development of the decision network in case more than one rule is applicable, or in
case a dead-point is reached where no further conclusion can be made. In this case back-
tracking will be required.

PERFORMING ILL-DEFINED-SYSTEM SIMULATION

At the University of Ghent, Belgium a simulation environment is being developed to
study different methodologies for building computer models of ill-defined systems.
Figure 2 shows the lay-out of the simulation environment currently under installation.
Highly interactive workstations -"intelligent terminals"- are joined to each other and
to network servers. The network server provides user-access to specialized software
systems, described in previous sections, to a fast central processing unit and to the
process measurements. It is the intention to implement the functional tasks displayed in
figure 1 in a distributed fashion between server and workstations.

Very high processing speeds are achieved using the Applied Dynamics International
AD-10 as a peripheral array processor designed for the simulation of large time-critical
dynamic systems. It is the purpose to employ the AD-10 as a high performance back up for
computer models too computation-intensive to be handled by an individual workstation or
by the network server. The I/O processor connects the process (in this case a fermentation process) to the simulator.

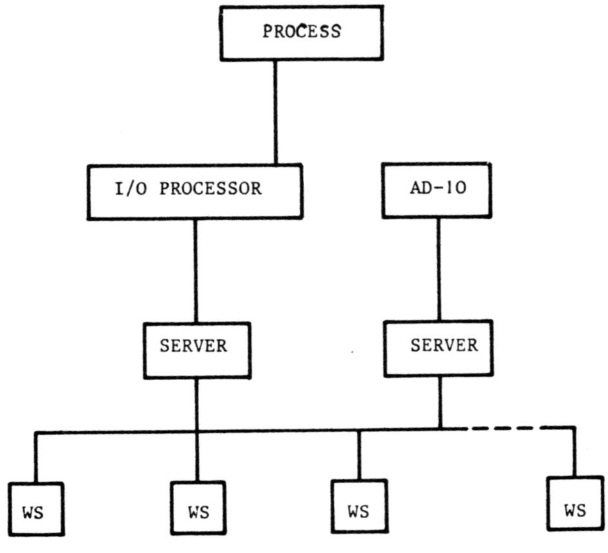

Figure 2. Simulation environment lay-out

ACKNOWLEDGEMENT

The author is gratefull to Bart Dobbelaere for his collaboration in providing the simulation results.

CONCLUSION

The purpose of the paper is on one hand to underline the almost unlimited capabilities of simulation in solving complex large-scale problems ; with on the other hand the instructional power to improve phenomena understanding and à priori knowledge.

REFERENCES

[1] Doherty, W.J. : The commercial significance of man-computer interaction. In : Proc. of the Conference on Man/Computer Communication, Amsterdam, 1978.

[2] DELAND, E.C. : Conceptual models in physiology, where are we ?. In : Proc. of the IFIP Working Conference on Modelling and Data Analysis in Biotechnology and Medical Engineering, Ghent, Belgium, North-Holland Publ. Co., 1983.

[3] Vansteenkiste, G.C., J.A. Spriet and S. Akutagawa : On the precision of modeling techniques for ecological parameter identification. In : Proc. of the IASTED Symp. on Simulation, Modeling and Development, 1981, Cairo, Egypt and of the INRIA International Conference on Analysis and Optimization of Systems, Versailles, France, dec. 1982.

[4] Vansteenkiste, G.C. and J.A. Spriet : Structure characterization for ill-defined systems. In : Simulation and Model-based Methodologies : an integrative view, NATO Advanced Study Institute, Ottawa, 1982.

[5] Spriet, J.A. and G.C. Vansteenkiste : Analysis techniques for the evaluation of the representative value of stochastic models for ill-defined systems. In : Transactions of the Society for Computer Simulation, San Diego, U.S.A. (in press).

[6] Oren, T.I. and B.P. Zeigler : Concepts for advanced simulation methodologies. In : Simulation, 32, nr.3, 1979, 69-82.

[7] Spriet, J.A. and G.C. Vansteenkiste : Computer-Aided Modelling and Simulation. International Lectures Notes in Computer Science, Academic Press, London, 1982.

Methodological Basis of the Simulation System SIMPLEX-II

Bernd Schmidt[1]

Abstract

SIMPLEX-II is a simulation system which provides extensive support in model studies. The most important areas of support are the description and construction of models and experimentation with these models.

The individual steps in the investigation of a model are analysed in detail in order to show the nature of the support provided for the user by SIMPLEX-II. Special attention is given to the conceptual model. Finally the way in which the basic concepts and ideas are incorporated in the structure and framework of SIMPLEX-II is shown.

1 The framework for model studies

Scientific model studies generally follow the same pattern independent of the area of application. This pattern is briefly described.

1.1 Model construction

The human being finds himself faced with a motley collection of elements belonging to the real world. He is always surrounded by an infinite number of different things.

From this myriad of differing elements found in the real world, the human perception cuts out a certain section to be analysed.

This section of reality is defined here as the real system.

One should note that only the input-output behaviour of the real system is known. The internal structure cannot yet be specified in any way. It is similar to a black box whose internal workings remain hidden.

However, it is possible to carry out experiments with the real system. This means that a given input is applied to the real system. This input excites the internal mechanisms of the real system and causes it to assume a new internal state. The new internal state becomes apparent through the resulting observable output.

All information and the measurements which can be collected on the real system together form the system data. They represent the empirically determined behaviour of the real system.

Systems analysis now attempts to design a conceptual model based on the observed system data. The conceptual model is a mental image of the internal structure and behaviour of the real system, constructed by the human mind. For example, one can try to explain the observed input-output

---

,[1]IMMD IV, Universität Erlangen-Nürnberg
Martensstraße 1, 8520 Erlangen, West Germany

behaviour of the atomic nucleus by assuming that, hidden within the black box there is a mechanism consisting of protons and neutrons attracted and held together by forces.

1.2 Model behaviour

The conclusions about the behaviour of the conceptual model can be derived in two different ways: One differentiates between analytical methods and methods involving simulation models.

If the model is described using a formal language, it is possible to come to conclusions about the behaviour of the conceptual model using the deductive rules defined in the language, i.e. certain propositions can be proved.

Example:

- Differential equations are used to represent abstract models which change continuously with time and where the rates of change are used to describe the model. The exact analytical solution of the differential equations can be derived mathematically in simple cases.

Analytical solutions normally provide general results for a whole class of abstract models. Results to a specific model are obtained by giving particular values to the variable coefficients in the general solution .

There is another way of obtaining conclusions about the behaviour of a conceptual model. One can attempt to find or construct a real system whose behaviour can be described by the conceptual model.

The behaviour of the conceptual model can be simulated by the behaviour of the real system. Real systems which serve as models in this way are called simulation models.

An experiment using a simulation model provides only a particular solution. Only the behaviour of that model which has been constructed is observable. Information about the general behaviour of the conceptual model is not provided. However simulation produces results which cannot be obtained analytically.

The results which describe the behaviour of a conceptual model can be derived analytically or by using a simulation model. If an analytical solution and a simulation model exist for a conceptual model, the results obtained from the two differing procedures are comparable. They are not identical as they have been derived using differing procedures. The difference depends on the errors made during model construction and the experimental phase.

The evidence that the conceptual model really does satisfactorily represent the behaviour of the system to be studied is particularly important. This is known as validation.

In figure 1.1 it is important to note that it is not possible to directly compare the results from a real system with those of a conceptual model. A validation can only be based on a comparison of system data and model data.

Figure 1.1 shows a diagram of the scientific methodology.

The most significant points are be summarized here:

- The system to be studied presents itself to the human perception initially as unstructured system data.

- Our thought process provides a representation of the system as a conceptual model.

- The behaviour of a conceptual model can be studied in two ways. The analytical method provides comparable results to simulation.

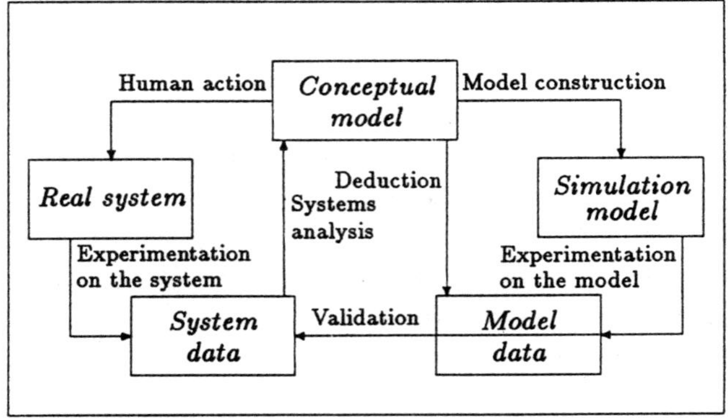

Fig. 1.1: Scientific methodology is based on making a conceptual model whose behaviour can then be studied. The results from this can then be applied to the real system.

2 Functions of the simulation system SIMPLEX-II

The simulation system SIMPLEX-II supports the user in the following ways:

- *Conceptual model*
  The model description language SIMPLEX-MDL is used to describe the conceptual model.

- *Model construction*
  A simulation model is produced from a description of the abstract model.

- *Simulation model*
  A model bank is available which stores and administers model components and the different model versions.

- *Experimentation on the model*
  Different experiments can be carried out on the models.

- *Model data*
  The results of the model experiments can be analysed, prepared for presentation, and archived.

- *Validation*
  The process of validation is supported by a number of procedures which help to compare results from the model and the original system.

A detailed description of the functions provided by the simulation system SIMPLEX-II is found in /1/.

2.1 Representation of the conceptual model

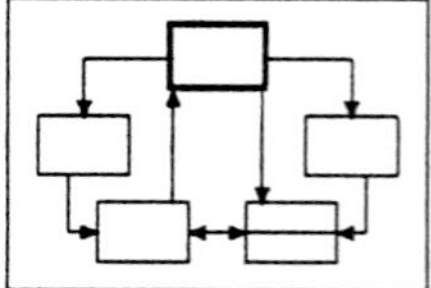

The model description language SIMPEX-MDL is part of the simulation system SIMPLEX-II.

SIMPLEX-MDL permits a descriptive, non-procedural representation of the conceptual model. It is based on the general model theory concepts.

A model consists of components and connections and has a hierarchical structure.

A component on the lowest level is called a basic component. It includes a description of the dynamic behaviour.

Model components are subject to the principles of object oriented programming.

- *Modular structure*
  A component has control of its own state transitions. It cannot alter the state variables of other components. The interaction between components is realized by communication via the connections. A component can thus be induced to change its behaviour by an external factor.
  Every component is an autonomous entity and can run independently.

- *Class concept*
  All components belonging to a specific component class has the same model description. Each has, however, its own parameter values.

The model description language SIMPLEX-MDL provides a general framework for dealing with continuous, event oriented, and transaction oriented models in the same way.

SIMPLEX-MDL recognizes only components and connections. A description of the dynamic behaviour within a component makes continuous and discrete state transitions possible. Communication takes place via the connections. A message is transferred by the exchange of mobile elements. Also transactions as used in GPSS can be represented using mobile elements.

A detailed description of SIMPLEX-MDL is given in /2/ and /3/.

2.2 Model construction

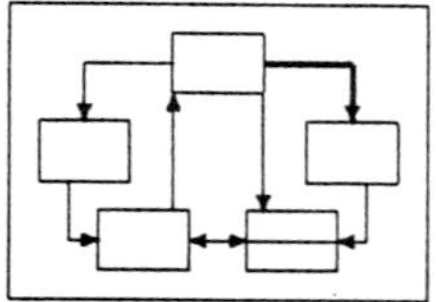

An executable simulation model is produced from the model description. This is done by first translating the model description into C-code.

The resulting object code is linked to the run time system providing an executable model.

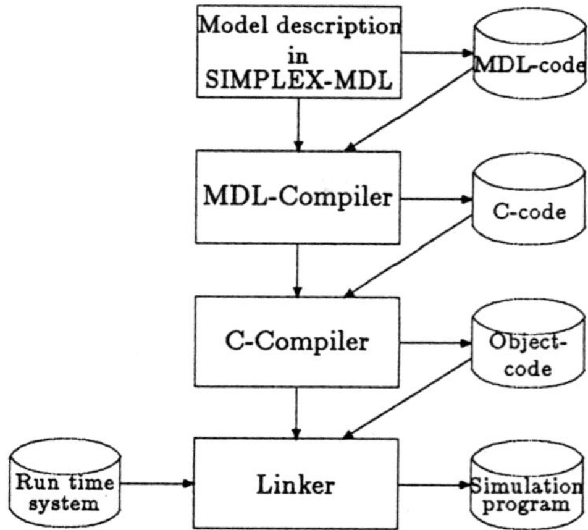

Fig. 2.1:   C-code initially results from the model description. A translation of the C-code
together with the run time system finally produces an executable simulation program.

2.3   The simulation model

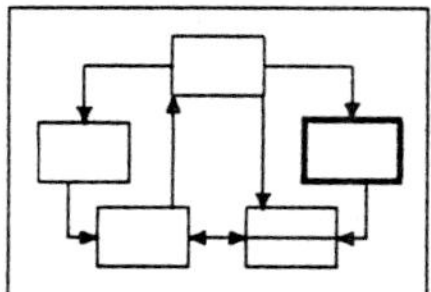

SIMPLEX-II provides the user with model banks and a version concept to assist the user in working
with simulation models.

Models and components belonging to a specific application area or to a particular model class can
be held in a model bank. Commands exist for creating, deleting, and renaming components, models
and model banks. Models or components can also be copied from one bank to another.

Different versions of a component can be exist within a model bank. The user can specify which
version of the component is to be integrated into the model.

The version concept has proved itself as useful in the following cases:

- *The investigation of modifications*
  The influence of one or more modifications to a lower level component on the overall behaviour
  of the model is to be investigated.

- *Rapid prototyping*
  One starts initially with only a rough formulation of a model component. The exact and
  accurate representation of the details follows later.

In both cases various versions are constructed which can be integrated into the model without
having to translate the entire model every time.

The individual components are relatively autonomous in SIMPLEX-II because of its hierarchical
structure. The user must therefore make sure that a modification to a lower level component has
no effect on the model as a whole.

Example:

- A modification in a lower level component causes a state variable to be eliminated. Another component has access to this variable.

Similarly it must be ensured that no error occurs when one component version is replaced by another.

SIMPLEX-II ensures that all components within a model bank are consistent with one another.

2.4 Experimentation on the model

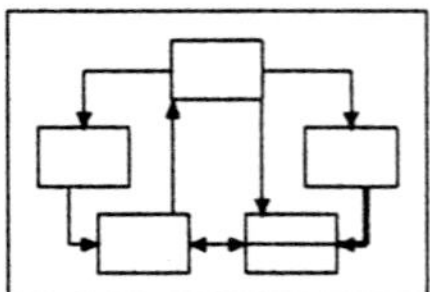

Different experiments are possible on models once they have been constructed. These can be grouped in SIMPLEX-II into a number of differing types:

- *Single run*

- *Sequential analysis*
  A number of runs with different parameter values are carried out.

- *Sensitivity analysis*
  One determines to what extent the model behaviour is affected by the modification of certain variables.

- *Optimization*
  A search is made for a set of parameter values which best satisfy the given aim. This is known as goal directed simulation.

The models and results of an experiment are stored under the name of the experiment. All experiments can thus be repeated. It is particularly important that the the experimental results can be reconstructed at any time.

2.5 Model data

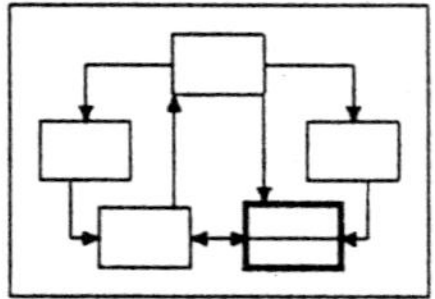

The state variables to be monitored during a simulation run can be specified. This does not depend on the model description. The values of these variables are stored in a file and are available for further processing.

SIMPLEX-II offers numerous mathematical and statistical methods for the evaluation of simulation run data including:

- Calculation of averages, variance, and the confidence interval

- Estimation of length of time needed to arrive at steady state

- Fourier analysis

2.6 Validation

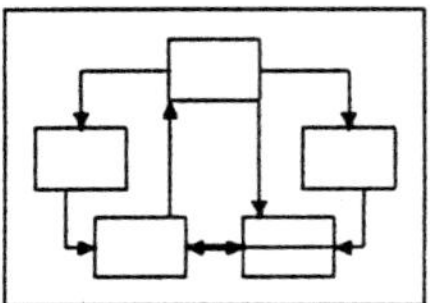

Validation can be understood to be a confirmation or proof that the model data derived from the behaviour of the conceptual model do in fact correspond within certain tolerance limits to the system data. In such a case, one can assume that the conceptual model is useable representation of a real system.

Validation is a very intricate and complex procedure, requiring judgement and experience in the user.

Depending on the model's field of application, the user has to decide on the acceptable tolerance limits, depending on the application area of the model. This decision is related to the availability of alternative models.

Validation can be supported by allowing a comparison of model data with system data, and by providing indicators for the degree of agreement.

3   The structure of the simulation system SIMPLEX-II

Simplex-II is based on the UNIX V operating system. It makes use of the user functions provided by this system. SIMPLEX-II is programmed in C. This ensures that it does not depend on the hardware and can run on any system supporting UNIX V and C.

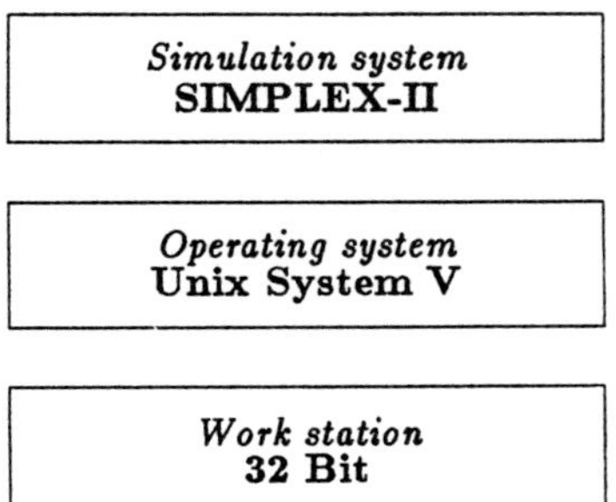

Fig. 3.1:   SIMPLEX-II makes use of the UNIX V operating system. It is thus independent of the hardware

3.1 The processes

SIMPLEX-II consists of 4 independent but interrelated processes:

- *Computer terminal process*
  This process manages the alphanumeric terminal screen. User commands are relayed to the operating process and messages are sent back to the screen.

- *Graphic process*
  This process is in charge of graphical representation on the screen in colour and is used for graphic model description and presentation of the results.

- *Operating process*
  The operating process decodes the commands received from the terminal process and executes.
  Executable simulation programs are written to the appropriate file where they remain available to the simulation process.
  The simulation results, which are written by the simulation process to a special file during the run, are available to the operating process for mathematical analysis and result presentation. The operating process can also make use of external data in the same way.

- *Simulation process*
  A simulation run is executed by the simulation process. The simulation process informs the operating process of the names being used in the model so that it can organize the interactive simulation operation of the model.

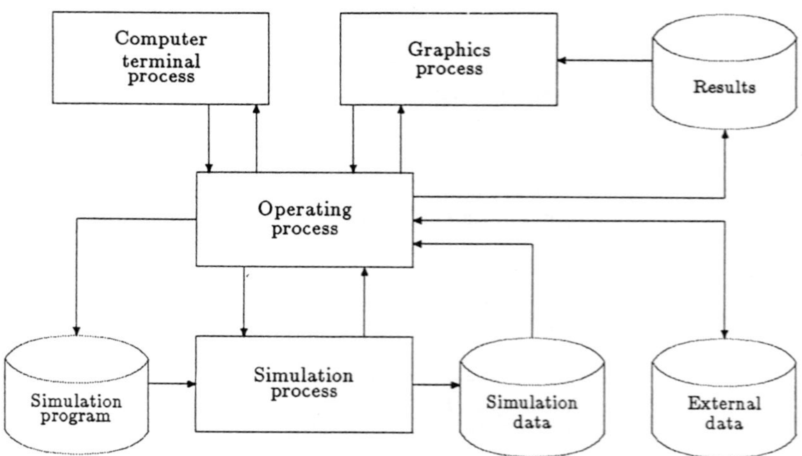

Fig. 3.2: SIMPLEX-II consists of 4 interrelated processes

3.2 The operating process commands

The operating process interprets and processes the commands received form the screen.
Related commands comprise a menu.

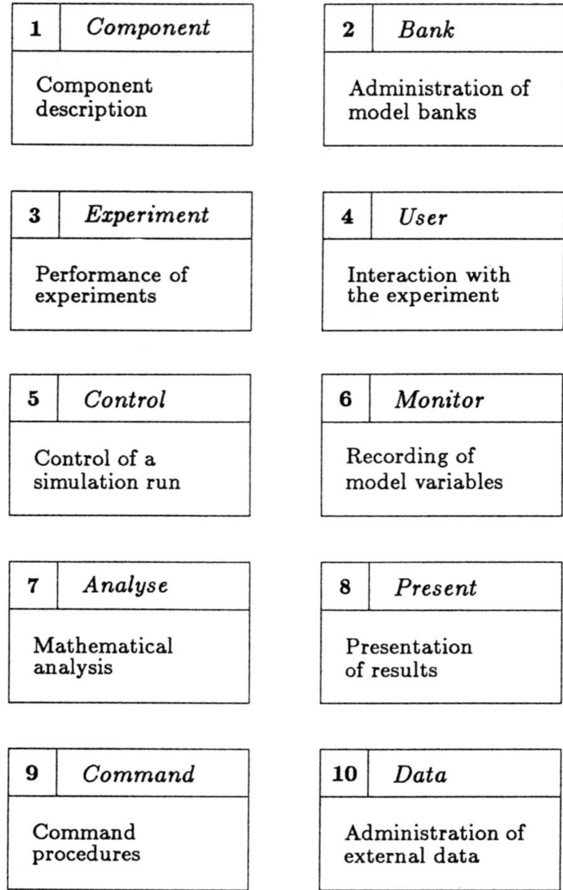

Fig. 3.3:   Interrelated commands make up a menu. SIMPLEX-II offers 10 different menus.

An example of this is given by a short description of the commands found in menu one. This menu
provides the commands needed to create and manage components in a model bank.

The simulation system SIMPLEX-II is a contribution to the application of simulation to new
application areas. It will help to make simulation accepted as a technique for design, planning and
decision making.

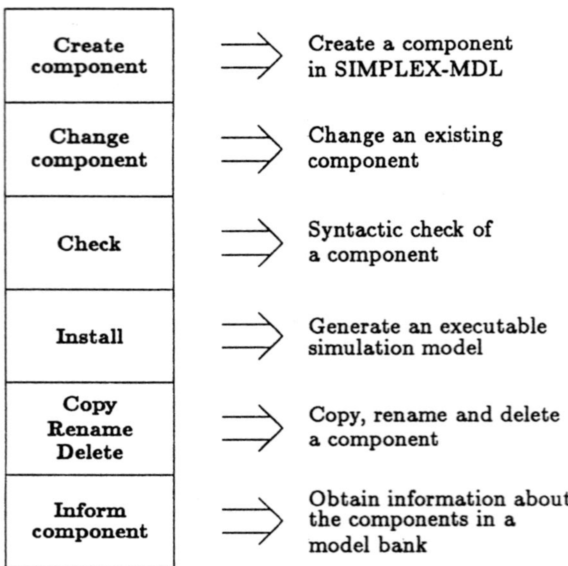

Fig. 3.4: The commands of the menu *Component* are used to create and organize components
and models

In this respect it joins the common efforts of others e.g. /4/.

References

[1] Eschenbacher, P., Schmidt, B.; The Model Description Language SIMPLEX-MDL; Discrete
Event Simulation and Simulation and Operations Research; Proceedings of the European Si-
mulation Multiconference, Vienna 1987; ed by Adelsberger H. and Broeckx; The Society for
Computer Simulation (SCS), 1987

[2] Eschenbacher, P.; An Approach to Transaction-Oriented Modelling Based on Systems Theory;
Discrete Event Simulation and Simulation and Operations Research; Proceedings of the European Simulation Multiconference, Vienna 1987; ed by Adelsberger H. and Broeckx; The Society
for Computer Simulation (SCS), 1987

[3] Langer, K.-J., Schmidt, B.; The Simulation System, SIMPLEX-II – A Model Construction
and Experimentation Environment; Discrete Event Simulation and Simulation and Operations Research; Proceedings of the European Simulation Multiconference, Vienna 1987; ed by
Adelsberger H. and Broeckx; The Society for Computer Simulation (SCS), 1987

[4] Zeigler, B.; Hierarchical, modular discrete-event modelling in an object-oriented environment;
Simulation 49, 5; 1987

Real-Time Simulation of Non-Linear Quadratic Gaussian Adaptive Control System

Andrzej Dzieliński[*]

SUMMARY

Real-time simulation of a non-linear quadratic gaussian adaptive control system is described. The piecewise linearization of the non-linear plant is used. Linear stochastic parameters identification methods are used. Unknown state variables of the plant are estimated by the well-checked Kalman Filter algorithm. Control law is a state feedback whose parameters are computed from the solution of the Algebraic Riccati Equation.

1. INTRODUCTION

There is a lot of work done in the field of linear adaptive control e.g. [1,2,3], but there are still many problems concerning the non-linear adaptive systems. There were many trials to solve these problems and there were some results obtained in the field of identification of non-linear systems. Main disadvantage of these methods however is the fact that they apply to only very narrow class of non-linear systems. There was rather not very much work done in general case The results in the domain of suboptimal control e.g. [4] as well as lack of the efficient identification method for non-linear case show that one of the reasonable ways to solve the problem of adaptive control is the method of piecewise linearization of the non-linear plant and, following from that, use of linear technique methods. This approach is especially advantageous if the non-linear plant linerization results in the linear time-variant model whose parameters change slowly in comparison with the identification algorithm.

2. PLANT DESCRIPTION AND IDENTIFICATION

Let us consider the following discrete-time non-linear plant model described by the state-space equations:

$$x_1(k+1) = x_1(k) + T_i x_2(k) + \zeta_1(k) \tag{1}$$
$$x_2(k+1) = x_2(k) + T_i b(u - W_0 - W_1 x_2(k) - W_2 x_2^2(k)) + \zeta_2(k) \tag{2}$$
$$y(k) = x_1(k) + \nu(k) \tag{3}$$

---

[*] Institute of Control and Industrial Electronics, Warsaw University of Technology, ul. Koszykowa 75, 00-662 Warszawa, Poland

where: x_1 - distance (m)

x_2 - velocity (m/s)

$b = \frac{1}{m}$ m - mass of the vehicle

u - driving force (N)

T_i - sampling time

ζ_1, ζ_2, ν - gaussian white noises

W_o - resistance factor due to ground shape

W_1, W_2 - resistance factors due to velocity

Driving force is restrained in the following way:

$$-U_H \leq u \leq U_R$$

where: U_H - maximum braking force ; U_R - maximum propelling force

Initial and final values of state variables are stated as follows:

$$x_1(0) = 0; \ x_2(0) = 0; \ x_1(t_F) = x_F; \ x_2(t_F) = 0; \ t_F - \text{motion time}$$

These non-linear discrete-time system is now piecewise linearized to obtain a linear time-variant system of the form:

$$x(k+1) = A^*(k)x(k) + B^*(k)u(k) + \zeta(k) \tag{4}$$

$$y(k) = C^*(k)x(k) + \nu(k) \tag{5}$$

where: $x(k) = \begin{bmatrix} x_1 \\ x_2 \end{bmatrix}$, $\zeta(k) = \begin{bmatrix} \zeta_1 \\ \zeta_2 \end{bmatrix}$

$$A^*(k) = \begin{bmatrix} 1 & T_i \\ 0 & 1 + T_i b(-W_1 - 2W_2 x_2(k)) \end{bmatrix} \tag{6}$$

$$B^*(k) = \begin{bmatrix} 0 \\ T_i b \end{bmatrix} \qquad C^*(k) = [1 \quad 0] \tag{7}$$

and the same system's matrices in canonical observable form:

$$A_o(k) = \begin{bmatrix} 0 & 1 \\ -(1 + T_i b(-W_1 - 2W_2 x_2(k))) & 2 + T_i b(-W_1 - 2W_2 x_2(k)) \end{bmatrix} \tag{8}$$

$$B_o(k) = \begin{bmatrix} T_i b \\ T_i^2 b \end{bmatrix} , \qquad C_o(k) = [1 \quad 0] \tag{9}$$

This description has to be changed into an input-output ARMA-type relation:

$$y(k) + A_1(k)y(k-1) + A_2(k)y(k-2) = B_2(k)u(k-2) + B_1(k)u(k-1) + \\ + e(k) + C_1(k)e(k-1) + C_2(k)e(k-2) \tag{10}$$

where $A_i(k)$, $B_i(k)$ and $C_i(k)$ depend on T_i, b, W_1, W_2 and $x_2(k)$.
Let us denote:

$$\gamma^T(k) = [\; A_1(k) \;\; A_2(k) \;\; B_1(k) \;\; B_2(k) \;\; C_1(k) \;\; C_2(k) \;] \tag{11}$$
$$\phi^T(k) = [-y(k-1) \;\; -y(k-2) \;\; u(k-1) \;\; u(k-2) \;\; \varepsilon(k-1) \;\; \varepsilon(k-2)] \tag{12}$$

Then we may introduce one of the modifications of the Recursive Least Squares method the so called Recursive Extended Least Squares method algorithm of which is as follows [6]:

$y(k+1)$ and $u(k+1)$ measurement and $\phi(k+1)$ updating

$$\varepsilon(k+1) = y(k+1) - \hat{\gamma}^T(k)\phi(k+1) \tag{13}$$
$$\lambda(k+1) = \lambda_o \lambda(k) + (1-\lambda_o) \tag{14}$$
$$M(k+1) = P(k)\phi(k+1)[\lambda(k+1) + \phi(k+1)^T P(k)\phi(k+1)]^{-1} \tag{15}$$
$$\hat{\gamma}(k+1) = \hat{\gamma}(k) + M(k+1)\varepsilon(k+1) \tag{16}$$
$$P(k+1) = \frac{1}{\lambda(k+1)} [I - M(k+1)\phi(k+1)]P(k) \tag{17}$$

- initial value of covariance matrix $P(0) = \text{diag}[10000]$
- initial value of parameters estimates vector $\hat{\gamma}(0) = [000000]$

Having the parameters vector $\hat{\gamma}(k)$ estimated the parameters of the matrices $A_o(k)$ and $B(k)$ of the canonical form of the state-space description can be easily determined. So the parameters of the matrices $A^*(k)$ and $B^*(k)$ can also be found.

3. KALMAN FILTER ALGORITHM.

Let us consider the linear time-variant system given by the state-space equations (4) and (5) then the optimal filtering algorithm (Kalman Filter) has the following form:

$$\hat{x}(k+1) = A^*(k)\hat{x}(k) + B^*(k)u(k) + K(k+1)[y(k+1) - C^*(k)(A^*(k)\hat{x}(k) + B^*(k)u(k)) \tag{18}$$
$$K(k+1) = P(k+1)C^*(k)^T[C^*(k)P(k+1)C^*(k)^T + V]^{-1} \tag{19}$$
$$P(k+1) = A^*(k)T(k)A^*(k)^T + W \tag{20}$$
$$T(k) = (I - K(k)C^*(k))P(k) \tag{21}$$

with initial conditions $\hat{x}(0) = \bar{x}_o$, $P(0) = \bar{P}_o$

where: W – covariance matrix of state disturbances

V – covariance matrix of output disturbances

4. LINEAR QUADRATIC OPTIMAL CONTROL

For the linear time-variant system given by the state-space equations (4) and (5) and the cost function given in the form:

$$J = \sum_{k=0}^{n-1} [(x_n - x(k+1))^T Q(x_n - x(k+1)) + u^T(k)Ru(k)] \tag{22}$$

where: Q – positive semidefinite matrix

R - positive definite matrix

x_n - final state of the system

Optimal control as a state feedback may be computed from the following set of equations:

$$u^o(k) = -G(k)x(k) + L(k) \tag{23}$$

$$G(k) = (R + B^*(k)^T H(k+1) B^*(k))^{-1} B^*(k)^T H(k+1) A^*(k) \tag{24}$$

$$L(k) = (R + B^*(k)^T H(k+1) B^*(k))^{-1} B^*(k)^T V(k+1) \tag{25}$$

$$H(k) = A^*(k)^T H(k+1) A^*(k) - A^*(k)^T H(k+1) B^*(k) G(k) + Q \tag{26}$$

$$V(k) = A^*(k)^T V(k+1) - A^*(k)^T H(k+1) B^*(k) L(k) + Q x_n \tag{27}$$

Equations (26) and (27) are the Algebraic Riccati Equations and they are solved backwards from the final conditions $H(n) = Q$ for equation (26) and $V(n) = Q x_n$ for equation (27).

5. SIMULATION, CONCLUSIONS AND FINAL REMARKS.

Simulation of an adaptive, non-linear control system was performed on SM 1420 minicomputer (PDP 11 compatibile) using the software package written in FORTRAN 77.

System parameters and state variables change slowly, so that the Recursive Extended Least Squares and Kalman Filter algorithms can follow up with determining the valuable estimates.

The target of control was to reach the final state of the system $x_n = (1000, 0)$, starting from the initial state $x_o = (0, 0)$ minimizing the cost function given by (30) where $Q = \begin{bmatrix} 1 & 0 \\ 0 & 1 \end{bmatrix}$ and $R = 1$.

The quality of control in adaptive case is slightly worse (value of the cost function is greater) than in optimal one, but in adaptive system results are still within reason although the system parameters and state variables were unknown at the beginning.

REFERENCES

[1]. K. J. Äström -"Theory and applications of adaptive control - A survey", **Automatica**, 19 (1983) 5, **pp. 471**

[2]. A. Dzieliński -"Real-time simulation of Linear Quadratic Gaussian adaptive control system", **Proceedings of European Congress on Simulation, Prague, Sept. 21-25, 1987**

[3]. C. Samson -"An adaptive LQ controller for non-minimum phase systems.", Int. J. Control, 35, (1982) 1, pp. 1,

[4]. V. Strejc -"Least squares in identification theory.", **Kybernetika, 13 (1977) 2, pp. 83,**

[5]. R. E. White, G. Cook -"Use of piecewise linearization for suboptimal control of non-linear systems.", Int. J. Control, 18, (1973) 3, pp. 385

Simulation of Complex Real Systems: Theory and Practice

Vlatko Čerić[*]

1. Introduction

Discrete event simulation modelling has a solid methodological and computational foundation that has been created through more than thirty years.
It was heavily used for solving various problems in real complex systems.
Nevertheless, there appears to be a "grey zone" of problems that are
situated between theory and practice. For those problems there is not
enough interest on both the theoreticians' and the practitioners' side.

The author, being mainly a practitioner dealing with simulation modelling
and analysis of complex real systems (postal centres, airport terminal
buildings and big hospitals), identified through his own work several
problems for which first attempts toward the precise description and
solution were done. Some of these problems are also relevant to the other
operations research methods. Here we shall present some of these problems
and new approaches and give the references that describe the issue in
more details.

2. Input data for queueing systems design

One of the primary applications of simulation modelling is in the system
design fiels, where the value of some of the system parameters have to be
determined in such a way as to ensure the proper system functioning with
minimum total system construction cost. The convenient method of system
design is system modelling. Significant element of the model is its input
data. Input data are most often stochastic variables whose characteristics can be obtained through measurement and analysis of the relevant
real processes. The question is which statistical characteristics of the
input data have to be known in order to ensure the proper system design.

System design practitioners usually treat model input data with not
enough care, i.e. they make insufficient number of measurements and
inadequate data analysis. On the other hand, the texts about modelling
treate input data analysis procedures independently of the final aim of
the system study. However, input data analysis has to be related to the
system design requirements both from the theoretical point of view and
for the aim of improving the system design practice.

---

[*]University Computing Centre, Engelsova bb, 41000 Zagreb, Yugoslavia

In the queueing systems design the usual goals are: selection of the
appropriate number of servers and the necessary storage space needed for
customers, and ensurance of the appropriate treatment (e.g. waiting
time) of customers. The estimation of the system performances is usually
done through the percentiles of the appropriate variables. It was
shown [2] that the percentiles for a quite wide range of distributions
which are significant in simulation are some function of variances.

As an example of a simplified and analytically tractable model of input-
output relationships we selected the M/G/1 queueing system. We have
shown that the variances of dependent variables in this system are proportional to the third moments of independent variables. Since percentiles are functions of variances, it follows that for the queueing systems'
design it is necessary to obtain the estimation of third moments of the
independent variables. The relation between the required sample size,
measurement error and its probalility was also found [2].

3. System representation for simulation modelling purposes

One of the first tasks in the system modelling study is the system representation task. Although a wide variety of representation methods exists,
as far as we know there is no comparative analysis which would enable the
practitioner to select the representation method appropriate for a specific simulation modelling study and the resources available. In our
research (which is in progress) we have identified the need for such
comparative analysis and structured the investigation along the following
lines.

The first step will be to enumerate the existing representation techniques with their main characteristics and typical applications' classes.
Thus, the activity cycle diagrams, simulation graphs, function nets,
event graphs, schema representation method, conical method and production
system modelling will be investigated. In the comparative analysis of
these representation techniques the criteria for comparison will be:
simplicity (small number of simple tools and concepts), selfunderstan-
ding, unambiguity, modularity, expressive power (which system characteristics can be represented and how naturally), ease of modelling and
programming (with possible automatic program generation), and overall
effectiveness of the representation (taking into account the whole modelling process).

On the basis of these investigations, and of the classification of complex real systems (structure, elements, interaction etc.) and simulation
study goals (system design, analysis, operation control etc.), the taxonomy and criteria for selection of appropriate representation technique.

4. Identification, representation and modelling of discrete system
   operation control

In order to enable the proper functioning of the system and use the
system resources as efficiently as possible, operation of most complex
systems have to be controlled. While the control theory of continuous
dynamic systems is a well recognized field of investigation with developed modelling and design methods, control of discrete queueing systems
has not been investigated to a larger extent.

In the simulation study of the parcel flow and sorting system in postal
centre [3] we made the identification, representation and modelling of
the system operation control. We identified two types of the parcel
processing operation control. First one is the closed loop control of the
parcel distribution (sorting) over the departure directions, and the
second one is the open loop control of the shipment of sorted parcels for
departure directions.

We developed the representation of the system operation control using the
representation scheme of the continuous control circuits, and integrated
it in the overall system representation. The simulation program, built in
the GPSS language, enables the experimentation with the alternative control strategies.

5. Influence of the system operation control on system design

Our experience in simulation modelling of complex systems with control
provides the evidence that the system operation control strategies have
influence on system design [1]. In this context, simulation modelling is
significant because of its ability to enable the modelling of complex
systems and quantitative analysis of their behaviour.

Simulation experiments for the design of the parcel processing postal
centre have shown [3] that the alternative strategies of parcel distribution control have significant influence on the parcel storage conveyer
area. Comparison of two alternative control strategies with different
structure of parcel distribution time subperiods indicates e.g. that the
parcel storage conveyer area can be reduced by approximately 10 percents
merely by adequate selection of the appropriate control strategy.

Similar type of influence has been obtained in the simulation studies of
the airport terminal building and the automatic guided vehicle transportation network in hospital.

6. Simulation modelling with arbitrary precision

In development of the simulation model determination of the level of
approximation in the model is connected both with the model preciseness
and the computer resources expenditure. The usual modelling approach is
to select the fixed approximation level which seems most appropriate for
the problem solution needs and for the available computer resources.

We have developed the alternative approach [3] in which the modelling
approximation can be changed during the experiment execution time. The
dynamic entities of the model can be aggregated in various sizes (by
simply changing the program input data). All the consequences of such
aggregation are built into a simulation model. Results of experiments
with various levels of approximation gave the dependencies of both the
model accuracy (for the most significant dependent variables) and the
required computer resources upon the aggregation factor.

Such an approach was implemented in the simulation model of a postal
centre for parcel processing. Simulation experiments have shown that, for
example, for the increase of aggregation factor from 1 to 10 the execution time decreased 8 times, while the maximum increase of the system
performances relative error was always under 4 percent. This result
encourages the usage of models with arbitrary precision in designing
cheaper and faster simulation experiments with controlled precision.

References

[1] Čerić, V.: Queueing Systems Design, Control and Simulation : the
    Interrelationships. Cybernetics and Systems'86, Wien, 1986, 261-265.

[2] Čerić, V. and L. Lakatos: Input Data for Queueing Systems Design.
    Proceedings of IX International Symposium "Computer at the Univer-
    sity", Cavtat, 1987, 6R.05.

[3] Čerić, V.: Simulation of Parcel Flow and Sorting System Control in
    Postal Centre for Parcel Processing. Proceedings of European Congress
    on Simulation, Vol. A, Prague, 1987, 54-58.

A Systolic Extrapolation Design

D.J. Evans & G.M. Megson
Department of Computer Studies
Loughborough University of Technology
Loughborough, Leicestershire,
U.K.

ABSTRACT

A systolic array for the Aitken extrapolation formula is presented.

1. INTRODUCTION

The use of extrapolation formulas and systolic arrays for the efficient construction of extrapolation tables has been previously studied in [1],[2]. In this paper we study Aitken's extrapolation formula and develop an algorithmic form suitable for systolic implementation of extrapolation and general table algorithms.

2. THE AITKEN EXTRAPOLATION FORMULA

Let $U_0(k_m)$ be starting values for the extrapolation procedure, that is, the approximations of some required functions using different point spacing as shown in Fig.1. Obvious spacing sequences are,

$$k_m : \quad \{k_0, k_0/2, k_0/3, k_0/4, k_0/5, \ldots\}$$
$$k_m : \quad \{k_0, k_0/2, k_0/4, k_0/5, k_0/6, \ldots\}$$

where k_m is the spacing between points with k_0 the starting spacing, i.e., the largest spacing, also $m > 0$. The successive extrapolations are expressed as elements in the table, i.e. columns $U_i(k_m)$ with i denoting column i, and the largest i represents the best extrapolated result. Aitken's formula [3] uses three current estimates from this table to produce an improved one. If these estimates are $U_{i-1}(k_{m+2})$, $U_{i-1}(k_{m+1})$, and $U_{i-1}(k_m)$ the formula is given by,

$$U_i(k_m) = U_{i-1}(k_{m+2}) \frac{[U_{i-1}(k_{m+2}) - U_{i-1}(k_{m+1})]^2}{[U_{i-1}(k_{m+2}) - 2U_{i-1}(k_{m+1}) + U_{i-1}(k_m)]} \tag{2.1}$$

where $k_m = k_0/2^m$.

3. THE SYSTOLIC DESIGN

The basic cell (Fig.2) can be constructed to compute formula (2.1) using the equivalent of three ips cells (where we consider the two subtracters to the left and the delay registers denoted Ω as a single ips).

A linearly connected array of n cells can be used to construct the first n columns of the extrapolation table (Fig.1), with each cell generating a single extrapolation column. Figure 3 indicates the operation of a single Aitken extrapolation cell, from which we observe from Table 1 that each cell produces a single output five cycles after the first input and then one on every cycle afterwards. Thus, as the nth cell produces the best extrapolated value, there is only a single value to be output hence after $T = 5n$ the final result is output, and the whole table has been constructed.

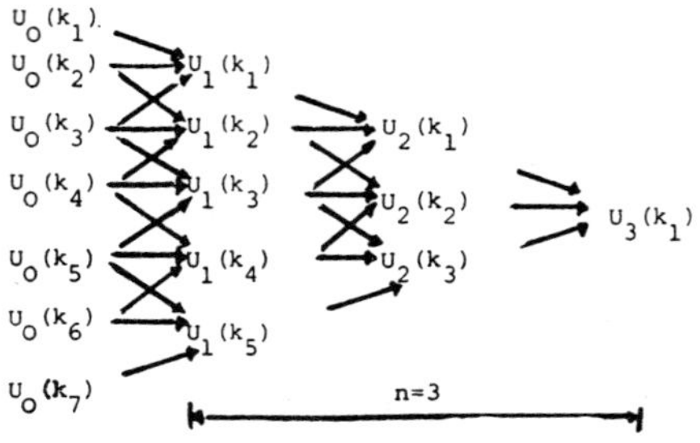

FIGURE 1: Extrapolation Table

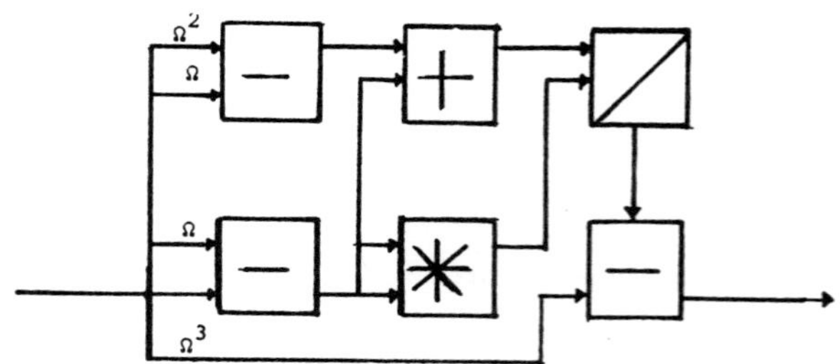

FIGURE 2: Aitken Extrapolation Cell

$$t=3 \qquad \begin{aligned} t_0^{(0)} &= u_0(k_1)-u_0(k_2) \\ t_1^{(1)} &= u_0(k_3)-u_0(k_2) \end{aligned}$$

$$t=4 \qquad \begin{aligned} t_0^{(1)} &= u_0(k_2)-u_0(k_3), & t_2^{(0)} &= t_0^{(0)}+t_1^{(0)} \\ t_1^{(1)} &= u_0(k_4)-u_0(k_3), & t_3^{(0)} &= [t_1^{(0)}]^2 \end{aligned}$$

$$t=5 \qquad \begin{aligned} t_0^{(2)} &= u_0(k_3)-u_0(k_4), & t_2^{(1)} &= t_0^{(1)}+t_1^{(1)}, & t_4^{(0)} &= t_3^{(0)}/t_0^{(0)} \\ t_1^{(2)} &= u_0(k_5)-u_0(k_4), & t_3^{(1)} &= [t_1^{(1)}]^2, \end{aligned}$$

$$t=6 \qquad \begin{aligned} t_0^{(3)} &= u_0(k_4)-u_0(k_5), & t_2^{(2)} &= t_0^{(2)}+t_1^{(2)}, & t_4^{(1)} &= t_3^{(1)}/t_0^{(1)}, & u_1(k_1) &= u_1(k_3)-t_4^{(0)} \\ t_1^{(3)} &= u_0(k_6)-u_0(k_5), & t_3^{(2)} &= [t_1^{(2)}]^2, \end{aligned}$$

$$t=7 \qquad \begin{aligned} t_0^{(4)} &= u_0(k_5)-u_0(k_6), & t_2^{(3)} &= t_0^{(3)}+t_1^{(3)}, & t_4^{(2)} &= t_3^{(2)}/t_0^{(2)}, & u_1(k_2)-u_1(k_4)-t_4^{(1)} \\ t_1^{(4)} &= u_0(k_7)-u_0(k_6), & t_3^{(3)} &= [t_1^{(3)}]^2, \end{aligned}$$

TABLE 1: Partial Results for Aitken Cell

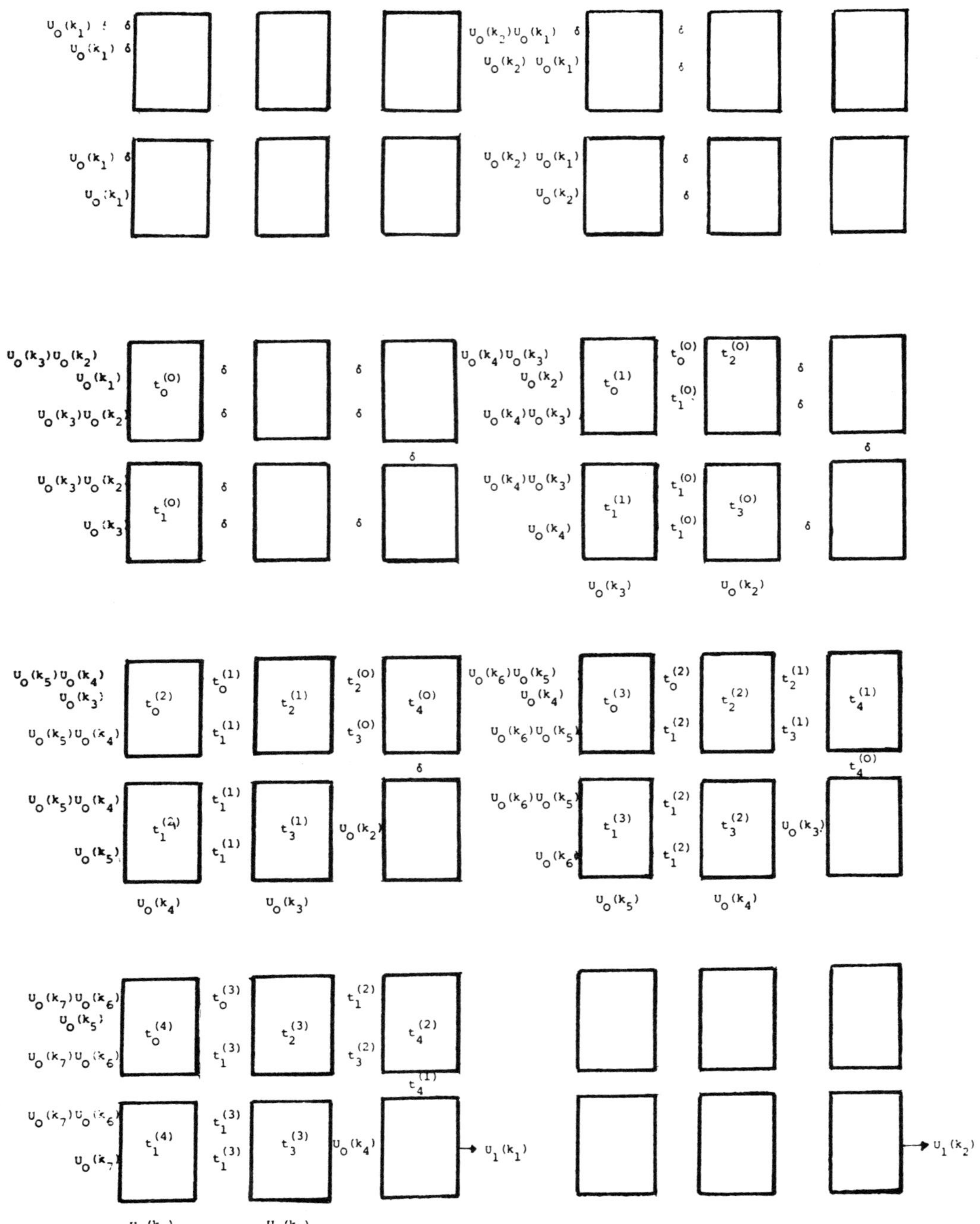

FIGURE 3: Data Flow for Aitken Cell

Similar to other designs like the Romberg Table construction [1] a systolic array can be folded into a ring structure. The ring requires $z=\lceil m/5 \rceil$ cells as when the last starting value is input the data has entered the zth cell and may be ready to output, as no data will now enter the first cell the results of the zth cell can be wrapped around to the first. Hence in the systolic ring Aitken's method requires $3z=3\lceil \frac{m}{5} \rceil$ inner product cells. It also follows that there must be $n=(m-1)/2$ cells in a linear systolic ring as the successive extrapolation columns decrease by 2 elements for each successive extrapolation level. Hence, $m=2n+1$ and the systolic ring requires $3\lceil \frac{2n+1}{5} \rceil$ of the original linear systolic array ips cells which is approximately $\frac{6n}{5}$ as opposed to 3n cells in the linear case, hence the systolic ring is area efficient as we save more than half the cells.

REFERENCES

[1]   Evans, D.J. & Megson, G.M., "Romberg Integration Using Systolic Arrays, Parallel Computing 3, 289-304, 1986.

[2]   Evans, D.J. & Megson, G.M., "Construction of Extrapolation Tables by Systolic Arrays for Solving Ordinary Differential Equations", Parallel Computing 4, 33-48, 1987.

[3]   Lapidus, L. & Pinder, G.F., "Numerical Solution of Partial Differential Equations", Acad.Press, 1984.

SLA – A Language for Simulational Evaluation
of Concurrent Systems Performance

Stanislaw Szejko [1]

Abstract. The paper introduces a language for
specification of concurrent systems.   Due to the
conceptual model features it enables achieving a row of
desired evaluations by means of simulation.

1. Introduction

Performance evaluation and correctness analysis have won the respected
place in system design process. They are practically obligatory for
concurrent and real time systems. However these two classes present the
most complex and environment-dependent areas what makes the methods of
analytic modeling and the associated methods of formal specification be
hardly applicable.

Instead, simulation methods can be used validly in order to achieve de
sired evaluations (Fig. 1). It can be seen [1,2,5,8] that world views of
the (languages) models are moreoften derived from analytical modeling. In
case of complex systems this reduces system specifications to formal in
limited aspects (structured) ones, representing systems as collections of
static and transient entities. And no matter how strange it seems, this
is not sufficient in case of dynamic correctness and quality metrics
evaluations [4].

The objective of this paper is to bring a method and a language which
can improve evaluation of the concurrent designs.

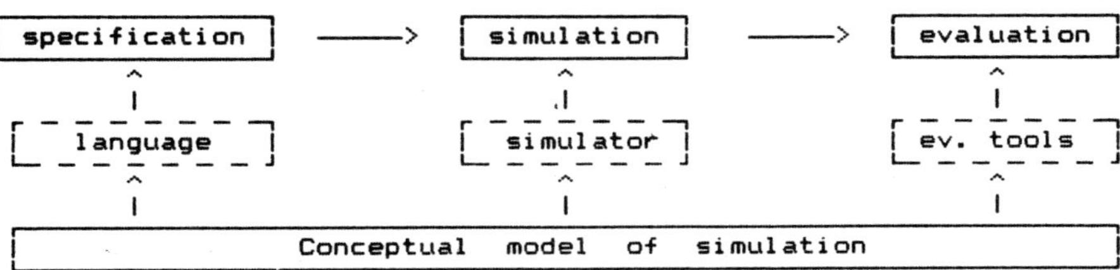

Fig. 1. Main components of the system design process supported by
means of simulation (refinement loops are not marked)

2. Simulation conceptual model

The method is based on the model which has been proposed in [6]. Accor
ding to the model, a system is viewed (Fig. 2) as a set of processors

---

[1]
  Institute of Computer Science, Gdansk TU,
  Majakowskiego 11/12 80-952 Gdansk, Poland

mapped to system <u>structure</u> <u>hierarchies</u>. Each hierarchy is a set of <u>levels</u> partially ordered by the closure of <u>direct</u> <u>support</u> <u>relation</u> There is a set of <u>instructions</u> available at any higher level and a set of <u>modules</u> either. Each module is a sequence of the level instructions established to follow system algorithms; each higher-level instruction is a call for <u>a process</u> of the lower-level module (or for <u>an operation</u>) - according to the support relation. Any process may be <u>marked</u> as handling events or as raising the event at the moment of its termination. Active processes of the hierarchy and ready ones (due to calls, raised events or just started by other processes) compete for the procesor at

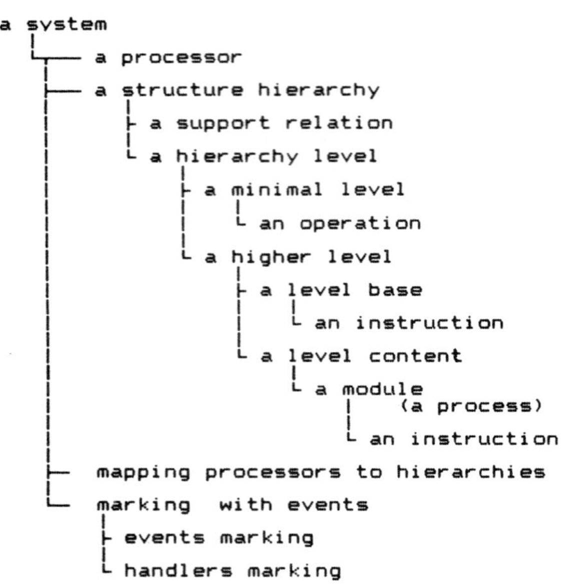

Fig. 2. The morphology of the product
of the model objects

every its cycle i.e. after the stepdown identification of the supports and the execution of the minimal level operation.

A row of attributes has been bound to the model objects while specifying and implementing the model (like states, timers or selection rules).

3. SLA - language

Extended BNF is used to present SLA syntax. The specification subject can be defined as follows:

```
<system specification>  :=  <lumped model specification>
                      <simulation parameters>  <required evaluations>
```

The way of <lumped model specification> allows counting SLA among descriptive scenarios languages [1]:

```
<lumped model specification> :=  'SYSTEM'  <system structure>  <process
            algorithm description>  [<process algorithm description>]*
```

The description of <system structure> is done in accordance with the established model objects and their attributes:

```
<system structure> := <processor> [<processor>]* <hierarchy>
            [<hierarchy>]* <mapping processors to hierarchies>

<processor> := <pr_identifier> <operation execution_time>
      [<operation execution_time>]* <instruction identification_time>

<hierarchy> := <h_identifier> <level> [<level>]* <direct_support
      relation> <minimal_level operation> [<minimal_level operation>]*
```

Level bases result from the declared supports, hence only level contents
must be specified:

        <level (content)> := <module> [<module>]*

          <module> := <m_identifier> <process state (initial value)>
                      <process time (initial value)>
                      <process priority (initial value)>
                      <process instruction counter (initial value)>
                      <call handling rule> <initial events_marking >
                      <process algorithm description>

          <process state> := ACTIVE / BUSY / IDLE / READY
                               / IN_QUEUE / SUSPENDED

          <call handling rule> := FIFO / LIFO / RR

Encapsulating of a single process in a module makes a clear link between
system static and dynamic structure. The process algorithm is specified
in the way which links process function with its run-time behaviour:

          <process algorithm description> := 'BEGIN'
      [<SLA passive instruction> <instruction behaviour equivalent>]*
        <SLA active instruction> <instruction behaviour equivalent>
      [<SLA active instruction> <instruction behaviour equivalent>]* '.'

          <instruction behaviour equivalent> :=
                          <m_identifier> [<m_identifier>]*

The simulation run is driven by the <instruction behaviour equivalent>.
Then, the simulation time is suspended and the SLA_instruction
corresponding to the modelled sequence is interpreted:

          <SLA passive instruction> := <declaration of a constant>
              / <declaration of a type> / <declaration of a variable>

Declarations are organized Pascal-like. Variables can be declared as
global or local ones [1]. Active instructions are of two kinds: typical,
high-level structured instructions acting variables and the ones directed
to the model objects or their attributes:

          <SLA active instruction> := <instruction acting variables>
                                      <instruction acting objects>

          <instruction acting variables> := CASE / IF / JUMP
                          / FETCH / LOOP / READ / WRITE / WITH

          <instruction acting objects> :=

    ACTIVATE   - change process state to READY (after process creation)
  / AWAITED   - check if there are processes suspended till the event
  / CLOCK   - set the process clock
  / CREATE   - create a new object
  / DELETE   - delete an object
  / DELEVENT   - delete the process marking
  / EVENTAS   - mark the process as handling the event
  / GOTOPROCESS   - terminate process, activating the other one
  / MASK   - mask events handled in the hierarchy
  / NOOP   - no operation
  / PRIOR   - set process priority
  / RETURN   - terminate process with or without an event
  / STARTPR   - allocate processor to a process/hierarchy
  / STARTPS   - change process IDLE state to READY one
  / STOP   - halt the processor's run
  / T   - increment process own timer
  / WAIT   - suspend process till the event raises or till the time
  / call   - a call for a standard procedure

In fact, each of instructions acts attributes of the model objects, at
least setting the new value of the process instruction pointer . And the
simulation run is resumed.

<simulation parameters> include maximal simulation time, level of detail of the output results, etc.

<required evaluations> consist of a row of declarations as only the history of run is always produced and deadlock detected. There can be required: testing the synchronization between any declaired pair of processes, critical sections observing, process tracing, evaluation of response, execution and delay times, evaluation of queues and starvations, processors' activity and workload.

4. Notes on implementation

The language is to play its role in a package which is being developed to model and evaluate concurrent systems [7]. Its primary version was written in Pascal and run on IBM PC under DOS [3]. This has enabled modeling of selected aspects of a distributed data-exchange. The beneath fragment of specification of a SUBSCRIBER process is taken therefrom; some interesting results delt with the time of connection of two subscribers and with times and degrees of processors' workload.

```
Process 44 - ASUBS1 - Select process initial state (Active/Ready/Idle/Busy/Waitng) - IDLE
Process 44          - Select process initial priority ............................. - 6
Process 44          - Timer initial value ......................................... - 0
Process 44          - Instruction pointer initial value ........................... - 0
Process 44          - Call handling rule (FIFO/LIFO/RR) ........................... - FIFO
Process 44          - Processes marked as handling its termination event ......... - ; (no)
Process 44          - Algorithm description:

    /function/          /behaviour/                 /comments/

BEGIN
WAIT (ASUBS1, SENDBS1)  WAITS1;                     ;waiting till completition of an address by MANAGER process
                                                     WAITS1 models the wait_function of the system kernel
NOOP                    MACHS1, READDBS1, 3xMACHS1,  ;checking services, categories, defining rates and address
                        READDBS1, 3XMACHS1, READDBS1, translation. MACHS1 represents host machine instructions,
                        MACHS1, READDBS1, 3XMACHS1;   READDBS1 mimics read_from_database procedure
NOOP                    SEXTAS1;                     ;kernel_send to the processor of the B-subscriber
WAIT (ASUBS1, SEXTAS2)  WAITS1;                      ;waiting for a response
IF S2OK <> TRUE THEN    MACHS1;                      ;connection allowed ?
RETURN                  3xMACHS1, WRITEDBS1, DELETES1; ;write_database procedure, kernel_delete function
ELSE                    MACHS1;
NOOP                    SEXTBK1;                     ;kernel_send to the commuting processor
WAIT (ASUBS1, SEXTBK1)  WAITS1;                      ;wait till the commutation is realized
IF KOK = TRUE THEN      MACHS1;                      ;connection done?
WAIT (ASUBS1, CORRESP)  WAITS1;                      ;wait till the end of the correspondence
```


Bibliography

[1] Beilner H.: On the construction of computing system simulators. In Experimental Computer Performance Evaluation, North-Holland 1981, 1-32.
[2] Dahl O.-J.: Languages for descrete event systems simulation. In Programming Languages, Academic Press, London and New York 1968.
[3] Flisek A. and Wichorowska J.: A Design of a Specification Language and Its Interpreter for Simulation of Systems (in Polish). Graduation work. Gdansk TU, 1987.
[4] Knuth E.: Analysis of specifications. Proc. EWICS'83, Graz, Austria.
[5] McDougall M. H.: System level simulation. In Digital System Design Automation. Pitman, USA 1978.
[6] Szejko S.: Structured model of real-time systems. Proc. of IFAC/IMACS Symposium on Control Systems, Vienna Austria 1986.
[7] Szejko S.: SLAS - A package for simulational evaluation of concurrent systems performance. Preprints of MIMI'88 Conf., Spain 1988.
[8] Zeigler B.P.: Theory of Modelling and Simulation. J.Wiley & Sons 1976

----
1)
Not all planned TYPES and INSTRUCTIONS are implemented yet [3].

The Off-Line Motion Planning via the Computer Graphics Simulation System

Kesheng Wang and Øyvind Bjørke *

ABSTRACT

This paper describes the off-line robot motion planning via the computer graphics simulation system for robot manipulators (CGSROB) and its environment. The general structure and the mathematical theories of a generalized computer program developed for this purpose are proposed. Approachs based on B-splines are emploied for the robot trajectory planning here. The motion is specified by a sequence of cartesian knots, i.e. positions and orientations of the end effector of robot manipulators. For a 6-joint manipulator, these cartesian knots are transformed into six sets of joint variables, with each set corresponding to a joint. Splines, represented as linear combinations of B-splines, are used to fit the sequence of joint variables for each of the six joints. This approach offers flexibility, computational efficiency, and compact representation and fits the jerk constraint reguired.

1. INTRODUCTION

In applications such as weilding and painting, it is necessary for the robot manipulator to follow the shape of the object on which it is working. In other applications, it may be necessary for the robot manipulator to avoid obstacles between goal points. This motion is specified by a sequence of cartesian knots, i.e. positions and orientations of the end effector of robot manipulator. The robot motion planning is to develop the necessary joint-level trajectories for a robot to perform a given motion.

In most commercial application, the motion planning is performed by the programmer who use a teacher pendent, a joystick, or a high-level programming language to facilite the teaching or programming of the robot. Once the teaching or programming is completed for a task, the robot hardware is used in playback mode to test the program and the

* Production Engineering Laboratory,NTH-SINTEF,N-7034 Trondheim,Norway

motion of the robot. The debugging is accomplished through a
teache-playback or edit-compile-play loop. This can be a very tedious
and unsafe procedure, requiring task-planning modifications that would
result, for example, in rearrangment of worktables or design changes in
fixtures and part-presenting devices. Furthermore, errors in moiton
planning or programming may cause the robot to collide with fixture or a
worktable. In addition, the robot programmers usually decided on robot
trajectories in between positions in ad hoc programtic manner. These
trajectories may be far from optimal and may result in longer cycle time
or decreased accuracy during motion performance. A totally different
approach to robot motion planning and programming is to use computer
graphics simulation system rather than the actural hardware and to
perform the motion planning and programming in an off-line manner. Such
an approach would allow the user to plan motion strategies and to roughly
debug programs without needing access to the robot or other devices in
its environment.

In this paper we present the structure and the mathematical theories
of the CGSROB which was developed to aid the design of industrial robot
system, the planning of robot motion task, and the testing of robot
programs (Wang and Balchen 1987) and a robot joint trajectory planning
based on B-splines.

2. FACILITIES OF THE SYSTEM

2.1 Hardware
The hardware arrangements are shown in Fig. 1. The host computers are
VAX 11/750 or VAX 11/780 which can communicate with each other. User
application programs are created, compiled, linked and executed on these
host computers. The graphics display devices are a Westward station, VAX
stations, and Tektronix tubes.

2.2 Software
The software of the CGSROB system is divided into three main levels,
namely the VAX-VMS operating system; the graphics system which includes
the standard graphics packages such as GPGS-F, GKS, and other CAD
geometric models; and an application system which is written in FORTRAN.

2.3 Geometric model
A 3D wireframe geometric model is used in the CGSROB system to represent
robots and the environments The description for a link or an object is
input to the program either by specifying the coordinates of several
points on an object to form a wireframe graphic representation of the
object in the GPGS-F package or by using the database generated in
external CAD geometric models.

3. OVERVIEW OF THE SYSTEM

The structure of the computer graphics simulation system developed is shown in Fig. 2. The whole system consists of the following basic modules: user inputs; geometric model; kinematic calculation; trajectory planning; motion programming; and graphics display. The configuration of the robot and world environment can be created in the user inputs and geometric modules Robot motion strategies are entered via an manipulator level language in robot programming. At this stage, the CGSROB system fully uses interactive computer graphics techniques to provide an efficient interface between the user and the program. The motion data is computed and manipulated in the kinematic calculation and simulator modules. Finally, all results are visualized on screen through the graphics display module.

4. MODEL BUILDING

4.1 Robot modelling
The graphic simulation of robot kinematics utilizes the technique of geometric transformations, which has been used to describe robot kinematics and transform objects in graphic space (Newman and Sproull 1979). A coordinate frame is rigidly fixed to each robot link to describe the position and orientation of the link. The Denavit-Hartenberg representation (Denavit and Hartenberg 1955) is used for assigning the coordinate frames. First, we create an object for each robot link. Next, we use the GPGS subroutine calls to draw the robot links one by one on the graphics terminal. We continue this procedure to draw all the robot links from the base to the end effector.

4.2 World modelling
Objects such as the conveyer, working table, workpieces, etc. are similarly created for general geometric bodies. First the coordinate frame of working table and coordinate frame of the part are located. Then the object can be created in every coordinate frame as picture segments. By rotating and translating the coordinate frames in which the object is fixed, the world environment can be conveniently created.

If subroutines for drawing and transforming objects are available, which is quite common in most graphics systems, a similar version of the statements can be directly used for simulation program development. If not, these statements can still be used although the programmer must develop his own routines for drawing and transforming objects.

5. MATHEMATIC EQUATIONS FOR KINEMATICS

In many robot applications such as assembly, material handling, etc, we
describe the robot configuration in terms of the location of the end
effector (i.e. position and orientation). In order to display the robot
configuration corresponding to the location of the end effector, joint
variables have to be solved. This can be done by using algebraic,
geometric, or iterative approaches to find the solution of the inverse
kinematics problems. The kinematics equation for a six-degree-of-freedom
robot manipulator can be written in terms of 4X4 homogeneous
transformation matrix as

$$\,^0_6T = A_1 \ A_2 \ A_3 \ A_4 \ A_5 \ A_6$$

where

$\,^0_6T$ — matrix representing the transformation of the coordinate
frame of the end effector with respect to the base
coordinate frame (XYZ)0.

A_i — matrix representing the transformation of the coordinate
frame fixed to the ith link, (XYZ)i, with respect to the
coordinate frame attached to the (i-1)th link, (XYZ)i-1.

CGSROB uses the closed form algorithm (Wang and Lien 1987a) to solve the
inverse kinematic equation for the joint variables. For example, there
are eight solutions for the PUMA robot. In actual robot simulation, one
solution must be chosen based on practical constraints and other
conditions. After the joint variables are obtained, they are
incorporated in the simulation program to create the robot configuration
corresponding to the specified position and orientation of the end
effector.

6. TRAJECTORY PLANNING

One approach to robot trajectory generation is to use polynomial
trajectories (Bollinger and Duffie 1979; Paul 1981; Ho and Cook 1982;
Luh, Lin and Chang 1983; Derby 1983; Hornick and Ravani 1986).

All of them have not considered the jerk constraint. Here we propose
a new approach in which splines, represented as linear combinations of
B-spline, are used to generate the robot trajectories. Such profiles are
especially suitable for off-line motion planning and programming.

6.1 Interpolation by B-splines

We will use B-splines as building blocks to construct curves that
interpolate to given points(De Boor 1978). B-splines of order k are
piecewise polynomials of degree k-1 in one variable. They are k-2 times
continuously differentiable, positive on exactly k consecutive intervals
and zero everywhere else. Inside their support they are bellshaped.

Let $t_1 \leqslant t_2 \leqslant \ldots \leqslant t_{n+k}$ be a nondecreasing sequence of real numbers, where $n \geqslant k \geqslant 1$ The points $t_1, \ldots, t_{n+k}$ are called knots. In 1972 both De Boor (1972) and Cox (1972) discovered a stable recurrence formula for computing B-spline values.

The simplest B-spline, that of order 1, is given by

$$B_{i,\,1}(t) = \begin{cases} 1, & t_i \leqslant t < t_{i+1} \\ 0, & \text{otherwise} \end{cases} \tag{1}$$

The B-splines of higher order are recursively determined by

$$B_{i,\,j}(t) = \frac{t - t_i}{t_{i+j-1} - t_i} B_{i,\,j-1}(t) + \frac{t_{i+j} - t}{t_{i+j} - t_{i+1}} B_{i+1,\,j-1}(t) \tag{2}$$
$$i = 1, 2, \ldots, n \qquad j \geqslant 2$$

The derivative of a B-spline is given by the following formula (de Boor 1978)

$$B'_{i,\,j}(t) = (j - 1)\left(\frac{B_{i,\,j-1}(t)}{t_{i+j-1} - t_i} - \frac{B_{i+1,\,j-1}(t)}{t_{i+j} - t_{i+1}} \right) \tag{3}$$

A spline of order k is a piecewise polynomial of degree less than k which is k-2 times continuously differentiable. Schoenberg (1946) originally introduced the splines and showed together with Curry (1966), that on (t_k, t_{n+1}) splines and linear combinations of B-splines are the same thing. The "B" in B-splines stands for basis. The B-splines constitute a basis for the vector space of kth order splines.

If I = [a,b] is the global parameter-domain of interest, then by choosing

$$t_1 = t_2 = \ldots = t_k = a, \; t_{n+1} = t_{n+2} = \ldots = t_{n+k} = b \tag{4}$$

any spline of order k on I is equal to some linear combination of B-splines on the whole of I.

Let m > 1. In our case usually m = 6. A parametric spline curve of order k in $\mathbb{R}^m$ can now be defined by

$$s(t) = \sum_{j=1}^{n} c_j \cdot B_{j,\,k}(t), \qquad t \text{ in } I \tag{5}$$

where the c_j's are vectors in $\mathbb{R}^m$.

Let $\tau_1 \leqslant \tau_2 \leqslant \ldots \leqslant \tau_n$ be n abscissas in I and set $d_i = \underset{\tau_{i-j} = \tau_i}{\text{Max}\, j}$.

If $P_1, P_2, \ldots, P_n$ are n points in $\mathbb{R}^m$, then we say that the curve (5) Hermite interpolates to these points at $\tau_1, \tau_2, \ldots, \tau_n$ respectively if

$$s^{(d_i)}(\tau_i) = P_i, \qquad i = 1, 2, \ldots, n \tag{6}$$

Let us relate this to ordinary scalar spline Hermite interpolation. Let $P_{l,\,i}$ be the component of P_i and $c_{l,\,i}$ similar for c_i . By (5), the lth component of (6) is then equal to

$$\sum_{j=1}^{n} c_{l,j} B_{j,k}^{(d_i)}(\tau_i) = P_{l,i}, \qquad i = 1, 2, \ldots, n \tag{7}$$

This is an ordinary linear system of n equations in n unknowns $c_{l,1}, c_{l,2}, \ldots, c_{l,n}$. Hence the interpolation condition (6) is equivalent to m different systems of linear equations. The system (7) has a unique solution if the matrix

$$M = (B_{j,k}^{(d_i)}(\tau_i))_{i=1 \quad j=1}^{n \quad n} \tag{8}$$

is non-singular. This is the case if there is at least one interpolation point in each (t_i, t_{i+k}), i=1,2,...,n (Schoenberg and Whitney, 1953).

The curve (5) which solves (6) can now be constructed in the following steps:

Step 1. Construct the matrix (8) by using the recurrence formulae (2) and (3).

Step 2. Solve the linear system (7) for each l=1,...,m.

Step 3. For each evaluation point t calculate the right side of (5) by using the recurrence relation (2).

If there are many t´s for which the sum (5) has to be calculated, it is more economic to first find the different polynomial pieces of each component of s and then use Horner´s scheme. This then replaces Step 3 above.

In practice only the vectors $P_1, P_2, \ldots, P_n$ are given so that the interval I, the knots $t_1, t_2, \ldots, t_{n+k}$ and the interpolation points $\tau_1, \tau_2, \ldots, \tau_n$ can be chosen by the user. The first k and last k knots have to be chosen according to (4), and Schoenberg-Whitney condition has to be fulfilled. Since the interpolation points are ordered in an increasing sequence, the following inequalities must hold:

$$t_i < \tau_i < t_{i+k}, \qquad i = 1, 2, \ldots, n \tag{9}$$

where the left inequality may be weakened if k knots coalasce. This leaves quite a bit of freedom. If we want the abscissa τ_i to have multiplicity d, i.e. $\tau_{i-1} < \tau_i = \ldots = \tau_{i+d-1} < \tau_{i+d}$, the difference $\tau_{i+d} - \tau_i$ and the distance between P_i and P_{i+d} ought to be proportional. a reasonable means of selecting the knots is the following (De Boor, 1978):

$$t_i = (\tau_{i-1} + \tau_{i-2} + \ldots + \tau_{i-k+1})/(k-1) \qquad i = k+1, \ldots, n. \tag{10}$$

In the next section we will look at some numerical examples for robot manipulators.

6.2 Numerical Example

In order to illustrate how to use B-splines to construct joint trajectories for robot manipulators, a PUMA 600 robot manipulator with six joints is considered as a numerical example.

Ten knot points from a Cartesian path of the end effector of the robot manipulator are chosen. Using the closed form solution for the inverse kinematics (Wang and Lien, 1987b), the joint variables are solved for these knots and shown in Table 1.

Table 1   Joint variables for the PUMA 600 robot manipulator

Knot joint	1	2	3	4	5	6	7	8	9	10
1	15.0	30.0	50.0	90.0	130.0	90.0	45.0	-10.0	-30.0	-50.0
2	10.0	25.0	30.0	15.0	-20.0	-55.0	-70.0	-20.0	0.0	10.0
3	50.0	70.0	150.0	200.0	120.0	35.0	-10.0	50.0	60.0	50.0
4	15.0	20.0	40.0	80.0	80.0	40.0	-60.0	-100.0	-60.0	-30.0
5	10.0	30.0	10.0	-40.0	-60.0	10.0	50.0	-40.0	-20.0	10.0
6	6.0	20.0	40.0	80.0	70.0	10.0	-10.0	15.0	30.0	20.0

The robot is at rest at the starting point, and comes to a full stop at the end point.

The velocity, acceleration and jerk constraints are given in Table 2.

Table 2   Joint constraints of the PUMA 600 robot manipulator

Joint constraint	1	2	3	4	5	6
Velocity (degree/sec)	100.0	95.0	100.0	150.0	130.0	110.0
Acceleration (degree/sec^2)	45.0	40.0	75.0	70.0	90.0	80.0
Jerk (degree/sec^3)	60.0	60.0	55.0	70.0	75.0	70.0

Since the velocity and acceleration have to be zero at the start and end points, the spline to be constructed must Hermite interpolate to the following points:

$$P_1 = (15.0, \ 10.0, \ 50.0, \ 15.0, \ 10.0, \ 6.0)^T$$
$$P_2 = (0.0, \ 0.0, \ 0.0, \ 0.0, \ 0.0, \ 0.0)^T$$
$$P_3 = (0.0, \ 0.0, \ 0.0, \ 0.0, \ 0.0, \ 0.0)^T$$
$$P_4 = (30.0, \ 25.0, \ 70.0, \ 20.0, \ 30.0, \ 20.0)^T$$
$$\cdots$$
$$P_{11} = (-30.0, \ 0.0, \ 60.0, \ -60.0, \ -20.0, \ 30.0)^T$$
$$P_{12} = (0.0, \ 0.0, \ 0.0, \ 0.0, \ 0.0, \ 0.0)^T$$
$$P_{13} = (0.0, \ 0.0, \ 0.0, \ 0.0, \ 0.0, \ 0.0)^T$$
$$P_{14} = (-50.0, \ 10.0, \ 50.0, \ -30.0, \ 10.0, \ 20.0)^T$$

where $\tau_1 = \tau_2 = \tau_3$ and $\tau_{12} = \tau_{13} = \tau_{14}$ and the rest of the τ's are simple ($d_i = 0$).

We want the spline curve to be three times continuously differentiable. Hence we cannot use cubic splines, k=4, since they are only twice continuously differentiable. But quartic (k=5), quintic (k=6) and higher degree splines can be used. We choose quartic splines, since piecewise 4th degree polynomials are cheaper to evaluate, needing only 4 multiplications per point.

We want to find interpolation abscissas and knots (the dividing points between subintervals) on an interval $[0, T]$, where the spline curve Hermite interpolates to the data and satisfies all the joint constraints, and where the total time T is as short as possible.

Let us observe that we can work with a fixed interval, for instance $[0,20]$, and scale this interval at the end of the calculations.

If we find interpolation abscissas τ_i and knots t_i on $[0, 20]$ for which all the constraints are satisfied, that means we can use a linear scaling to a smaller interval $[0, T]$. The new abscissas and knots will be $(T/20)\tau_i$ and $(T/20)t_i$.

If s_j was the joint j spline on $[0, 20]$, then the corresponding scaled spline on $[0, T]$ will be $s_j((20/T)t)$. Let us denote the constraint for joint j and the ith derivate by Mji. By differentiation, the following inequalities have to hold

$$\left(\frac{20}{T}\right)^i \| s_j^{(i)} \|_\infty \leq M_{ji}, \qquad \begin{matrix} i = 1, 2, 3 \\ j = 1, 2, \dots, 6 \end{matrix} \qquad (11)$$

where $\| \ \|_\infty$ denotes the sup-norm. By (11), the least total time T we can have and still satisfy all the constraints, is the following

$$T = 20 \ \underset{1 \leq j \leq 6}{\text{Max}} \ \underset{1 \leq i \leq 3}{\text{Max}} \left(\frac{\| s_j^{(i)} \|_\infty}{M_{ji}} \right)^{1/i} \qquad (12)$$

This T is a function of the abscissas τ_i and knots t_i on $[0, 20]$. Here, $\tau_1 = \tau_2 = \tau_3 = 0$ and $\tau_{12} = \tau_{13} = \tau_{14} = 20$ and n=14.

The knots have to satisfy (4), so $t_1 = t_2 = \dots = t_5 = 0$ and $t_{15} = t_{16} = \dots = t_{19} = 20$. Hence T given by (12) is a function of the 8 internal abscissas $\tau_4, \dots, \tau_{11}$ and the 9 internal knots $t_6, \dots, t_{14}$.

For each abscissa and knot set the six different interpolating splines were constructed and transformed to piecewise polynomial form by some subroutines written by Carl de Boor and modified by D. E. Amos, see the SLATEC library of FORTRAN subroutines and (de Boor 1977).

Our FORTRAN program needs a set of starting values for abscissas and knots. Let us now only consider those vectors P_i that correspondto spline values and not to derivatives. Set $P_1^* = P_1, P_2^* = P_4, P_3^* = P_5, \dots, P_{10}^* = P_{12}$ and $\tau_1^* = \tau_1, \tau_2^* = \tau_4, \dots, \tau_{10}^* = \tau_{12}$. It seemed reasonable that the difference between two abscissaas, $\tau_{i+1}^* - \tau_i^*$, should be proportional to the distance between P_i^* and P_{i+1}^*, $\| P_{i+1}^* - P_i^* \|_\infty$. We tried with a weighted proportionality,

$$\tau_{i+1}^* - \tau_i^* = 20\,\frac{W_i \|P_{i+1}^* - P_i^*\|_\infty}{\displaystyle\sum_{j=1}^{9} W_j \|P_{j+1}^* - P_j^*\|_\infty}\,, \qquad i = 1, 2, \ldots, 9 \qquad\qquad (13)$$

where the W ´s are some chosen positive weights.   It   is   reasonable   to have higher weights at the end points since the robot manipulator starts and stops with zero velocity and acceleration.   Therefore   it · must   have more   time   there.   So let $W_2' = W_3' = \ldots = W_8' = 1$ and $W_1' = W_9'$. The knots were then chosen according to (10) in the start.

We have tried different W_1 and found that the last experiment where $W_1 =$ 2.3 gave the least total time.   The interpolation abscissas and knots on [0,20] in this case are shown in Table 4. The plots are given in Figure 3.

Table 4   Interpolation points and knots ($W_1 =$ 2.3)

i	τ_i	t_i
1	0.000	0.000
2	0.000	0.000
3	0.000	0.000
4	2.186	0.000
5	3.868	0.000
6	5.508	1.468
7	8.006	2.748
8	10.659	4.731
9	13.780	7.028
10	16.590	9.486
11	17.840	12.258
12	20.000	14.718
13	20.000	17.052
14	20.000	18.607

7. ROBOT PROGRAMMING

The user can create a motion program via manipulator level commands interactively.   At the manipulator level, robot motions are implicitly defined by specifying motion operations that are to occur between the objects being manipulated. Input commands are defined with reference to VAL, where some commands are listed as follows (Pai and Leu 1986; User,s guide 1980):

```
SET --------  initial starting point
MOVE_J -----  move hand in joint interpolation
MOVE_S -----  move hand in a straight line
APPROACH_J -  moves hand to the location and an offset
              along the z axis in joint interpolation
APPROACH_S -  'moves hand to the location and an offset
              along the z axis in a straight line
DEPART_J ---  move hand to the location along the current
              z axis in joint interpolation
```

          DEPART_S ---   move hand to the location along the current
                         z axis in a straight line
          CLOSE ------   close hand
          OPEN -------   open hand

The simulation program accepts input commands and uses the language
processing module to interpret them to generate the required motion data
to the respective joints.

The CGSROB system can display a multiple view of the robot and its
environment, (Fig. 8), enabling the programmer to plan the layout of the
robot system and to visually detect any potential interference between
the robot and objects in its environment at the planning stage.

8. CONCLUSIONS

CGSROB - an interactive computer graphics simulation system for
industrial robots has been developed. This system can be used for the
off-line robot motion planning.

A scheme for generation of B-splines joint trajectories which is
widely used in the field of computer graphics for connecting data points
with smooth curve has also been presented. These curves, known as
B-apline functions, to provide good inherent behaviors, such as to
prevent the loss of accuracy, to reduce greatly the amount of
computation, to help the convergence analysis. When this approach is
used to establish the robot joint trajectories, it will offer
flexibility, computation efficiency, and compact representation and can
fit the jerk constraint required.

9. REFERENCES

1. Bollinger, J. and Duffie, N. (1979): Computer algorithms for
    high speed continuous path robot manipulator, Ann. CIRP 28(1),
    PP. 391-395.
2. Cox, M. G. (1972): The numerical evaluation of B-splines, J.
    Inst. Math. Applic. 10, PP. 134-149.
3. Curry, H. B. and Schoenberg, I. J. (1966): On polya frequency
    functions IV: The Fundamental spline functions and their limits,
    J. d´Analyse Math. 17, PP. 71-107
4. De Boor, C. (1972): On calculating with B-splines, J.
    Approximation Theory 6, PP. 50-62.
5. De Boor, C. (1978): A practical guide to splines, New York,
    Springer Verlag, P. 219 & P. 138.

6. Denavit, J., Hartenberg, R.S.   (1955):A kinematic notation for lower pair mechanisms based on matrices. J. of Applied Mechanics, Trans. ASME, Vol. 77,pp. 215-221.

7. Derby, S.J. (1981):Kinematic elasto-dynamic analysis and computer graphics simulation of general purpose robot manipulators. Ph.D Thesis, Rensselaer Polytechnic Institute, Troy, NY.

8. Ho, C.Y. and Cook, C.C. (1982):The application of spline functions trajectory generation for computer controlled manipulators, Digital Systems for Industrial Automation, 1(4),PP. 325-333.

9. Hornick, M.L., Ravani, B. (1986):Computer-aided off-line planning and programming of robot motion, The International Journal of Robotics Research, Vol. 4, No. 4., PP. 18-31.

10. Luh, J. Y. S., Lin, C. S. and Chang P. R. (1983): Formulation and optimization of cubic polynomial joint trajectory for industrial robots, IEEE Trans. Automatic Control, AC-28(12), PP. 1066-1074.

11. Newman, W.M., Sproull, R.F. (1979): Principles of interactive computer graphics. 2nd edition, McGraw-Hill, New York,

12. Pai, D.K., Leu, M.C.(1986): Ineffabelle - an environment for interactive computer graphic simulation of robotic applications. Preceedings 1986 IEEE International Conference on Robotics and Automation, USA, PP. 897-903.

13. Paul, R. C. (1981): Robot manipulators: mathematics, programming and control, Cambridge: MIT press.

14. Schoenberg, I. J. (1946): Contribution to the problem of approximation of equidistant data by analytic functions, Quart. Appl. math. 4, PP. 45-49 & 112-141.

15. Schoenberg, I. J. and Whitney, A.(1953): On polya frequency functions III: The positivity of translation determinant with applications to the interpolation problem by spline curves, Trans. Amer. Math. Soc. 74, PP. 246-259.

16. Wang, K., and Lien, T.K. (1987a): The structure design and kinematics of robot manipulators, The Proceedings of International Conference on The Robotics, Paper No.9, Yugoslavia.

17. Wang, K., and Lien, T.K. (1987b): The solution with closed form for the inverse kinematics of PUMA robot manipulators, The Proceedings of International Conference on The Robotics, Paper No.10, Yugoslavia.

18. Wang, K. and Balchen, J. G., (1987): Computer graphics simulation system for industrial robots. The Proceedings of 17th ISATA Conference, No. 87168, Munich, West Germany.

19. User´s Guide to VAL (1980):Unimation, Inc., Danbury CT.

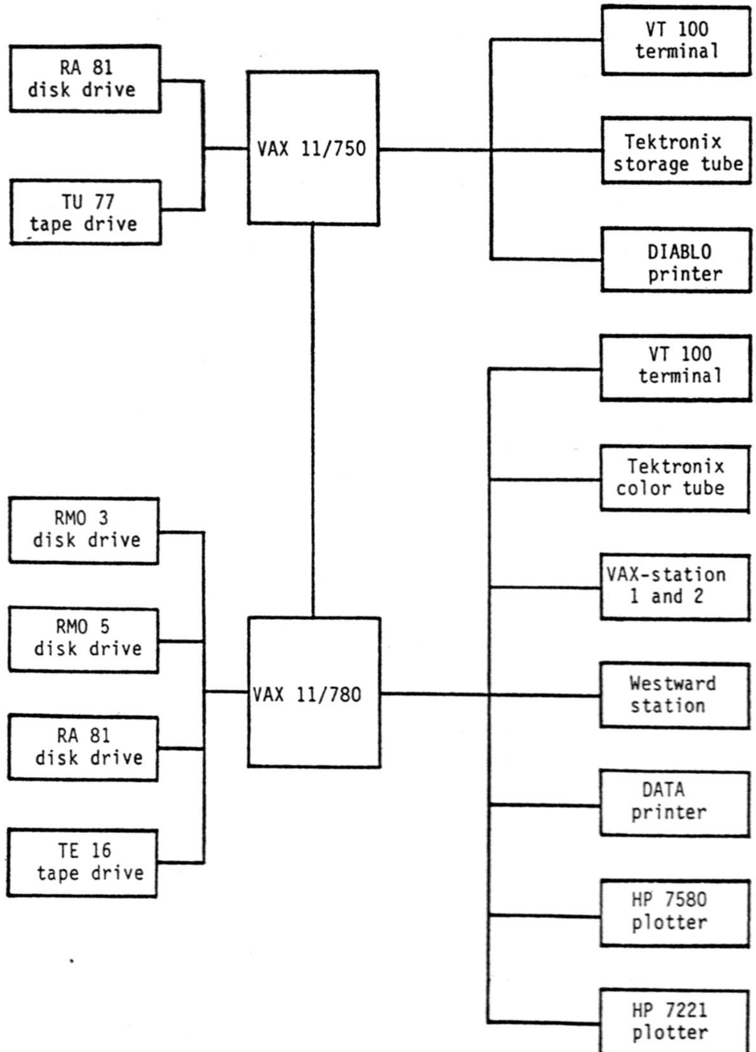

Fig. 1 The computer graphics simulation facilities

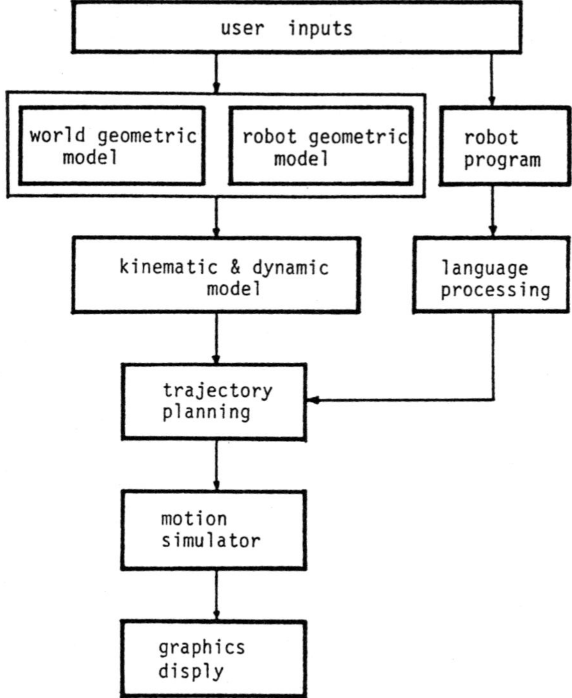

Fig. 2 The structure of the CGSROB system

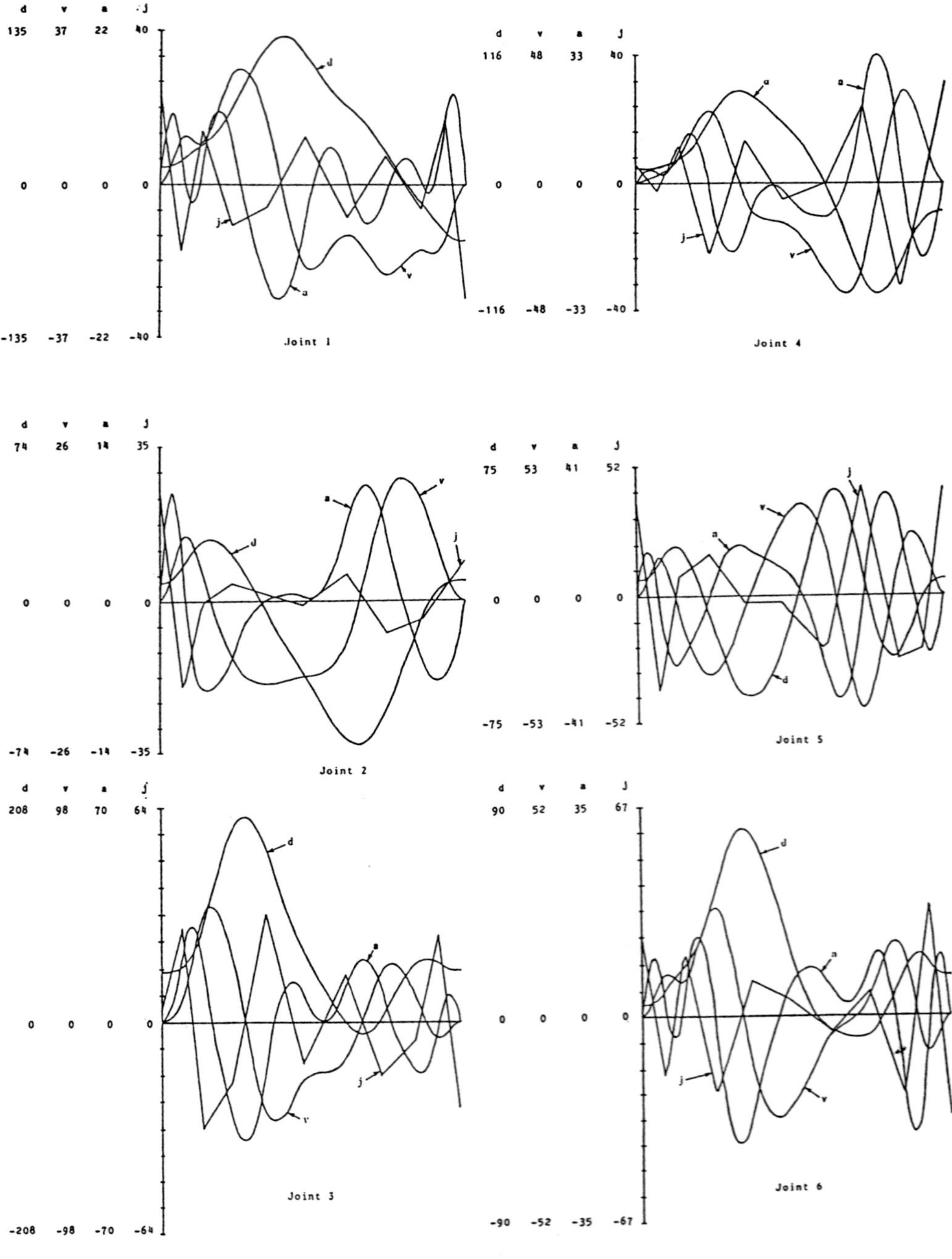

d=displacement, v=velocity, a=acceleration, j=jerk

Fig 3. Joint trajectories for the example

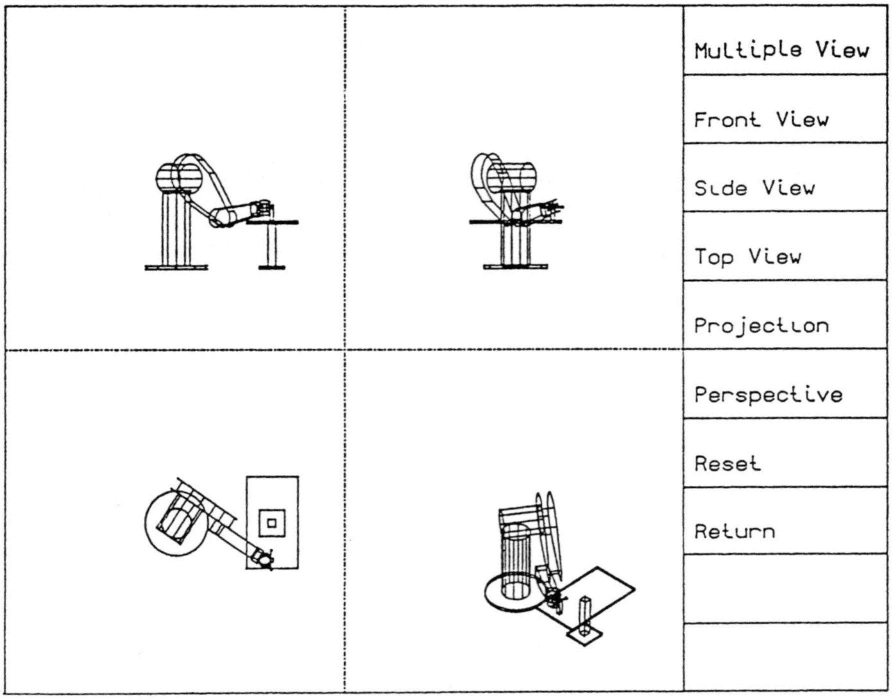

Fig. 4. Multiple views for interactive motion programming

Mini-Supercomputers: New Perspectives in Scientific Computation and Simulation

E.J.H.KERCKHOFFS *)

ABSTRACT

Mini-supercomputers and their significance in scientific computation and systems simulation are discussed. These mini-supercomputers with a considerably favorable cost/performance ratio appeared on the market recently. In this paper we consider in some detail two specific examples of these commercial products: the NCube/4+ parallel computer and the Convex C-1 vector processor, both available at the author's institute (Delft University of Technology).

1. INTRODUCTION

The simulation of (complex) continuous systems has been particularly influential in the evolution of special-purpose computer systems to attain high computing speeds and performance. In this respect two major computer architecture innovations have permitted the circumventing of the von Neumann bottleneck in order to attain high speeds: parallelism and pipelining. The implementation of these techniques has resulted in some distinct families of high speed digital computers including supercomputers, peripheral array processors and multiprocessors [FADD84], [HOCK81], [KARP84], [KARP87], [LOUI81], [SPRI82].

A particularly important phenomenon in the evolution of simulation-oriented digital computer systems was the appearance of mini-supercomputers in the mid 1980's: commercially available multiprocessing systems and relatively inexpensive vector processors, all with a very favorable cost/performance ratio. An important subset of these mini-supercomputers shows the successful implementation of the MIMD architecture concept, implying a (smaller or larger) network of identical processing elements (PEs). Examples are [KARP87]:

        N-CUBE, NCUBE   (max. 1024 PEs)
        INTEL, iPSC   (max. 128 PEs)
        ALLIANT, FX/8   (max. 8   PEs)
        AMETEK, SYSTEM 14   (max. 256 PEs)
        ENCORE, MULTI-MAX   (max. 20   PEs)
        FLEXIBLE, FLEX/32   (max. 20   PEs).

Another subset of the present mini-supercomputers is based upon pipelining, and is therefore more similar in organization to the well-known CRAY computers and other pipelined supercomputers. Computers of this category include:

        CONVEX COMPUTER CORPORATION, C-1
        FLOATING POINT SYSTEMS, INC., FPS 164,264 and 364
        SCIENTIFIC COMPUTER SYSTEMS CORPORATION, SCS-40.

Most systems employ a Unix-based operating system and are committed to providing compilers for the FORTRAN 77 and the C languages, and sometimes for PASCAL.

In the following sections we discuss in some detail the first mentioned in both above lists of computers: the N-Cube parallel computers (especially the NCube/4+) and the vector processor Convex C-1, both available at Delft University of Technology.

2. THE NCUBE/4+ PARALLEL COMPUTER

2.1 The hypercube architectures

An n-dimensional hypercube computer, also known as a binary n-cube computer, is a multicomputer characterized by $N=2^n$ processors interconnected as an n-dimensional binary cube. Each processor forms a node (or vertex) of the cube and has its own CPU and local memory. Each processor has direct, hardware implemented communication paths to n other processors (its neighbors); these paths correspond to the edges of the cube. The conventional method to address the nodes in a hypercube structure is to use the so-called Hamming Codes; whenever the nodes are neighbors, their addresses differ only in one bit. An n-dimensional cube can be constructed from two (n-1)-dimensional cubes according to a simple rule. A feature of a hypercube structure is that the maximum distance between two processors grows linearly with respect to the dimension of the hypercube. In an n-dimensional hypercube (with 2^n processors) the maximum distance between two processors is n.

Proposals to build large hypercube computers have been made for more than 20 years. The early hypercube designs were not practical because of the large number of combinatorial logic units and processors required. The situation changed rapidly after the introduction of the VLSI technologies. The first working hypercube has been demonstrated at Caltech University (USA) in 1983. Each node consisted of a single-board microcomputer containing a 8086 microprocessor and a 8087 floating point processor. In 1985, INTEL announced the development of hypercube machines with the 80286/287 CPU as a node processor. The first commercially successful hypercube was introduced by N-Cube: the NCube/10 system with a (theoretical maximum) performance of 500 MFLOPS. Other recently developed hypercube-style machines with supercomputing potential include among others the Connection Machine, transputer-based hypercubes, INTEL iPSC and Caltech's MARK III [HAYE86].

2.2 The NCube/4+ architecture

The NCube/4+ system is a mini-supercomputer with the size of a microcomputer; its maximum performance is 8 MFLOPS. It uses up to 16 identical high-speed processors connected in a 4-dimensional hypercube network to deal with a single problem. The NCube/4+ (the + is denoting an extended amount of memory, namely 512 KByte instead of 128 KByte in the smaller model NCube/4) is the smallest system of the NCube family; other members are the NCube/7 and the NCube/10 systems, consisting of a maximum of 128 and 1024 processors respectively.

The NCube/4+ consists of 2 main components: an IBM AT (80286) functioning as host and a 16-processor (4-dimensional) hypercube. The hypercube is linked to the host computer by 4 I/O-channels in such a way, that 4 processors are directly connected to the host. The IBM system has a 72 MByte hard disk, from which 65 MByte is available for AXIS, the Unix-like operating system of the NCube. At Delft University of Technology we have connected two CIT-101 terminals to the NCube, as well as a Canon inkjet printer, a Modem and the DUneT entrance (Delft University of Technology network system).

The NCube node processor is a 32-bit processor with 512 KByte local memory. It is a single VLSI chip, including a full floating point instruction set and the combinatorial logic needed for memory management and interprocessor communication. An entire NCube node only contains 7 chips. This compactness has prompted the introduction of a 4-node IBM AT board; four of such boards can be combined to construct a 16-node hypercube. Each processor has 11 full-duplex communication channels (allowing the maximum configuration of the 10-dimensional hypercube NCube/10). A node processor can communicate with other nodes by means of asynchronous DMA operations. Once a processor has initiated a "send" or "receive" operation, it can continue with other operations while the DMA channel completes the internode communication operation.

2.3 The operating systems and languages supported

For the NCube/4+, the operating system running on the IBM AT host is called "AXIS", the nucleus (a small operating system that supports the communication between nodes by an efficient routing algorithm) running on the nodes is called "VERTEX" the languages provided are Fortran 77 and C.

AXIS, the N-Cube's version of the Unix-family, embodies the features common to the Unix main dialects. Two features of AXIS greatly facilitate program development:
-- the ability to share files
-- the ability to manage the main cube and subcubes.

The AXIS environment supplies the user with editors, compilers, cross compilers, a symbolic debugger, text-formatting tools, etc. As in Unix, all devices are treated as files. AXIS controls a hypercube of nodes simply as a device, which is a file. AXIS permits the user to allocate subcubes with the appropriate size for his application. Consequently, several users with smaller problems can use the hypercube simultaneously, which considerably increases the system's efficiency and gives the hypercube supercomputer significant advantage over other (mini-) supercomputers. Partitioning the main hypercube into subcubes is simplified by the fact that each subcube is protected from access by another one.

VERTEX is a small nucleus (less than 4KBytes) resident in each of the nodes; it serves as an operating system. Its primary function is to support the communication between the nodes in order not to bother the user about routing. This is achieved with "send" and "receive" system calls which transfer messages between any two nodes in the hypercube by activating the DMA interfaces. The maximum size of a message is 32 KBytes. It is possible to assign a type to a message, distinguishing different messages and admitting a selective receipt at a destination node.

The current NCube application languages (apart from the host languages and node assembly language) are Fortran 77 and C, chosen because of the users' interest in technological and scientific problems. Compilers for other languages, such as OCCAM, are presently being developed.

3. THE CONVEX C-1 VECTOR PROCESSOR

3.1 Architectural features

The Convex C-1 from Convex Computer Corporation is a high speed scientific 64-bit computer with integrated vector processing. It consists of a tightly coupled asymmetric multiprocessor that comprises four functional subsystems: Central Processing Unit (CPU), Memory Subsystem, Input/Output Subsystem and Service Processor Subsystem. The CPU portion has the same architecture as the Cray supercomputers. It performs the computational functions; these functions include arithmetic and logical operations on scalar and vector data. The Memory Subsystem provides the main memory. It consists of one Memory Control Unit and from one to eight Memory Array Units (each of 16 MBytes capacity, access cycle time: 400 nsec.). The Input/Output System consists of distributed intelligent I/O-processors. The Service Processor Unit is microprocessor based and controls the operation of the diagnostic programs. The theoretical maximum performance of the Convex C-1 is 60 MFLOPS.

3.2 The operating system

The operating system consists of a modern Unix version. Convex Unix is giving the users a number of features:
-- Convex Unix is providing a maximum of 4 GigaBytes of virtual memory, allowing programmers to concentrate on their applications rather than on memory limitations of the programs.
-- Subroutines written in C, Fortran or assembly language can be linked together to make one program.
-- The Unix operating system is distributed over the CPU and all I/O-processors individually, allowing more computational power to be devoted to solving application problems.
-- In addition to several text and screen editors, the so-called Convex Consultant is available for profiling program execution (run time analysis) and debugging.
-- Users are allowed to manage the work load on the system efficiently, permitting interactive users to remain productive even when large batch processing jobs are running on the system.

3.3 Languages supported

One of the languages supported is C, associated with the Unix operating system and presently extensively used for systems programming. The Convex C compiler is an extended implementation of the C language. The C compiler is supported by an additional utility program called "lint", which can check programs for correctness. The in C written subroutines can be linked with other subroutines, written in Fortran or assembly language.

Another language supported is Fortran 77. The Convex Fortran compiler is a

multipass, optimizing and vectorizing compiler. Existing standard Fortran programs can be processed by this compiler without any modifications to the source. The following features are making this language the most powerful high level language to use on the Convex C-1:
- The source programs are optimized and vectorized (a parallel executable code is produced that utilizes efficiently the vector processing capabilities).
- The Fortran compiler provides the user an easy to read vectorization summary.
- No special syntax is needed for vectorization. The user writes the program in standard Fortran.
- A run time performance summary is provided, giving information of the execution time of each included subroutine and the number of times, each subroutine is called.

3.4 Vectorization

Although the Fortran compiler provides optimization and vectorization facilities, the user can still create better performing programs when he is programming according to some vectorization rules. Vector processing is applied on program loops, which frequently occur in scientific and technical application programs. The percentage of vectorization depends on 1) the ability of the programmer to write his source in a code easily executed in a vector processor, and 2) the efficiency of the compiler's automatic vectorization possibilities.

3.4.1 Programming vectorization rules

At the programming level a lot can be done to make the program running faster on the specific hardware, without the necessity of programming in machine language. Let us consider a simple example, in which a call to a scalar function subroutine is done:

```
      DO 10 i=1,n
         CALL dist(x(i),y(i),hyp(i))
   10 CONTINUE

      SUBROUTINE dist(x,y,hyp)
      hyp=sqrt(x**2+y**2)
      RETURN
      END
```

This source program cannot be vectorized. Scalar code is generated as the vectorizer cannot vectorize subroutine calls. The following code can be easily vectorized:

```
      CALL vdist(n,x,y,hyp)

      SUBROUTINE vdist(n,x,y,hyp)
      DIMENSION x(n),y(n),hyp(n)
      DO 10 i=1,10
         hyp(i)=sqrt(x(i)**2+y(i)**2)
   10 CONTINUE
      RETURN
      END
```

The two vectors x and y are loaded completely and the operations are executed pipelined and concurrently. The given example demonstrates one of the numerous ways to optimize a program at the programmers level.

3.4.2 The compiler's abilities to vectorize

The vectorizing features of the compiler are the following:
- IF-statements in DO-loops can be vectorized.
- Vectorization is done across all nested loops within a program.
- Loop interchange can be performed in nested Do-loops to ensure that a contiguous vector (i.e., the vector elements have been stored in adjacent memory locations) comes first; in addition, it is advantageous to nest the longer vector in the inner loop.

-- Partial vectorization of loops is possible: some loops are divided in a part that can be vectorized and a part that is calculated sequentially.

The following example illustrates the compiler's methods of vectorization:

```
      DO 20  i=1,n
        b(i,1)=0
        DO  10   j=1,m
          a(i,j)=a(i,j)+b(i,j)*c(i,j)
   10     CONTINUE
        d(i)=e(i)+a(i,1)
   20 CONTINUE
```

In the first step the compiler uses its ability to vectorize across all nested loops and creates the following loops:

```
      DO 20a  i=1,n
        b(i,1)=0
   20a CONTINUE

      DO 20b  i=1,n
        DO 10  j=1,m
          a(i,j)=a(i,j)+b(i,j)*c(i,j)
   10     CONTINUE
   20b CONTINUE

      DO 20c  i=1,n
        d(i)=e(i)+a(i,1)
   20c CONTINUE
```

The original outer loop is distributed over three loops 20a, 20b and 20c. The loops 20a and 20c and the inner loop 10 can be vectorized directly. In the next step the compiler uses its ability to interchange loops on the second DO-loop (20b):

```
      DO 10  j=1,m
        DO 20b  i=1,n
          a(i,j)=a(i,j)+b(i,j)*c(i,j)
   20b    CONTINUE
   10    CONTINUE
```

The loops are interchanged to ensure processing on a contiguous vector. In the given example only a few of the compiler's abilities are illustrated. The programmer is informed about the compiler's operations in a vectorization summary, showing exactly which loops are and which loops are not vectorized by the compiler.

4. PARALLEL SIMULATION

With the term "parallel simulation" is meant simulation on a multicomputer system. Users of scientific digital computers are motivated to demand high processing speeds for two distinct reasons: the increasingly detailed representations required for more and more complex distributed parameter systems, characterized by PDEs, and real- time (or faster than real- time) computation of complex lumped parameter systems described by ODEs.

In the early 1980's there was talk of a centralization of supercomputer facility (such as the Cyber 205 and CRAY-1), used among others for challenging complex simulations of systems characterized by PDEs and accessed remotely by scientists and engineers with large simulation problems to solve. The advent of mini-supercomputers may affect a fundamental change in this picture, since presently these are generally marketed at a cost less than 10% of that of supercomputers (and therefore may well be within the reach of many research groups and institutions, and may open the possibility of owning a dedicated and decentralized computer, which seems to permit a more attractive mode of interactive scientific computing).

Rather than the performance measures in MIPS (million instructions per

second) and MFLOPS (million floating-point operations per second) provided by
the vendors, realistic performance comparisons and predictions are to be
based more reliably upon benchmark problems. For the simulation of systems
characterized by PDEs the treatment of sufficiently large sets of simultaneous linear algebraic equations (preferably with sparse characterizing
matrices) on different computers may give some performance indications;
recent studies (however, for dense matrices) showed some mini-supercomputers
to reach 5% - 20% of the performance of supercomputers, resulting in a
favorable cost/performance ratio [DONG85]. No comprehensive benchmark studies
have been as yet published dealing with solving ODEs on mini-supercomputers.A
general remark is that, based on the special characteristics of the mini-
supercomputer and the size of the simulation problem (the number of state
equations, the number of finite difference or finite element points, the
types of nonlinearities to be included), the user must be enabled to provide
various directives to allow the compilers to recognize parallelizable or
vectorizable code in order to take maximal advantage of the mini-supercomputers' potentialities.

In addition to the afore-mentioned reasons to use multicomputers in (real-
time) simulations of complex distributed or lumped parameter systems, there
can be other motivations to parallel simulation. In methodology-based
interactive simulation the specific architecture of such multicomputer
systems might be well exploited. Focussing the attention on MIMD arrays of
processing elements, examples of this in model implementing and experimenting
are [KERC85]:
-- the exploitation of the one-to-one analogy realizable between a model
   structure and its physical implementation on the multicomputer;
-- model composition by assembling components running in different processing
   elements (configuring excitable units);
-- interactive experimentation on model bases (for instance, multimodel
   output analysis after one single run).

REFERENCES

[DONG85] DONGARRA, J.J.: Performance of various computers using standard
         linear equations software in a Fortran environment. Mathematics and
         Computer Science Division, Technical Memorandum No. 23, Argonne
         National Laboratory, Argonne, IL., December 1985.
[FADD84] FADDEN, E.J.: The System 10 Plus: Broader horizons. In: W.Karplus
         (Ed.): Peripheral Array Processors. Simulation Series, Vol.14 No.2.
         Simulation Councils, Inc. (Society for Computer Simulation), San
         Diego/California, 1984.
[HAYE86] HAYES, J.P., T. MUDGE, Q.F. STOUT: A microprocessor- based hypercube
         supercomputer. IEEE Micro, October 1986, pp.6-17.
[HOCK81] HOCKNEY, R.W., C.R. JESSHOPE: Parallel Computers. Adam Hilger,Ltd.,
         Bristol, UK, 1981.
[KARP84] KARPLUS, W.J. (Ed): Peripheral Array Processors. Simulation Series,
         Vol.14 No.2. Simulation Council, Inc. (Society for Computer Simula-
         tion), San Diego/California, 1984.
[KARP87] KARPLUS, W.J. (Ed.): Multiprocessors and Array Processors. Simula-
         tion Series, Vol.18 No 2, Simulation Council, Inc. (Society for
         Computer Simulation), San Diego/California, 1987.
[KERC85] KERCKHOFFS, E.J.H., S.W. BROK: The Delft Parallel Proces-
         sor:Properties and utilization in simulation and related fields.
         Systems Analysis, Modelling and Simulation (Journal of Mathematical
         Modelling and Simulation in Systems Analysis). Akademieverlag,
         Berlin (GDR), Vol.2 (1985); pp. 175-208.
[LOUI81] LOUIE,T. Array Processor: A selected bibliography. Computer, Septem-
         ber 1981, pp.53-57.
[SPRI82] SPRIET,J.A., G.C.VANSTEENKISTE: Computer-Aided Modelling and Simula-
         tion / International Lecture Notes in Computer Science. Academic
         Press, London 1982.

*) Delft University of Technology
   Faculty of Mathematics and Informatics
   Julianalaan 132
   2628 BL DELFT.

The Simulated Performance of a Real-Time Interprocessor Synchronization Algorithm Based on Event-Driven Method

Duan   Ping[1]

Abstract

The stochastic simulation to the performance of a real-time synchronization algorithm for interprocessor communications is discussed. The results of the simulation are consistent with that of the mathematical analysis. This work is a successful example of the simulation to the performance of distributed systems.

1.   Introduction

Duan Ping and Cai Xiyao [1987]   present a real-time synchronization algorithm for interprocessor communications in distributed computer systems   (without shared memory among processors) with relatively great transmission power (not less than 2), for example, tree-structured , cube-structured and fully interconnected systems , etc. [1]. Transmission power indicates the number of messages that can be simultaneously transmitted in the system. This algorithm is not appropriate for application in bus-structured system and star-structured system because their transmission power are respectively 1 and less than 2.

Some probability assumptions about the system behaviour are involved when analyzing the algorithm's performance. Based on these assumptions , the upper and lower bounds of the mean response time of the algorithm are derived in [1]. One of these assumptions is not very practical in most real systems. What will happen without this assumption? When the difference between the upper and lower bounds is relatively large , the analyzed results have little meaning. In addition, it is difficult to derive some algorithm's performances by mathematical analysis. This is why we discuss the simulation of the algorithm's performance in this paper.

2. A Brief Review of the Real-time Synchronization Algorithm

The algorithm employs buffered communication and considers only communications between adjacent nodes.   The communication between non-adjacent nodes will be transformed into the communication of some pairs of adjacentnodes by the distributed path choice algorithm.

A synchronization management module is set for every node. Every module exchanges its status information by reading and writing some boolean flag variables in order to implement the synchronization of interprocessor communications. Synchronization management modules of different nodes have the same structure. The only difference among them is the values of some parameters , which depend on the node code (1,2,...,N). In each time phase of operation, each module will perform one of the following two actions according to the values of the status variables " bufferfull" and "send" , which respectively indicate whether the communication buffer of this node be full and whether this node needs to send a message to another node at present:

(1) Asking phase: try to send a message to an adjacent node.
(2) Responding phase: accept messages from adjacent nodes.
(3) Closing phase: neither send nor accept the message from or to the adjacent node.

In the analysis and the simulation of distributed systems , the following assumptions are frequently involved:

Assumption 1. When a message is to be sent, the probability of every node in the system as source node is equal and the probability of one of the others as target node is equal.

Assumption 2.   The frequence at which a node sends messages follows negative exponent distribution.

Because only communications between adjacent nodes are considered , it is implicit that assumption 1 is changed into the following assumption to make time analysis easier in [1]:

Assumption 3. When a message is to be sent, the probability of every node in the system as source node is equal and the probability of one of its adjacent nodes as target node is equal.

The following assumption is also involved in [1] to make time analysis easier:

Assumption 4. The frequence at which a node sends messages is low.

----

1) Department of Computer Science, Northwest Telecommunication Engineering Institute, Xi'an, Shaanxi, P.R. of China

3. The Design Idea of the Simulation Program

Traditional discrete-event simulations have two forms : one is called event-driven and the other is called time-driven. This paper discusses event-driven simulation , i.e. , the simulation clock is advanced after simulation of an event to the time of the next event.

The basic idea of simulation is as follows:

(1) Input the interconnection structure of the distributed system for which the algorithm is used and some parameters. Set the initialization status and the initialization time for each node;

(2) Fetch the node which current time is the minimum;

(3) Simulate the running of this node for one or more phase(s) and update the data structure;

(4) If the number of messages sent is not adequate then go to (2), otherwise output the simulation result.

Fig.1 is the status transformation diagram of the node. In Fig.1, S indicates that the node will want to send a message in the next phase; succ indicates that the communication request is successfully responded to; $\bar{S}$ and $\overline{succ}$ respectively indicate the non-logic of S and succ.

The simulation is based on assumptions 1 and 3. Therefore, two kinds of stochastic variables are involved in the simulation system: one of them follows uniform distribution and the other follows negative exponent distribution.

The function $X(i+1)=aX(i)+b$ (mod m) is used to generate a pseudo stochastic number.

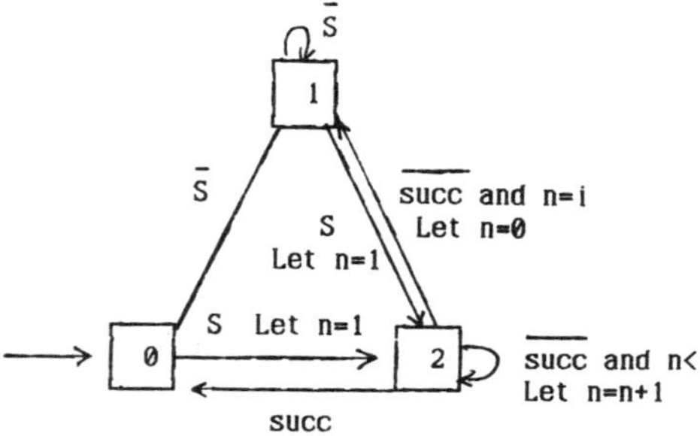

Fig.1　The Status Transformation Diagram of the Node

About this function, the following theorem is known:

Theorem 1. The period of the pseudo stochastic number generator $X(i+1)\equiv aX(i)+b$ (mod m) is m, iff

(1) b (mod m)$\neq$0 and m (mod b)$\neq$0;

(2) For any prime number p such that m (mod p)=0, (a-1) mod p=0;

(3) If m (mod 4)=0, then (a-1) (mod 4)=0.

In generally, $m=2^k$, where k is a positive integer. In this paper, let m=2048, a=817, b=431, and X(0)=626.

The function $\delta =max \{X (n) , 1 \}/m$ is used to generate the stochastic real number which follows uniform distribution in the interval (0 , 1). From this , we can get:

(1) Negative exponent distribution with mathematical expected value $1/\lambda$: $\eta =-(\ln\delta)/\lambda$;

(2) Uniform distribution in the set $\{1,2,...,k\}$: $l=\lfloor k*\delta \rfloor$.

4. On the Realization of the Simulation Program

4.1 The Basic Idea

Let λi be the joint probability of the following probabilities:

(1) the probability that node i sends a message to another node per phase, and

(2) the probability that the communication buffer of node i is full.

Let $R= \lceil Rmax/Rmin \rceil$, where Rmax and Rmin respectively indicate the upper and lower bounds of the node's speeds, 0<Rmin$\leqslant$Rmax.

In order to make the simulation easier, we assume that:

(1) The ratio of node i's speed to Rmin is $1+i*(R-1)/N$, and it does not change in the running process.

(2) The value of λi is the same, where $1\leqslant i\leqslant N$.

(3) The capacity of the communication buffer is large enough , i.e. , the communication buffer will never be full. So no closing phases are involved in the simulation system. The above assumptions have no inherent affect on the validity of the simulation results.

Because the node's speeds are different , the upper bound of one phase's length of nodes will be different. Let uppl(i) be the upper bound of one phase's length of node i.

Preseted below is a description of the simulation algorithm:

Algorithm for Simulation

```
status(S): 0..2;          the current status of node S
curtime(S): Real;         the current time of node S
Begin
    initialization;
    input the interconnection structure of the distributed system, R and λ ;
```

```
do while (the number of messages sent is not enough)
  choice the node S which current time is the minimum;
    do case
      case status(S)=0
        comment: simulate the enviornment in which the algorithm works
        according to negative exponent distribution, compute the length Δt of the
        interval between curtime(S) and the time at which node S wants to send the
        next message;
        according to uniform distribution, compute the target node's code;
        curtime(S)=max{curtime(S), curtime(S)+Δt-the response time of node S's
                    last communication};
        record related information;   status(S)=1
      case status(S)=1
        comment: simulate the responding phase
        n(S)=0;  status(S)=2
      case status(S)=2
        comment: simulate the asking phase
        if n(S)=S then  status(S)=1
        else
          begin
            if (the communication request has been successfully responded) then
                begin  compute and record related information;   status(S)=0   end
            else n(S)=n(S)+1
          end;
        if status(S)=0 then
          curtime(S):=the time of node S at which the communication is established
        else  curtime(S):=curtime(S)+uppl(S)
    endcase
  enddo;
  output the simulation results
end.
```

4.2 On Simulation Precision

There are two major factors which will affect the simulation precision:
  (1) How to judge whether the communication request will be successfully responded to, and
  (2) Because of the response time, it is difficult to precisely simulate the frequence of
sending messages which follows negative exponent distribution.
  We give an example to explain the first problem. At a certain time in the simulation
process , fetch node i which has minimum current time and is in an asking phase. Assume that
node j is the target node and curtime(j)-curtime(i) is less than uppl(i). Let t1=curtime(j)
and t2=curtime(i)+uppl(i).   If and only if in the interval (curtime(i), t1) node j is in
status 1 for enough time , we think that the communication request is successfully responded.
If the communication request cannot be successfully responded before curtime (j), we think it
will not in (t1, t2]. We have enough reasons to believe that this is true in most cases.
  This brings us to the second problem. In a real system , a node will not request to
communicate with another node before the last message to be sent of this node has been sent.
But when we simulate the frequence of sending messages, this case may happen.
  The method for solving this problem in this paper is as follows : let T be the response
time of this communication. Assume that the time at which the next communication request will
occur will be T1 according to negative exponent distribution. Let T'=max{0, T1-T} be the time
at which the next communication request will occur.Because assumption 4 is NOT involved in
the simulation system, this method will cause three problems:
  (1) The simulated frequence of sending messages is smaller than the value of input λ .
  (2) The reason of involving assumption 4 in the time analysis is: under this assumption,
most communication requests will be successfully responded to in one phase. But this is NOT
true in the simulation system.
  (3) The simulated frequence cannot exactly follow negative exponent distribution.
  In order to solve the first two problems , let λsimu=the total number of asking phases in
the simulation/the total number of phases in the simulation. It corresponds to (λ1+λ2+...+
λN)/N   under assumption 4   because in this case most communication requests will be
successfully responded to in one phase. When the simulated performance is compared with the
analyzed performance , the upper bound of the mean response time in phase is computed
according to λsimu rather than λ .
  Because sometimes a successful communication needs more than one communication request , it
is impossible that the simulated frequence of sending mesages exactly follows negative
exponent distribution.   When λ's value is relatively small (<0.15) , T1's value will be

greater than 0 in most cases , so it will roughly follow negative exponent distribution. When λ's value is relatively large ($>$0.25) , T1's value will be 0 in most cases , so it cannot follow negative exponent distribution. It seems impossible to find a perfect solution to this problem.

5. The Simulated Performance and the Discussion of It

The simulation program is written in PASCAL and runs on VAX-11/750. Five distributed computer systems with different interconnection structures are considered. The simulated and analyzed performance presented in Table 1 is for cube structured systems which have eight nodes with R=2.

Table 1. The Simulated and Analyzed Performance of the Algorithm
Simu.T: the simulated response time in phase
Simu.N: the simulated number of messages sent per phase
Low.T (Upp.T): the lower (upper) bound of the mean response time in phase

Table 1 demonstrates the following:

(1) The simulated response time of the algorithm is always between the analyzed upper and lower bounds of the mean response time, which means that, to a certain degree, the results of the mathematical analysis of the response time are reasonable even without assumption 4 (the definition of joint probability λ, of course, should be changed into: the probability that the node is in an asking phase).

(2) The joint probability λ has certain saturation value which depends on the interconnection structure and R.

(3) When the value of λ is too large, the algorithm's performance will slow down.

λ	λ simu	Simu.T	Simu.N	Low.T	Upp.T
0.01	0.0102	1.5146	0.0101	1.0100	2.0414
0.1	0.1240	1.7865	0.1002	1.1000	2.6063
0.2	0.3104	2.3612	0.1871	1.2000	4.2057
0.25	0.4369	2.9791	0.2073	1.2500	6.3075
0.3	0.5525	3.8350	0.2113	1.3000	9.9872
0.4	0.6820	4.9322	0.2043	1.4000	19.778
0.5	0.7479	5.8436	0.1931	1.5000	31.469
0.6	0.7793	6.3599	0.1887	1.6000	41.661
0.7	0.7918	6.7127	0.1879	1.7000	46.139
0.8	0.8033	7.1710	0.1849	1.8000	51.692

5. Conclusion

This paper, in certain degree, demonstrates that the simulation is an effective method for deriving the performance of distributed computer systems. Through this simulation system , we can clearly see the effect of various kinds of factors on the algorithm's performance. It is very difficult to derive the number of messages sent per phase , the distribution of the response time, and the saturation value of parameter λ by mathematical analysis. We have also simulated the performance of the algorithm presented in [2], and the result is very ideal.

6. Acknowledgments

The author wishes to thank Prof. Cai Xiyao for his suggestion on this topic and Mr. James Yale (from the Department of English at the University of Notre Pame, U.S.A) for his help in proof reading this paper.

References

[1] Duan Ping and Cai Xiyao : A Real-time Interprocessor Synchronization Algorithm for Communications in Distributed Computer Systems. Journal of Computer Science and Technology, 2(1987)4, 292-302.

[2] Duan Ping: A Real-time Synchronization Algorithm for Interprocessor Communications Based on Duplex System. Proceedings of IEEE Asian Electronics Conference 1987, Sept. 1987, 473-475.

[3] Duan Ping: Communication Mechanisms and Some Real-time Synchronization Algorithms for Interprocessor Communications in Distributed Systems (in Chinese). Master Thesis, The Department of Computer Science, Northwest Telecommunications Engineering Institute, Nov. 1985.

[4] Jin Lan and Zheng Weimin : Stochastic Simulation of A Distributed Multiprocessor System. Chinese Journal of computers, 7(1984)2, 100-107.

Simulation Environment in SONCHES

Flechsig,M. , Matthaus,E. , Wenzel,V. [1]

Abstract: The simulation environment of the branch specific simulation system SONCHES for ecosystems is introduced, including facilities for model use and result processing, model change, model documentation and library services.

Keywords: simulation system, simulation environment, model use, model handling

1. Introduction

In recent years, simulation systems have seen a drastic improvement of their capabilities, mainly influenced by the development of interactive computer systems. Many of the enhancements are concerned with

- the process of modelling, which is mostly supported now by a model description language and
- the process of model use and handling, performed by flexible and user friendly tools, forming the environment of the simulation system /1/.

The simulation system **SONCHES** (S_imulation O_f N_onlinear C_omplex H_ierarchic E_cological S_ystems) is an interactive tool for computer aided model investigations especially for ecosystem models /2/. It is now implemented on personal computers. A model description language (/3/,/6/) forms the core of the user communication with the system. Model language concept is based on the use of processes as ecologically interpreted terms of difference or recurrence equations. In general processes are realized algorithmically. As a branch specific language SONCHES includes possibilities to map phenomenas in the fields of (agro-)ecosystems, water- and forest systems and pharmacocinetics by preformed modules.

Besides this modelling language SONCHES offers in its environment components for

- model documentation,
- model simulation and result processing,
- model change and
- library services.

They enable the user to concentrate on handling his model, independent of his knowledge in informatics or of the computer operating system. The user is guided through the simulation system by hierarchically ordered menues. During the process of entering a model, in a first step all the variables and manipulable parameters, which are used in the model must be declared. Additional rates and variable vectors can be derivated from the declared variables with the help of SONCHES modelling language prefixes and postfixes resp. All together they form a model dependent unique set of syntactical entities, used for model input, simulation preparation and

---

[1] Central Institute of Cybernetics and Information Processes, Academy of Sciences of the GDR, Kurstr. 33 Berlin 1086

result processing as well as for documentation and model change. The
first three activities are supported by language components, accessing to
the variable and parameter set of the model.

2. Model documentation

The more complex the structure of a model, the less it can be expected
from the user to keep in mind the actual model constellation. This is
especially important in the courses of interactive model change and
validation. SONCHES documentation is based on the model notation by means
of the syntax rules and the conventions of the modelling language. The
documentation menue offers possibilities for displaying any model detail
up to a total model overview.
Information can be displayed on terminal screen or can be stored in print
files, supplemented with headers. Switching between these output modes is
possible in all parts of the simulation system.

3. Model change

SONCHES model change service contains two different approaches. On the
one hand it offers tools for changing model details without modifying the
model structure, including

■ the recalibration of the model (initial value / manipulable parameter/
  table assignment (IPT) changes), performed temporarily (before running
  the model) or permanently

■ the change of process realizations (mathematical process descriptions), divided into

■ □ entering of a new process realization

■ □ refering to a predecessing process

■ □ editing a process realization by the help of a SONCHES implemented
  line editor.

On the other hand the change of the model structure (variable
interconnections, parameter assignments to processes, process sequence,
...) is a very important and recurrent step during the incremental (and
decremental) modelling process, facilitated by the process- and sub-
model- oriented model design in SONCHES.
Depending on the extend of the changing interference the SONCHES- and
operating system services, necessary for updating the model are performed
automatically.

4. Simulation experiments

Simulative case studies are the central technique for model verfication
and validation. A single model run results in a trajectory in the state
space. To get a comprehensive, user friendly overview on the model
dynamics dependent on IPT variations more complex experiments than a
single run are necessary. Preparing a simulation experiment in SONCHES,
the user can choose the following tasks /4/ :

(1) single run   (4) Monte-Carlo simulation
(2) behaviour analysis   (5) parameter adaptation (optimization)
(3) sensitivity analysis   (6) model debugging
                      (7) model interface

(2)-(5) are multi run tasks, performing automatically an ensemble of model runs dependent on user's request. Behaviour analysis comprises a model inspection based on deterministic IPT changes. The stochastic approach is supported by Monte-Carlo simulation in the sense of

- an error analysis, realized by pre-run IPT distributions or

- an uncertainty analysis, realized by time-stepwise fluctuating model variables.

Sensitivity analysis includes calculation of finite IPT sensitivity for different sensitivity functions. Parameter adaptation is based on adaptive stochastic optimization algorithms (/5/), while the debugging facility serves for interactive step-by-step simulation, checking the model numerically. The model interface gives the possibility to couple SONCHES models with other software tools.

After experiment selection the simulation preparation language guides the user to specify

- experiment control inputs for multi run tasks and

- model variables,

- □ to be stored time-stepwise,

- □ to be aggregated above time intervals (extrema, moments),

- □ to be aggregated over run ensembles (for Monte-Carlo simulation) during simulation.

For simulation experiments and result processing command files for repetitive tasks and interactivelly preparable batch mode files are available.

5. Result processing

Besides studying the model dynamics the main purpose of result processing is model - measurement comparison. Often the variables, representing the dynamics of the model don't match the measurable data of the real system, stored as tables in the SONCHES table library. The result processing language is devoted to this aspect and enables the calculation of output variables (functions) $g(t)$ to be derived from the state variables in the sense of systems theory:

$$g(t) = F ([Z(t)],[I],[p],[T(t)],t)$$

with   t   : time step
        F   : any user defined function, declared in FORTRAN
        [Z(t)]   : subset of model state variables
        [I]   : subset of initial values of state variables
        [p]   : subset of manipulable parameters
        [T(t)]   : subset of table functions at time step t

Declaration and interpretative calculation of output variables during result processing saves simulation time. Post processing of simulation

results by algorithmically defined output variables will be solved in
future (see also 6.2.).

Besides a number of service options SONCHES result processing menue
offers mainly presentation of

- output variables and/or their time aggregations (calculated during
  simulation or result processing) in tabular form
- output variables as graphic shapes
- □ versus time
- □ versus IPT realizations (behaviour analysis)
- □ as phase portraits
- □ as empirical model variables distributions (Monte-Carlo simulation)
  realized
- □ as pseudographic plots for print files,
- □ as 2-dimensional plots including lettering, hardcopy and storage
  possibilities for monocrome screen,
- □ as 2- or 3-dimensional plots including histograms with dyeing,
  colour editing and lettering for colour graphic display
- output variable time series to be stored in the table library.

6. Library services

Fig.1 shows the informational connections and pathways for simulation
experiments and result processing. Libraries serve for data supply and
effective result and model version management. They are directory oriented and model independent, that is different models can share a library.

6.1. Table library (data management)

The description of processes in an algorithmic manner enables an
approximation by table functions. External driving forces of the model
are mostly time series. Together with reference tables for model -
measurement - comparison and time series of output variables, stored
during result processing they form the table funds of the model.
Possibilities for storing (supported by an interface to other data funds)
and updating tables are realized. Before running the model, all assigned
tables are coupled to the model and during simulation missing values are
linearly interpolated.

6.2. Scenario library (result management)

Based on the external storage of results during simulation SONCHES
comprises also the management of simulation   results in a scenario
library. This allows the user

- to save experiments for later result processing (also in batch run
  mode),
- to supply other software packages with the dynamics of output
  variables (interface for external result processing) and
- to perform result processing over scenarios, stored in the library in
  the sense of a generalized behaviour analysis.

The latter topic offers the possibility to compare different models or
model versions of the same subject.

6.3. Model version library (constellation management)

Running a complex model under real conditions means specification of an IPT set adapted to the intended experiment. Studies of the model behaviour or model – measurement comparisons need steady readjustments of the above mentioned set during a simulation experiment. To support this task, to store and to manipulate these "sets of model constants" for adjusting a model to a real situation the user is provided with model initialization utilities, handling IPT sets together with their numerical values and process assignments resp. This set can be regarded as a new variable type, called version macro. These macros are the entities of the model version library. They can be displayed, edited, copied (to generate a similiar macro for another model) and updated, necessary if the model structure was changed. The version macro can be used as a supplementary variable (parameter) in the simulation experiments.

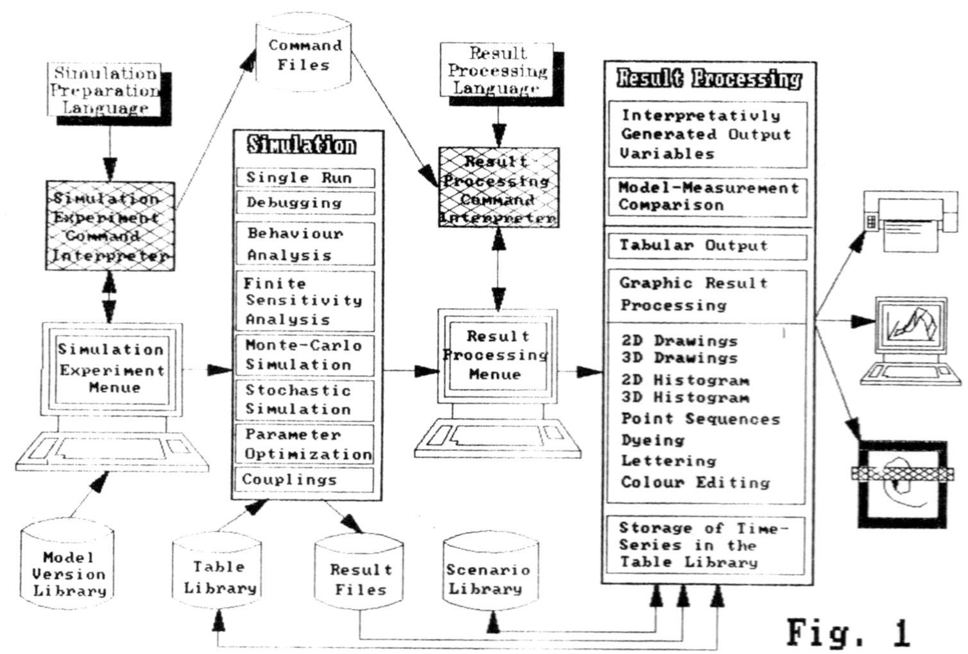

Fig. 1

<u>Literature</u>

/1/ Luker, P.A.: From CSSL to CSSE. In Proceedings of the Conference on Continuous Systems Languages, San Diego 1986, pp. 53-56
/2/ Knijnenburg, A. et al: Concept and Usage of the Interactive Simulation System for Ecosystems SONCHES. Ecol. Mod. 26(1984), pp. 51-76
/3/ Wenzel, V.: Problemlösungen bei Entwurf und Realisierung eines fachgebietsorientierten Simulationssystems. Diss. A, Berlin 1987
/4/ Matthäus, E. et al: Use of Task Facilities of the Simulator "SONCHES" for Agroecosystem Model Investigations and Control. In Proceedings of the 2nd Intern. Symposium on Systems Analysis and Simulation, Berlin 1985, Vol. 2, pp. 133-138
/5/ Born, J.: Adaptively Controlled Random Search – A Variance Function Approach. SAMS 2(1985), pp. 109-120
/6/ Wenzel, V. et al: Generic Modelling in SONCHES. In this proceedings

A Program Generator for a Model-Based Simulation System

Manfred Marx, Regine Czerner [1]

Mathematical software is often used for simulation studies in which a
frequent alteration of the algorithmen is needed especially by modell
searching. Model-based simulation systems enable the users to formulate their problems on a higher level of abstraction. Here a system
is considered that can be adapt to the concret models and methods of a
modeler at a relativ little expense. This capability can be obtained
if the model management system fulfils three essential demands:
(I)   Incorporation of user-specific models and   methods and a
         flexible definition of experimental frames and tasks,
(II)   A broad understanding of the user language by means of
         formulation the models and tasks in mathematical notation and
         with the help of elements of program languages,
(III) Coupling of models, methods and tasks for processing
         complex problems.
In following a realization of the program generator of this system is
considered. Components of the system joined with the program generator
are described.

1. Data bases of the system

The demands (I)-(III) cause the consequence that at the beginning
tasks, models and methods are not connected fixed. These relations
have to be defined additional. They are stored in a so-called abstraction base named like by Dolk [1]. The very important properties of the
data bases are described briefly.

Parameter base
All data to be found in models, tasks or methods have to be stored in
the parameter base. Belonging to models,tasks and methods is marked.
The declaration of attributs is possible.

----

1) Martin-Luther-Universitaet Halle-Wittenberg,   Organisations-
   und Rechenzentrum, Weinbergweg 17, 4010 Halle.

Model base

The model base contains user-specific models and models read for each user. A model describes objects and relations among them. It is formulated with linguistic elements which partly occur in other systems (s. Oeren [2]) in an analogues form. Im detail these are:

- a physical (INTEGER, REAL, ARRAY a.s.o.) or also logical (PARAMETER, VARIABLE, FUNCTION a.s.o.) declaration of objects and their kind of using (I,O),
- a description of mathematical relations in a DYNAMIC-construction,
- characteristics of relations by attributs,
- coupling among objects of submodels,
- allocation of names to objects,
- assignment of values fixed to objects,
- conversion of objects into the form needed in the model.

The models can be arranged hierarchically. The lowest model levels are the so-called basis models in which the objects and attributs are defined for a class of models. They present the interface to the methods which solve problems defined. The user forms his models by refering to basis models or also other models. Objects, attributs, values and coupling are leaved to the models of the higher level.

Procedure base

A procedure means an instruction to solve an actual task or an experimental frame. It is represented by a sequence of commands. Concerning the contents a procedure corresponds to a complex method, that on the contrary to a simple method is processed by an interpreter.

Following tasks may be solved by commands:

- adjustment of global system variables (numerical representation, working mode),
- modifying of values and object notations,
- input and output of objects,
- formation of control structures,
- method call by so-called activities.

Activities refer to a given model and cause together with the relevant basis model, that a corresponding method is called from a defined mathematical domain. On the other hand a method may be demanded explicitly.

Method base

By a method we understand a (methematical) algorithm for solving a special problem which is implemented in a higher programming language. It is processed in a compiled version and possesses one input and one output. The user interface of every method has been made available in a method describing part. The methods must be arranged corresponding to mathematical domains. All methods belonging to the same domain should possess a common interface with reference to data transmission. The classification into domains might be specified corresponding to users' aspects. In our system we prefer a classification into domains of numerical mathematics. Every domain must contain at least one basis model.

Abstraction base

First of all abstraction base is needed for storage of knowledge of problem solving process and for model information service. Therefore connection of models, activities and methods is its mean aspect. The connection contains the following logical details:

a) Characterization of activities

Every combination of activity and basis model is assigned to a mathematical domain.

b) Characterization of methods

Ordered by domains each existing method is characterized by nessecary and suitable conditions and possesses a priority. Conditions may be attributs or limitations for model quantities.

c) Characterization of domains

A table must be put to each defined combination of basis model and domain which contains all variables of the basis model by name. In this way notations in one domain are independent from those of another domain.

d) Charcterization of transformations

To every existing combination of basis model and method the name of the transformer must be stored which starting from DYNAMIC-section of the model prepares the objects demanded by the method. Further sequences of commands have to defined for working off the methods. These specifations would be identically if methods are standardized.

2. Working principle of program generator

Fundamentally programm generator has two functions:
a) Compilation of present command sequence into such a configuration
    which the used command interpreter can apply.
b) Determination of such a command sequence, that a method could be
    processed by specifation of activity and model.
In this paper we only shall consider determination of command sequence. First of all the mathematical domain is found by the help of
characterization of activities. It   contains a suitable solving method. An optimal method is selected by attention of following criterions: number of fulfilled nessecary conditions, number of fulfilled
suitable conditions and priority of method.
If the method is found out the objects get the names required by the
method . Following the competent transformer is activated. It verifies
all specific conditions of the domain and prepares interface conditions needed for working off the method. First of all the task of
transformer is to generate user functions and to transform data in
those special structures required by method.
Finally the symbolic sequence of commands for processing a method,
contained in characterization of transformations is completed by
specification of definit notations of commands and files. Essential
commands of this sequence are:
  - linkage of the method with user functions,
  - feed the method with data from parameter base by analyzing method
    describing part,
  - method call,
  - transfer of results of the method into parameter base.
At last the complete sequence of commands is worked out
interpretativly in the GO-step.

Remark: The system described here is realized on the UNIX-like
operating system VMX.

References:
[1] Dolk,D.R.: A Generalized Model Management System for Mathematical
    Programming, ACM Trans. on Math. Sofware 12(1986), P. 92-126.
[2] Oeren,T.I.: GEST - a Modelling and Simulation Language Based on
    System Theoretic Concepts, in Simulation and Model-Based Methodo-
    logies: An Integrative View, Berlin, 1984.

Software Package for Linear Nonstationary Systems Analysis and Simulation

Nicola E. Madjarov , Stoyan B. Maleshkov [1]

INTRODUCTION

In recent years a great interest has been paid to the computer-aided analysis and simulation of linear nonstationaty systems. This approach is very useful especially when the research requires a big computational effort. A new generation of software has been developed, which significantly improve the engineer's analysis and design environment. The engineer is able to concentrate on the engineering task rather than the detailed calculation and representation of the system performance. In this way different versions can be computed and compared in a short time and on this basis the optimal one can be chosen.

A wide variety of interaction styles are used in dialogue-oriented software. Without going into details for every software package, it can be concluded that there are two fundamental modes of operation: selection mode and command mode. The command mode of interaction [2] offers some advantages, but requires a good knowledge of the command language syntax and semantics. That is why the selection mode is chosen in the software package presented here, so it will suit the unfamiliar user as well and can be widely applied in education.

STATEMENT OF THE PROBLEM

Consider a linear nonstationary system of the form

$$\dot{x} = A(t)x + B(t)u \qquad (1)$$
$$y = C(t)x \quad ,$$

where $x \in \mathbb{R}^n$, $u \in \mathbb{R}^m$, $y \in \mathbb{R}^l$ and $A(t)$, $B(t)$, $C(t)$ are appropriately dimensioned matrices ($n \times n$, $n \times m$ and $l \times n$ respectively).

The transient matrix $\Phi(t, t_0)$ of this system is defined by [5]

$$\Phi(t, t_0) = X(t)X^{-1}(t_0) \quad , \qquad (2)$$

where $X(t)$ is the fundamental matrix of the homogeneous equation

$$\dot{x} = A(t)x \quad . \qquad (3)$$

The transient matrix is a solution of the matrix differential equation

$$\dot{\Phi}(t, t_0) = A(t)\Phi(t, t_0) \qquad (4)$$

with an initial condition of $\Phi(t_0, t_0) = I$.

Beside the expresions (2) and (4) the transient matrix can be described by an infinite matrix series of the form

$$\Phi(t, t_0) = I + \int_{t_0}^{t} A(\tau_1)d\tau_1 + \int_{t_0}^{t} A(\tau_1) \int_{t_0}^{\tau_1} A(\tau_2)d\tau_2 d\tau_1 + \ldots \qquad (5)$$

$$+ \int_{t_0}^{t} A(\tau_1) \int_{t_0}^{\tau_1} A(\tau_2) \ldots \int_{t_0}^{\tau_{k-1}} A(\tau_k) \, d\tau_k d\tau_{k-1} \ldots d\tau_1 + \ldots \quad ,$$

---

[1] Higher Institute of Mechanical and Electrical Engineering, 1156 Sofia, BULGARIA.

It can be proved [1] that the series (5) is absolute and uniformly convergent if the elements of the matrix $A(t)$ are continuous time functions in the interval of $[t_o, t]$. An algorithm for computing $\Phi(t, t_o)$ using (5), as well as a comparison of the three methos given above for computing the transient matrix are presented in [6].

The weighting matrix function of the nonstationary system (1) is determined by

$$W(t, t_o) = C(t)\Phi(t, t_o)B(t_o), \qquad t \geq t_o . \tag{6}$$

For sampled data control purposes the system is represented in a discrete form .

$$x(k+1) = F(k)x(k) + G(k)u(k) , \tag{7}$$
$$y(k) = C(k)x(k) .$$

For given matrices $B(t)$, $C(t)$ and $\Phi(t, t_o)$ of the continuous system (1), the matrices $F(k)$, $G(k)$ and $C(k)$ in (7) are calculated by

$$F(k) = \Phi[(k+1)T_o, kT_o] ,$$
$$G(k) = \int_{kT_o}^{(k+1)T_o} \Phi[(k+1)T_o, \tau] B(\tau) \, d\tau \tag{8}$$
$$C(k) = C(kT_o),$$

The linear nonstationary system given by (1) is controllable in the interval of $[t_o, t]$ if and only if the matrix $Q_x(t_o, t)$ which satisfies the matrix differential equation

$$\dot{Q}_x(t_o, t) = A(t)Q_x(t_o, t) + Q_x(t_o, t)A^T(t) + B(t)B^T(t) \tag{9}$$

with an initial condition of $Q_x(t_o, t_o) = 0$ is nonsingular.

The same system is observable in the interval of $[t_o, t]$ if and only if the matrix $S_x(t_o, t)$ which satisfies the matrix differential equation

$$\dot{S}(t_o, t) = -A^T(t)S(t_o, t) - S(t_o, t)A(t) + C^T(t)C(t) \tag{10}$$

with an initial condition of $S(t_o, t_o) = 0$ is nonsingular.

For discrete systems the approach for investigation of the controllability and the observability is analogous. The system given by (7) is controllable in the interval of $[k_o T_o, k_1 T_o]$ if and only if the matrix $Q_x(k_o, k_1)$, which is computed recurrently by

$$Q_x(k_o, k+1) = F(k)Q_x(k_o, k)F^T(k) + G(k)G^T(k) \tag{11}$$

with an initial condition of $Q_x(k_o, k_o) = 0$ is nonsingular.

The same system is observable in the interval of $[k_o T_o, k_1 T_o]$ if and only if the matrix $S(k_o, k_1)$, which is computed recurrently by

$$S(k_o, k+1) = [F^{-1}(k)]^T S(k_o, k)F^{-1}(k) + C^T(k+1)C(k+1) \tag{10}$$

with an initial condition of $S(k_o, k_o) = C(k_o)^T C(k_o)$ is nonsingular.

The matrices defined by (9), (10), (11) and (12) are symmetric, so they have only real eigenvalues. It is known [5] that these matrices are non-negatively defined. To analyze if the investigated matrix is nonsingular we can compute the eigenvalues and examine them if they are not equal to zero (i.e. they are positive).

SOFTWARE PACKAGE FUNCTIONS

The software package makes it possible to solve basic problems for nonstationary system analysis and simulation in interactive mode. The functions included in this software package involve:

1. Computing the transient matrix using different methods – solving the matrix differential equation (4), using the fundamental matrix in (2) or summing up matrix series members, according to (5).
2. System simulation applying various input signals.
3. Computing the weithting matrix function.
4. Determining the discrete description (7) and simulation.
5. Controllability and observability analysis. The problem is solved in the continuous case as well as in the discrete one.

Apart from the basic functions shown above, the software package includes powerful features for input–output operations (data input from keyboard and disk files, data output in numerical and graphical form). It is of great interest to see the data organization of the matrices $A(t)$, $B(t)$, and $C(t)$, whose elements can represent arbitrary functions of t. For this purpose an arithmetic expression interpreter is used which can analyze and calculate an arbitrary expression where arithmetic operations (+ – * / **) as well as trigonometric functions (sin, cos, asin, acos, tan, atan, ln, exp) are presented.

The software package consists of a subroutine library and of executable main programes, which realize the specified functions and organize the interaction with the user. All programes are written in Fortran 77. For user interface development, modules for different operations (windows management, editing, data input and output, menu organization and selection ect.) from a special toolkit are applied.

The software package is installed on microcomputer Pravez 16 (IBM PC compatible) and operates under DOS 3.10.

Only numerically stable computational algorithms are implemented in the software package. The matrix differential equations (4), (9) and (10) are solved by the Runge–Kutta–Felberg method [2] with some modifications. The computation of the eigenvalues of the matrices for controllability and observability analysis is realized by the programs described in [3].

EXAMPLES

Example 1. Simulate a nonstationary system with 2 inputs and 2 outputs given by (1), where the matrices $A(t)$, $B(t)$ and $C(t)$ are of the form

$$A(t) = \begin{bmatrix} 0 & 1 \\ -(2 + \dfrac{6}{t} + \dfrac{2}{t^2}) & -(3 + \dfrac{4}{t}) \end{bmatrix} \qquad (13)$$

$$B(t) = \begin{bmatrix} 1 & 0 \\ 0 & 1 \end{bmatrix}, \qquad C(t) = \begin{bmatrix} 1 & 0 \\ 0 & 1 \end{bmatrix} \; .$$

The state variables X_1, X_2 and the output variables Y_1, Y_2 are computed for 10 points, placed uniformly in the interval of [0.1, 1.0]. The problem is solved for an input vector $u(t) = (1, 0)$. The results are given in TABLE 1 and plotted in FIG. 1.

Example 2. The system, described in Example 1 is analyzed for controllability. The software package gives as a result

$$Q_x(t_0, t) = \begin{bmatrix} 0.34294 & -0.38842 \\ -0.38842 & 0.63319 \end{bmatrix}$$

Eigenvalues =   0.73421E-01   0.90271

which means that the system is controllable.

CONCLUSION

A dialogue-oriented software package for analysis and simulation of linear nonstationary control systems in the state space is described. Both continuous and discrete systems are considered. The user interface is menu-driven and suits the unfamiliar user as well. The software package is introduced in education. It is developed as a task within the framework of the great Sofia project which has for an objective to work out approaches, methods and tools to introduce advanced computer systems in education.

REFERENCES

1. Andreev, Y.: Finite-dimensional Linear Prosesses Control. Nauka, Moscow, 1976.
2. Forsythe, G., M. Malcolm, and C. Moler.: Computer Methods for Mathematical Computations. Prentice-Hall, Inc., 1977.
3. Garbow, B.S., J.M.Boyle, J.J.Dongarra and C.B.Moler: Matrix Eigensystem Routines- EISPACK Guide Extension. Springer-Verlag, Berlin, 1977.
4. Kalaikov, I., E. Garipov, S. Maleshkov, N. Sulemezova, and P. Ganev: Interactive System for Identification of Dynamical Models. Intern. Conf. "SYSTEM SCIENCE VIII", Wroclaw,(1983)p.66.
5. Madjarov, N.: Introduction to Modern Automatic Control Theory. Technica, Sofia, 1982.
6. Madjarov, N., and S. Maleshkov: Computing the Transient Matrix of Linear Nonstationary Systems. Automatica, Computers and Automated Systems, (1987), 12, pp. 7-13.

TABLE 1. RESULTS FOR EXAMPLE 1

t	X_1	X_2	Y_1	Y_2
0.10	0.00000E+00	0.00000E+00	0.00000E+00	0.00000E+00
0.20	0.88585E-01	-0.23312	0.88585E-01	-0.23312
0.30	0.15775	-0.36818	0.15775	-0.36818
0.40	0.21683	-0.44461	0.21683	-0.44461
0.50	0.26959	-0.49749	0.26959	-0.49749
0.60	0.31772	-0.53861	0.31772	-0.53861
0.70	0.36210	-0.57274	0.36210	-0.57274
0.80	0.40333	-0.60213	0.40333	-0.60213
0.90	0.44179	-0.62804	0.44179	-0.62804
1.00	0.47781	-0.65120	0.47781	-0.65120

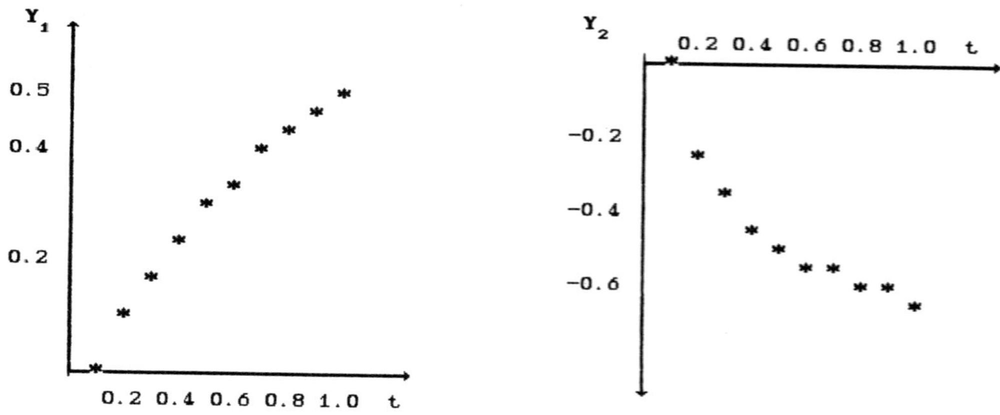

FIG. 1. Values of Y_1 and Y_2 for example 1.

SIMPC – An Implementation of GPSS for Personal Computer

Thomas Schulze [1]

1. Introduction

GPSS, the General Purpose Simulation System, which was developed by Geoffrey Gordon of the IBM Corporation around 1961, is the most important simulation language for discrete systems. Examples of such systems include flexible manufacturing systems, queueing networks, communication networks, etc.

PS SIMDIS is a powerful GDR implementation of GPSS-language for ESER computer. The user interface of this implementation has been improved since 1985. Release 3.0, an implementation for the TSO operating system, is offered for sale.

The increasing use of the personal computer allows us to use the simulation as a tool for problem solving in more areas. A new generation of simulation software can be developed using the new and better advantages of the personal computer. These include the use of multicoloured graphic display and the capability to simulate on your desk.

It is impossible to use the old PS SIMDIS software for applications on personal computers. The new generation brings for the user an essentialy higher user interface. The old known, prooved and existing simulation algorithms are the basis for the new software. Modelling power gets a new and better interface.

The Magdeburg University of Technology has started developing a GPSS oriented simulation system SIMPC for personal computer.

2. Targets

Starting demands for development were

- **GPSS oriented language;**
  The GPSS language is an old and known simulation language. This language is standard for description of discrete simulation models.

- **processing of GPSS programs;**
  The ability to process GPSS programs is one of the important requirements of a new system. The new system has to process old and tested simulation models.

----

[1] Technische Universitaet "Otto von Guericke" Magdeburg

- run on many personal computer

  The new system must be portable. That means, it must be written in
  a language wich can be processed by many computers.

The intention was not to develop a new simulation language. Targets
were to create an experimetal system based on the descriped points.
SIMPC allows us to research for
- syntax error free input of the simulation model;
- connection to known database systems for personal computer;
- userfriendly possibilities to observe the simulation process;
- animated graphic output of simulation results.

3. Available blocks

It isn't necessary for this experimental system to implement all known
GPSS blocks in the first version. The following blocks are avaible in
the first version running also on the 8-bit personal computer.

1. Creating and destroying transactions
GENERATE   TERMINATE

2. Assigning of parameter of transactions
ASSIGN   MARK   SELECT

3. Movement of transactions through a model
ADVANCE   GATE   TEST   TRANSFER

4. Facilities
SEIZE   RELEASE

5. Storage
ENTER   LEAVE

6. Logic switches
LOGIC

7. Functions
FUNCTION

8. Savevalues
SAVEVALUE

9. Queues
QUEUE   DEPART

10. User chains
LINK   UNLINK

These blocks allow us to simulate several queueing models.

4. Implementation

The SIMPC simulation system is written completely in TURBO-PASCAL and
runs under CP/M and MS-DOS operating systems. No statements were
imbedded valid for special units. TURBO-PASCAL compilers are avaible
on many personal computers. There are no problems for the portability
of this system. Another aspect, why this system was written in PASCAL
is the aid for further development by students. PASCAL will be learned
by all students.

The structure of SIMPC is offered by a PASCAL-like notation.

```
     PROGRAM   SIMPC ;
          PROCEDURE  DEFINITION ;
               (* This procedure defines the data structures of
               modelling entities and decides their quantity. *)

          PROCEDURE INITIAL ;
               (* This part initiale all entities  *)

          OVERLAY PROCEDURE COMPILE ;
               (* This procedure compiles the GPSS source file. All
               blocks and statements will be transformed in  internal
               block data structure.
               If errors occur the compiling process is aborted and
               messages are sent  to the terminal. *)

          OVERLAY PROCEDURE GO ;
               (* This procedure proceedes the internal block
               structure. Every block is interpreted by the procedure
               ACTION . The movement of transactions through the model
               is contolled by this procedure.
               If run-time errors occur, the run of the simulation is
               stopped. *)

     BEGIN
          DEFINITION ;
          INITIAL ;
          COMPILE;
          GO;
     END .
```

5. Example

The following example demonstrates the use of the SIMPC system. The
simulation was executed on an IBM AT/PC compatible computer. Six
trucks are loaded by two excalators standing on different places. Busy
trucks drive to the delivery station. Idle trucks drive back to
excalators. Simulation will be stopped if 1000 deliveries have been
executed.

```
*********************************************************************
*     3. INTERNATIONAL SYMPOSIUM ON SYSTEM ANALYSIS AND SIMULATION  *
*                  1988    BERLIN , GDR                             *
*                                                                   *
*          Application of SIMPC for truck delivery simulation       *
*                                                                   *
*********************************************************************
*
  LOTIM FUNCTION    RL1,D6        Load time for one truck
0.1,28/0.5,30/0.7,32/0.85,34/0.95,36/1.0,38
*
  UNLOT FUNCTION    RL1,D3        Unload time for one truck
0.1,13/0.9,15/1.0/17
*
*                               Drive time for one busy truck from
  BUSY1 FUNCTION    RL1,D7        excalator 1 to delivery station
0.2,28/0.7,30/0.9,32/0.95,34/0.98,36/0.99,38/1.0,40
*
*                               Drive time for one busy truck from
  BUSY2 FUNCTION    RL1,D7        excalator 2 to delivery station
0.2,38/0.7,40/0.9,42/0.95,44/0.98,46/0.99,48/1.0,50
*
*                               Drive  time  for  one  idle  truck  from
  IDLE1 FUNCTION    RL1,D5        delivery station to excalator 1
0.2,23/0.7,25/0.9,27/0.97,29/1.0,31
*
*                               Drive time for one idle truck from
  IDLE2 FUNCTION    RL1,D5        delivery station to excalator 2
0.2,33/0.7,35/0.9,37/0.97,39/1.0,41
*
          INITIAL    XH#COUNT,0
*
*
*              BLOCKS
*
          GENERATE    ,,,3              3 trucks for excavator 1
  EXCA1 QUEUE       EXCA1             Queue on excalator 1
          SEIZE       EXCA1             Start loading
          ADVANCE     FN#LOTIM          Loading time
          RELEASE     EXCA1             End loading
          DEPART      EXCA1
          ADVANCE     FN#BUSY1          Drive time
          TRANSFER    ,UNLO
*
          GENERATE    ,,,3              3 trucks for excavator 2
  EXCA2 QUEUE       EXCA2             Queue on excavator 2
          SEIZE       EXCA2             Start loading
          ADVANCE     FN#LOTIM          Loading time
          RELEASE     EXCA2             End loading
          DEPART      EXCA2
          ADVANCE     FN#BUSY2          Drive time
          TRANSFER    ,UNLO
*
  UNLO  QUEUE       UNLOD             Queue on delivery station
          SEIZE       UNLOD             Start unloading
          ADVANCE     FN#UNLOT          Time for unloading
          RELEASE     UNLOD             End unloading
          DEPART      UNLOD
          SAVEVALUE   COUNT+,1,XH       Drive counter
          TEST L      XH#COUNT,1000,END     End of simulation ?
```

```
        TEST LE    QN#EXCA1,QN#EXCA2,DRIV2    Drive to excavator 2
*
        ADVANCE    FN#IDL1                Drive time to excavator 1
        TRANSFER   ,EXCA1
*
 DRIV2  ADVANCE    FN#IDL2                Drive time to excavator 2
        TRANSFER   ,EXCA2
*
 END    TERMINATE  1
        START      1
```

RESULTS

****   BLOCK   REPORT   *****

Name	Aktuell	Gesamt	Name	Aktuell	Gesamt
1	0	1	UNLOD	1	1001
EXCA1	0	576	18	0	1000
3	0	576	19	0	1000
4	1	576	20	0	1000
5	0	575	21	0	1000
6	0	575	22	0	1000
7	1	575	23	0	1000
8	0	574	24	0	999
9	0	3	25	0	573
EXCA2	0	428	26	0	573
11	0	428	DRIV2	1	426
12	1	428	28	0	425
13	0	427	END	0	1
14	0	427			
15	0	427			
16	0	427			

**** FACILITIES   REPORT   ****

Name	mittlere Auslast.	mittlere Verweild.	Eintritte	belegender Aktivator
EXCA1	0.88	31.71	576	6
EXCA2	0.65	31.75	428	3
DELIV	0.72	15.00	1000	

***** QUEUE   REPORT   *****

Name	Aktuelle Laenge	Maximale Laenge	Eintritte gesamt	mittlere Laenge
EXAC1	1	3	576	1.25
EXAC2	1	3	428	0.82
DELIV	1	3	1001	0.84

Expert Systems in CIM Operations: Key to Productivity and Quality

SPYROS TZAFESTAS*

ABSTRACT: An overview of the principal manufacturing areas that provide a challenge for expert systems is presented. These are: dynamic simulation, CAD/CAM in terms of features,computer-aided process planning, and automated fault diagnosis. Due to space limitation a limited but sufficient set of references are included where several artificial intelligence/expert systems applications in computer integrated manufacturing (CIM) processes are described.

1. INTRODUCTION

Just recently the artificial intelligence (AI) discipline has arrived at the point where some of its applications have seen important practical results {1-3}. AI researchers attempt to construct computer programs that perform tasks for which, at the moment, people are better {4}. AI technology offers the tools that enable us to:

(i)   capture and retain expertise that was gained over many years of engineering

(ii) amplify expertise that is needed to successfully deploy new methods and applications

(iii) design systems that reason intelligently about nesessary actions to take in real time, thus freeing operational staff.

Manufacturing is one of the most attractive areas of application of AI techniques. However a considerable effort is required to capture and organize the accumulated knowledge of manufacturing engineers. There is a vast amount of·knowledge here, extending over many processes, diverse situations, and an infinite array of parts and products. Particular attention should be given for representing this knowledge, and adequately representing explicitly the characteristics of machines and processes {5-10}.

In this paper we discuss the principal areas of CIM operations where knowledge-based expert systems can play a major role for increasing pro-

* Control and Robotics Laboratory, Computer Engineering Division,
  National Technical University of Athens, Zografou, Athens, Greece

ductivity and quality. Specifically the following are considered:

- — Dynamic simulation of FMS
- — CAD/CAM in terms of features
- — Computer-aided process planning
- — Design for assembly
- — CAD in the CIM environment
- — Manufacturing systems diagnostics

2. DYNAMIC SIMULATION OF FLEXIBLE MANUFACTURING SYSTEMS (FMS)

To design "intelligent" manufacturing systems we need a large amount of structured information and powerful processing facility at the node level. Here is exactly where distributed database systems and knowledge-based expert systems are essential {11-15}. Usually, each processor of the distributed system needs different data types and has different data occurences. Therefore the local data bases are different, and to be able to communicate with them by means of the computer network, and handle each of them as a physical part of a "large" database, each $cell$ must have the appropriate system software and communication equipment. An FMS is a node in the LAN or in the distributed control system.

Dynamic FMS simulation uses a model of the system to study its behaviour and approximate its operation over a given time interval in order to evaluate alternative decision rules (for its design and operation). The simulation should be dynamic since the system behavior and the results obtained are time-dependent with deterministic or stochastic parameters. Dynamic FMS simulation techniques employ queueing models for the buffer stores or the automated storage and retrieval system (AS/RS) and calculate the required capacities for the various FMS modules. An FMS simulation model should be sufficiently complete, and take into account all previously discussed design rules and operation principles.

Some available FMS simulation models are:

- — Graphical models using CAD packages
- — Mathematical models (using functional expressions and appropriate simulation languages such as GPSS, DYNAMO, SLAM etc)
- — Scaled down microprocessor-controlled physical models (static and expensive).

These simulation tools combined with expert system tools and graphic/interactive systems make the dynamic simulation a key method for designing, diagnosing and running realistic models of large and complex multi-manufacturing systems.

3. CAD/CAM IN TERMS OF FEATURES

A key link between design and AI techniques in manufacturing is the
method of representing design geometry {16-18}. To obtain a true CIM
system, a single database of the design aspects, especially geometry,
should be accessible for all functions within the design/manufacturing
repertory. These functions include graphics, analysis, evaluation for
manufacturability, process design, and process planning.

The representation of geometry in terms of features is one solution
to the problem of obtaining a unified database. An identifiable geometric
shape or entity is considered as a feature of a design. Examples of such
features are 2D or 3D corners, slabs, boxes, holes etc. The importance
of this approach is that most of the knowledge required for evaluation
of manufacturability, design, and process planning is expressed in terms
of features or combinations of features. The creation of the desired
design representations in terms of features is a promising and active
area of CIM and knowledge engineering research. The link with the CAD
system is natural, since the information about features can be obtained
from the boundary representation of points, edges, and surfaces of objects that are provided by CAD systems. Of course another approach is to
design with features from the very beginning. Some appropriate combination of feature extraction and design-with features would give the most
general and successful solution. This can be best realized through AI/
expert systems techniques.

4. COMPUTER-AIDED PROCESS PLANNING

Process planning deals with the development of procedures and instructions for the manufacture of a designed part {19-34}. It is distinguished in *variant* and *generative* process planning. In the former a plan
from among a set of standard plans is selected. In the latter one uses
information on product or part features (provided for example by a CAD
system) for creating a process plan. Variant process planning is amenable
to the utilization of expert system techniques for classification and
selection (such as the one used in {35}). The use of knowledge-based
techniques for generative process planning is very attractive since machine tool capabilities are represented in the computer data-base. Ex-
aples of available conventional *computer-aided process planning* (CAPP)
systems are: AUTAP {19}, ARPL {20}, CAPP {21}, CAPSY {22}, etc. Some examples of knowledge-based systems for knowledge-acquisition/representa-
tion and process planning can be found in {23-30}.

A subclass of process planning problems is the planning of robot
work cell. This involves the selection of the robots and the auxiliary
equipment that best suits to the selected factory operations. Presently

there are available alternative systems (H/W plus S/W) for simulating
robot cells (e.g. PLACE and ROBOT-SIM) {31-32}. Newer systems for robot
work cell simulation are needed to include the representations of the
robot dynamics and controls. When the entire work cell is shown one can
easily check for interferences or collisions. Any simulation system must
be user-friendly such that to be usable by the manufacturing engineer,
the process planning staff, and the robot programmer on the shop floor.

5. DESIGN FOR ASSEMBLY

Design for assembly (DFA) belongs to the more general field of design
for manufacture (DFM) {36}. DFM deals with understanding how product design interacts with the other parts of the manufacturing system. It also
involves the problems of specifying product design alternatives which
help to globably optimize the manufacturing system as a whole. DFA is
concerned with the design of products with ease of assembly in mind. By
using DFA a product can be systematically designed to minimize the technological and financial efforts required for assembly and simultaneously
satisfy all constraints on the product functionality. To facilitate the
designers in this job, a decision support system is needed which will
take into account the knowledge of assembly requirements and help the
product engineer in his task (i.e. in designing for easing manual and
automated assembly, designing for assembly process planning anf for reducing reworking).Such a system should among others be able to:
— Analyse all decisions relevant to assembly and point out if there is
   some fault
— Indicate the critical feautures in the design
— Propose possible changes of the inappropriate design characteristics
— Suggest examples of design changes and check if a design change affects
   the product functionality.

Some available classical tools (not using expert systems techniques) for
DFA, that are not consistent with all the above objectives can be found
in {37-39}. These tools are tedious, costly and time consuming. Also they
do not guarantee success in all cases, since most of the knowledge on DFA
exists only as experience which brings irreproducible results. But most
of the DFA knowledge is available in a well formulated form, which means
that the greatest problem of knowledge acquisition is solved. Thus the DFA
field is an ideal field for the application of expert systems.

6. CAD IN THE CIM ENVIRONMENT

CAD systems under proper management can shorten substantially the
engineering development cycle in several ways {40-46}. Among the uses of
CAD geometric data we mention the development of finite element models,

creation of N/C tool paths, development of assembly drawings, development
of various simulation models, etc. Looking at the standard sequence of
operations and events in the development of a new product one can see
immediately the close relationship between CAD and CIM. The information
generated during the design phase (engineering drawing or CAD database)
plays a significant and unique role in almost all other product-related
activities of the manufacturing system. Very broadly any CIM model involves the following absolutely necessary functional sectors: *marketing,
design, material purchasing, manufacturing engineering, scheduling control, shop production, quality assurance* and *management*. In the center of
these sectors there is an information system which involves *database management, communications* and *distributed processing*.

At the one end is the product design function where CAD systems are
mostly used. The production related functions which constitute a good
portion of the CIM model, receive their orders in the form of engineering
drawings and specifications directly from the product design function. At
the other extreme is the management function that includes such things
as: planning and scheduling, organization, coordination, setting policies,
public relations, personnel, legal and financial problems, etc. Many of
these issues require a special study when they refer to interorganiza-
tional and multisupplier environments. In many (if not the most) existing
cases the link between CAD and CAM is not realized or is not satisfactory
to the system managers. Open questions that have yet to be answered include:

— Do existing CAD systems support downstream CAM/CIM functions?
  How much?
— Can we improve the costribution of CAD to CIM? What steps should be
  taken for this?

Looking at the existing manufacturing systems and the related technical literature, we are persuaded that, within a CIM environment, CAD
can and must do more than give a pure geometric database for production
operations {40,41}. The existing systems such as the *Initial Graphics
Exchange Specification* (IGES) {42}, the *Graphical Kernel System* (GKS)
{43} and the *Product Definition Data Interface* (PPDI) {44} provide only
part of the solution. These systems should be expanded using AI and
knowledge-based techniques that take into account product requirements
for interchangeable parts, time procurement, maintainability, logistics
support, ergonomics, fault detection and safety. The expert system(s) to
be developed   should include a *validation component*. The design engineer
should be able to observe the impact of various product design enhancements using a set of given measurable criteria. CAD has a strong effect
in many areas of the CIM environment. Therefore, the applications of AI/
expert system tools to the CAD generated product definition database can
increase substantially the beneficial issues of this impact both for the
manufacturer and the customer {45}.

7. MANUFACTURING SYSTEM DIAGNOSTICS

System fault (malfunction) diagnosis has become a standard area for
the application of knowledge engineering techniques {47-50}. Fault diagnosis or troubleshooting requires to pull apart the interactions of components in order to isolate the presence of a fault to a particular component or interconnection {47}. For example a robotic workcell involves
a variety of simple or complex components such as mechanical switches or
actuators or delicate end-effectors (grippers and tools) or sensing devices (visual, tactile, force, etc.).

Standard fault detection and correction techniques usually create
long down times of the system at hand, when no advice is given on how to
locate the malfunctioning component or part. Here is exactly where the
knowledge-based approach can help in improving the performance of diagnostic systems, especially as the knowledge and information about the mal-
fucntions is increased during the operation of the system. This is because
the AI techniques used in expert system development provide much better
tools for treating uncertain or incomplete knowledge.

The application of expert systems for manufacturing systems diagnosis
is presently very little. Much remains to be done in this area. An example
in this direction is the work described in {50}. Since technological systems parts can be exchanged or modules switched to locate a fault, the
fault diagnosis procedure of technological systems is much different than
medical diagnosis procedures. Tests are usually simpler and more effective than long reasoning chains. Also one has available special equipment
for easy diagnosis. But, due to the wide repertory of applications, components and options, the equipment under test is never the same. In {50}
much effort was spent on both cell level diagnosis (i.e. localize a particular equipment in the work cell) and equipment level diagnosis (i.e.
troubleshooting inside the equipment).

8. CONCLUDING REMARKS

Presently there are available several important techniques and tools
for designing and implementing CIM operations. Our purpose in this overview paper was to show that AI and knowledge-based expert systems techniques and tools provide a promising environment for further increasing
the productivity and quality of current manufacturing installations. In
particular, expert system tools can play a unique role in successfully
integrating such diverse aspects as managerial principles, market requirements, organizational goals, manufacturers interactions, logistic and
negotiation issues, and communication standards. The NTUA control and
robotics group has started work in this area with emphasis on knowledge-
based scheduling/capacity planning and FMS model building.

7. REFERENCES

1. A. Barr, A. Feigenbaum, P.R. Cohen (eds): The handbook of artificial intelligence, *William Kaufmann*, Los Altos, CA, 1981.

2. S. Tzafestas: AI techniques in control - An overview, in Kulikowski, C. and Ferrate, G. (eds.) "AI, Expert Systems and Languages in Modelling and Simulation" (IMACS Proc. 1987), *North Holland*, Amsterdam, 1988.

3. M. Gerencser, R. Smetek: Artificial intelligence technology and applications, *Militech Tech.*, No 6, p. 67, 1985.

4. E. Rich: Artificial intelligence, *McGraw-Hill*, Singapore, 1983.

5. M.S. King, S.L. Brooks, R.M. Schaefer: Knowledge-based systems - How will they affect manufacturing in 80's?, *Proc. 1985 ASME International Computers in Engineering Conf.*, Boston, MA, Vol. 2, pp. 383-390, 1985.

6. P.T. Rayson: A review of expert systems principles and their role in namufacturing systems, *Robotica*, Vol. 3, pt. 4, pp. 279-287, 1985.

7. E.L. Fisher: Expert systems can lay groundwork for intelligent CIM decision making, *Industr. Eng.*, Vol. 17, No 3, pp. 78-83, 1985.

8. R.A. Snyder: Artificial intelligence in the future factory - Test program, *Test Measure World*, Vol. 4, No 12, pp. 97-109, 1984.

9. H.-J. Schneider: Systems of the 5th generation and their impact on the qualification, training, and retraining of management, in T. Bernold (ed.) "Artificial Intelligence in Manufacturing", Elsevier Science Publ. (Nort-Holland), pp. 85-96, 1987.

10. E. Warman: AI in manufacturing: An organic approach to maufacturing cells, *Data Processing*, Vol. 27, No 4, pp. 31-34, 1985.

11. P. Ranky: Computer integrated manufacturing, *Prentice-Hall*, Englewood Cliffs, N.J., 1986.

12. P. Ranky, C.Y. Ho: Robot modelling - Control and Applications with software, IFS (Publications) Ltd., UK, 1985.

13. D.A. Bourne, M.S. Fox: Autonomous manufacturing - Automating the jobshop, *Computer*, pp. 76-79, 1984.

14. U. Rembold and R. Dillmann (eds.): Methods and tools for computer integrated manufacturing, *Springer-Verlag*, N.Y., 1984.

15. U. Rembold and R. Dillmann: Computer-aided design and manufacturing: Methods and tools, *Springer Verlag*, Berlin, 1986.

16. B.K. Choi, M.M. Barash and D.C. Anderson: Automatic recognition of machined surfaces from a 3D solid model, *Computer Aided Design*, Vol. 16, No 2, pp. 81-86, 1984.

17. M.R. Henderson and D.C. Anderson: Computer recognition and extraction of form features - A CAD/CAM link, *Comput. Indust.*, Vol. 5, No 4, pp. 329-339, 1984.

18. T.C. Woo: Feature extraction by volume decomposition, Proc. Conf. CAD/CAM Technology in Mechanical Engineering, Cambridge, MA, pp. 76-94, 1982.

19. W. Eversheim, H. Fuchs, K.H. Zons: Anwendung des Systems AUTAP zur Arbeitsplanerstellung, *Ind. Anz.*, H. 55, pp. 29-33, 1980.

20. K. Tuffentsammer, et. al.: Vorgabezeitermittlung und Arbeitsplanerstellung im Dialog mit dem Rechner. tz für Metallbearbeitung 71, H. 7, pp. 49-52, 1977.

21. C.H. Link, CAM-I, Automated Process Planning System (CAPP), *Techn. Paper*, Dearborn, Michigan, 1976.

22. G. Spur, E. Hein: Ergenbnisse zur rechnerunterstrtützten Prüfplannung, *Endbericht*, P6.4/28; B-PRi/2, KfK-BMFT, 1981.

23. M.S. Fox, S.F. Smith: ISIS-A knowledge-based system for factory sche-
    duling, *Expert Systems*, Vol. 1, 1, pp. 25-49, 1984.

24. F.G. Mill, S. Spraggett: Artificial intelligence for production plan-
    ning, *Computer-Aided Engineering J.*, Vol. 1, No 7, pp. 210-213, 1984.

25. S.Y. Nof: Expert system for planning/replanning programmable facili-
    ties, *Int. J. Prod. Res.*, Vol. 22, No 5, pp. 895-903, 1984.

26. M.C. Wu, C.R. Liu: Automated process planning and expert systems,
    Proc. IEEE Intl. Conf. on Robotics and Automation, St. Louis MO, pp.
    186-191, 1984.

27. C. Fellenstein et. al.: Prototype manufacturing knowledge base in
    Syllog, *IBM J. Res.*, Vol. 29, Nc 4, pp. 413-421, 1984.

28. F.G. Mill, S. Spraggett: An artificial intelligence approach to
    process planning and scheduling for flexible manufacturing systems,
    Proc. Intl. Conf. on Computer Aided Engrg., IEE London, 1984.

29. F. Gliviak et. al.: A manufacturing cell management system: CEMAS, *in
    Artificial Intelligence and Information-Control Systems of Robots,
    Proc. 3rd Intl. Conf.*, Smolenice, Czechoslovakia, I. Plander (ed.),
    North-Holland, 1984.

30. K. Szenes: An application of a parallel system planning language in
    decision support-production scheduling, in *Advances in production
    management systems*, IFIP Working Conf., Bordeaux, France, pp. 241-
    249, 1982.

31. R.N. Stauffer:Robot system simulation, *Robotics Today*, June, 1984.

32. J. Hansen: Use of simulation in planning of robot work cell, in
    "Robotics and Factories of the Future", S.N. Dwivedi (ed.), Springer
    Verlag, pp. 442-447, 1984.

33. S.M. Alexander, V. Jagannathan: Computer-aided process planning sys-
    tems: Current and future directions, *Proc. IEEE Intl. Conf. on Sys-
    tems, Man and Cybernetics*, N.Y., 1983.

34. E.D. Sacerdoti, A structure for plans and behavior, *Elsevier*, N.Y.
    1977.

35. S.G. Tzafestas, G. Tsihrintzis: ROBBAS: An expert system for choice
    or robots, in M. Singh, D. Salassa (eds.) Managerial Decision Sup-
    port Systems and Knowledge-Based Systems (IMACS/IFORS Proc.) North-
    Holland, Amsterdam, 1987.

36. H.W. Stoll: Design for manufacture - An Overview, Appl. Mech. Rev.,
    Vol. 39, No 9, pp. 1356-1364, 1986.

37. A. Gairola: Design for automatic assembly, in H.J. Warnecke and
    H.J. Burlinger (eds.) Factory of the Future, Springer, Berlin, 1985.

38. M.M. Andreasen, et. al.: Design for assembly, *IFS Publications Ltd*,
    Bedford, UK, 1982.

39. A. Gairola: Design for assembly: A challenge for expert systems,
    Robotics, Vol. 2, pp. 249-257, 1986.

40. A. Meister: The problems of using CAD-generated data for CAM, *CAD/CAM:
    Integration and Innovation*, SME Comput. and Automated Syst. Assoc.,
    Dearborn, MI, 1985.

41. A. Feder, K. Victor: Computer graphics benefits for manufacturing
    tasks that rely directly on the engineering design data base, *CAD/CAM:
    Meeting todays productivity challenge*, SME Comput. and Automated
    Syst. Assoc., Dearborn, MI, 1980.

42. B.M. Smith, J. Wellington: IGES - A key interface specification for
    CAD/CAM system integration, *CAD/CAM Integration and Innovation*, SME
    Comput. and Automated Syst. Assoc., Dearborn, MI, 1985.

43. N.N.: DIN-ISO 7942, Graphic Kernel System - A functional description,
    DIN Standard, 1982.

44. R. Carringer, Product definition data Interface (PPDI), *Proc. CIMTECH
    Conf.*, SME Comput. and Automated Syst. Assoc., Dearborn, MI, 1986.

45. K. Matsushima, N. Okada, T. Sata: The integration of CAD and CIM by application of artificial intelligence techniques, *Manufacturing Technology*, Techn. Rundschan, Berne, Switzerland, 1982.

46. D. Allen, W. Van Twelves: CAD in the CIM environment: Where do we go here?, *Appl. Mech. Rev.*, Vol. 39, No 9, pp. 1345-1349, 1986.

47. S.G. Tzafestas: Knowledge engineering approach to system modelling, diagnosis, supervision and control, *Proc. IFAC/IMACS Intl. Symp. on Simulation of Control Systems*, Vienna, Sept. 1986.

48. S.G. Tzafestas, M. Singh, G. Schmidt: System fault diagnostics, reliability and related knowledge-based approaches (Vol. 1,2), *D. Reidel*, Dordrecht, 1987.

49. S.G. Tzafestas, L. Palios: Improved diagnostic expert system based on Bayesian inference, *IMACS World Congress on Scientific Computation*, July, 1988.

50. M.Y. Chiu, E. Niedermayr: Knowledge-based diagnosis for manufacturing cells, *Siemens Forsch.-u. Entwichl.*, Vol. 14, No 5, pp. 230-237, 1985.

Construction of a Knowledge Base for Simulation and Control of Large Scale and Complex Systems. Applications

F. Stanciulescu [1]

Abstract. A methodology for the construction of a knowledge base of a large scale and complex system, with a view to generate the mathematical - heuristical model, for the simulation and control of such a system, is given. The knowledge base is composed of the system's structure, a data base, a mathematical models library, and a set of heuristical rules deducted from the expert knowledge. An application of a natural system is given.

Keywords: expert knowledge, knowledge representation, knowledge base, inference network, mathematical-heuristical model, knowledge-based simulation, completeness of knowledge, natural systems simulation.

1. KNOWLEDGE BASED SIMULATION

A tendency towards developing new methods for modelling, simulation and control of Large Scale and Complex Systems (LSCS), using both mathematical and heuristical methods, and elements of AI, has recently evolved, e.g. Davis and Lenat (1982), Reddy & Fox (1982), Kerckhoffs & Vansteenkiste (1985), Wonham and Ramadge (1986), Stănciulescu (1986) and others. What it is understood by knowledge - based simulation is the interpretation of a knowledge base in connexion with a standard simulation model.

The kernel of our Knowledge-Based Simulation System (KBSS) is the mathematical-heuristic model, composed of two main parts (see Stănciulescu 1987) :

. the heuristical model, which consists of a set of logical - linguistic rules derived from the expert knowledge and the experience (of the operator), concerning the system or process under analysis. These rules constitute the main part of a specialized knowledge base of the system ;

. the standard simulation model, which consists of a set of differential (or discrete - time) non-linear equations and interactions between subsystems; the simulation model also includes limits (or tolerance intervals) for each state, control and interaction variable.

Our methodology of the construction of the knowledge base of a LSCS is proposed with a view to generate the mathematical-heuristical model for simulation and control of such a system, suppose that the system is decomposable in several interconnected subsystems, so that the knowledge base be a reunion of n mini-knowledge base, one for each subsystem.

2. KNOWLEDGE REPRESENTATION

The process of developing an appropriate representation of the

---

[1] Research Institute for Computer Technique and Informatics,
Bd. Miciurin 8-10, Bucharest - Romania.

knowledge gets the main effort in any problem solving.However, it is
clear that no single representation scheme can hope to be universally
most powerful across all applications.

Besides these remarks, the knowledge representation is recommended
to have the following properties :

. knowledge representation should be both flexible and extendable,
i.e. modifications to existing knowledge in models do not necessarily
require substantial reorganisation of the model structure ;

. knowledge representation should be enbedded in the model (this
concerns the knowledge representation with representational semantics,
in general).

The mathematical-heuristical model is composed of conceptual objects
and relations (the conceptual knowledge), and of heuristical rules
(heuristical knowledge).

2.1.   Representation of the conceptual knowledge

2.1.1. The system's structure knowledge

Following the definition of the LSCS (see Stănciulescu 1987) these
systems can be decomposed into parts in any of a number of ways (see
e.g. Sydow 1982). One of these leads to a special decomposition, called
the tree diagram of the system. An example is given in Figure 1, where
the LSCS is decomposed into subsystems, and the subsystems are broken
into components, each of them being characterized by means of state variables and a set of parameters. To both states and parameters, one at-
taches tolerance intervals.

The tree structure shows how the system is decomposed into parts,
but it does not show the interactions among these parts. To represent
how the parts are connected to form the system, we use a graph; the
vertices (or the nodes) of the graph represent the part of the system,
and the arcs show the interaction between these. An example of graph is
given in Figure 2.

2.1.2. Data and formalized knowledge

To simulate and control a LSCS we need, for certain, a data base
and a set of (mathematical) models derived from the laws of nature. The
data base includes the values of the system parameters (which can be
constant or time series), initial values of the state variables, the
admissible tolerance intervals for both state variables and parameters,
a.s.o.

The formalized knowledge on the system is represented by means of a
models library, containing a set of standard modules, derived from the
laws, of nature and involved in the construction of the related models,
optional models, and other applications.

2.2.   Logical representation of the heuristical knowledge

2.2.1. Sequential Logic Principles

Sequential Logic is closely related to the concepts of moment, time interval, and sequence (see Stănciulescu 1965). Let

and
$$X = \left\{ x(0),\ x(1),\ \ldots,\ x(t-1),\ x(t),\ x(t+1),\ \ldots \right\}$$
$$Y = \left\{ y(0),\ y(1),\ \ldots,\ y(t-1),\ y(t),\ y(t+1),\ \ldots \right\}$$

be the sequences of the two logic variables $x(t)$ and $y(t)$, at various
moments t.

We call sequential logic sum the following logic function :

$$S_{t+1}(x,y) = x(t-1) \vee y(t).$$

We call sequential logic product :

$$P_{t+1}(x,y) = x(t-1) \wedge y(t).$$

We call sequential logic negation :

$$N_{t+1}(x) = \neg\, x(t).$$

The implication in the sequential logic is written as :

$$I(p(t-1),\ q(t)) \longleftrightarrow (p(t-1) \longrightarrow q(t)).$$

The assignation in the sequential logic is written as :

$$A(p(t-1),\ q(t)) \longleftrightarrow (q(t) \longleftarrow p(t-1)).$$

Other sequential logic functions and operators have been defined in
[5] .

2.2.2. Representation of the expert knowledge (heuristical rules).

The expert knowledge is represented by means of heuristical rules
in an analogue manner to the so-called (in AI) production rules. Mainly
there are two kinds of rules: behavioural (and interaction) rules and
control rules. The former describe the behaviour of a system under the
influence of parameters change (the inference between parameters and
states), while the latter describe the way thereby a system can be controlled (the inference between controls and states). From the control
rules we can deduce decision rules. Both behavioural and control rules
(as well as decision rules) have two parts: IF, which tests the applicability of the rule, and THEN, whose contents is executed, when the
rule in applicable.

The standard form of these rules is the following :

$$\langle\ \text{IF } C_i \text{ (condition), THEN } S_i \text{ (state)} \rangle\ , \tag{1}$$
(behavioural and interaction rule; $i = 1,2,\ldots,\ n$) ;

$$\langle\ \text{IF } C_i \text{ (condition) AND } S_i \text{ (state) AND } A_i \text{ (action),}$$
$$\text{THEN } R_i \text{ (result)} \rangle \tag{2}$$
(control rule; $i = 1,2,\ldots,\ m$) ;

From the last rule we can deduce :

$$\langle\ \text{IF } C_i \text{ (condition), THEN } A_i \text{ (action)} \rangle \tag{3}$$
(decision rule; $i = 1,2,\ldots,\ m$).

Using the Sequential Logic language, the heuristical rules (1) –
(3), can be rewritten as follows :

$$\langle C_i(\alpha_i(t)) \longrightarrow S_i(x_i(t+1))\rangle \tag{4}$$

(behavioural rule; i = 1,2,..., n) ;

$$\langle C_i(x_i(t), \alpha_i(t)) \wedge S_i(x_i(t)) \wedge (u_i(t+1) \longleftarrow u_i(t) +$$
$$+ \Delta u_i(t)) \longrightarrow R_i(x_i(t+2))\rangle \tag{5}$$

(control rule; i = 1,2,..., m) ;

and respectively :

$$\langle C_i(x_i(t), \alpha_i(t)) \longrightarrow (u_i(t+1) \longleftarrow u_i(t) + \Delta u_i(t))\rangle \tag{6}$$

(decision rule; i = 1,2,..., m).

Details concerning the representation of heuristic expert knowledge, using heuristical rules can be seen in [6].

3. KNOWLEDGE BASE CONSTRUCTION
3.1. The knowledge base structure

The essential step in the construction of a knowledge base, from the methodological point of view, is the setting out of its structure. This task is made easier by the facilities mentioned in the previous chapter, concerning the knowledge representation. On the other hand we take into consideration that the knowledge base is - in general - dedicated to a specific problem (e.g. simulation and control of large-scale ecological systems).

The experience acquired in the structuring of knowledge bases (see for example [2] and [4]) and our own experience, entitle us to propose the following definition :

By a knowledge base, created in order to answer "what if" question, we understand a collection of both conceptual and heuristical knowledge, which allow us to simulate the system behaviour and to control it.

Starting from this definition and from the knowledge representation (dealt with the 2 nd chapter), we can develop the structure of the knowledge base. The knowledge base is composed of two kinds of knowledge :

The conceptual knowledge. This component of the knowledge base includes the following knowledge :

. the system's structure knowledge, represented by both the tree diagram and the graph of the system (see the Figures 1 and 2). The tree diagram point out the components of the system, while the graph describes the system's interactions

. data and formalized knowledge of the system (the laws of nature). Besides data (generally contained in a data base), this component of the knowledge base includes a mathematical library, i.e. modules describing the laws of nature, and mathematical models representing the interactions betwen subsystems. The standard simulation model, included in the knowledge base is a discrete-time, non-linear model composed of :
(1) The state equations :

$$x_i(t+1) = A_i x_i(t) + B_i u_i(t) + f_i(x_i,\alpha_i) + v_i(x) ; \qquad (7)$$

$$(i = 1,2,\ldots, n)$$

(2) The initial conditions: $x_i(0) = x_{io}$; $\qquad$ **(8)**

(3) The interactions description $\quad v_i(x) = \displaystyle\sum_{j=1, j\neq i}^{n} g_{ij}(x_j) ;$ $\qquad$ (9)

(4) The admissible tolerance intervals for both states and interactions: $[x_{i1}, x_{i2}], [v_{i1}, v_{i2}]$.

(5) The constraints for the control variables: $u_{i1} \leq u_i \leq u_{i2}$.

The <u>heuristical knowledge</u>. The heuristical rules deducted from the
expert knowledge constitute the other component of the knowledge base.
The heuristical knowledge is composed of knowledge based on inference,
tendencies, rules of behaviour, and generally on all which can be called, in brief, heuristics. We do not include in the knowledge base all
heuristics, but only the heuristics which can be transformed into
heuristical rules, useful in practice (i.e. in simulation experiments
and system control). We remind that there are two kind of heuristical
rules :

. the behavioural rule, the standard form of which is:

$$\langle C_{i1}(\alpha_1(t)) \wedge \ldots \wedge C_{ip}(\alpha_p(t)) \longrightarrow S_i(x_i(t+1)) \rangle ; \qquad (10)$$

$$(i = 1,2,\ldots, n)$$

. the control rule, of the standard form :

$$\langle C_i(x_i(t), \alpha_1(t),\ldots, \alpha_p(t)) \wedge S_i(x_i(t)) \wedge (u_i(t+1)\longleftarrow$$
$$\longleftarrow u_i(t) + \Delta u_i(t)) \longrightarrow R_i(x_i(t+2)) \rangle ; \qquad (i = 1,2,\ldots,n) \qquad (11)$$

From the last rule we can deduce the so-called decision rule, of
the standard form :

$$\langle C_i(x_i(t), \alpha_1(t),\ldots, \alpha_p(t) \longrightarrow (u_i(t+1)\longleftarrow u_i(t) + \Delta u_i(t)) \rangle \qquad (12)$$

$$(i = 1,2,\ldots, n).$$

Figure 3 shows the typological structure of the knowledge base.

<u>Remark</u>: The set of inference procedures that operate on the knowl-
edge base, usually called inference engine, such as the inference network, are considered separately from the knowledge base. We are not
treating this subject matter in this paper. We simply show the inference rules, which can be used in the inference process :

. modus ponens : $\quad \dfrac{p \longrightarrow q,\ p}{q}$ $\qquad\qquad\qquad (13)$

(if p implies q, and p is true/then q is true) ;

. modus tollens: $\quad \dfrac{p \longrightarrow q,\ \neg q}{\neg p}$ $\qquad\qquad\qquad (14)$

(if p implies q, and q is false/then p is false).

3.2. Knowledge base validation

The central problem of validation is to validate the (conceptual
and **heuristical**) knowledge in relation to the tasks and the activities

it involves. The fundamental question will be to verify if the elicited and/or the acquired knowledge will allow the expert to succesfully complete the tasks. In essence we propose the knowledge base validation by means of one or several simulation experiments, and by comparing the simulation results with the experimental one (if possible), or to submit these results to the judgement of the experts (especially, for prediction purposes). In this case the conceptual knowledge must provide models, useful to correctly characterize the system's behaviour, while the heuristical knowledge must **cover** all the decision situation.

3.3. Methodological aspects of the construction of a knowledge base.

To accomplish a knowledge base it will be, in our oppinion, necessary and sufficient to pass through the following stages :

1 st stage: knowledge elicitation and acquisition,
2 nd stage: construction of the tree diagram,
3 rd stage: construction of the graph of the system interactions,
4 th stage: construction of the data base,
5 th stage: construction of the mathematical models library
6 th stage: construction of the heuristical knowledge base,
7 th stage: experimentation of the knowledge base,
8 th stage: validation of the knowledge base.

4. APPLICATION: THE KNOWLEDGE BASE FOR SIMULATION AND CONTROL OF A LARGE-SCALE HYDROLOGICAL SYSTEM

In this chapter we present a knowledge-base for simulation and control of a large-scale hydrological system, composed of six large lakes, interconnected by a network of canals and channels, of the Danube Delta.

4.1. The conceptual knowledge

4.1.1. The system's structure knowledge

The system's structure knowledge is represented by means of the diagrams of Figure no.1 and 2. The first represents the tree diagram of the analysed system, and emphasizes the six lakes, the state variables and their admissible tolerance intervals, the involved parameters. The second represents the graph of the system, which describes the interaction between the lakes, by canals and channels.

4.1.2. The formalized knowledge

This kind of knowledge is composed of both the data base and the mathematical models, derived from the applied physics.

The state equations :

$$\frac{dV_k}{dt} = (p-e)\,A_k + \sum_{i=1}^{n} Q_{ik} + Q_{rk} + \sum_{e=1}^{m} Q_{ke} + u_k \;,$$

$$V_k(0) = V_{ko} \quad (k = 1,2,\ldots, 6) \;,$$

$$Q_{ik} = \rho_{ik}\,A_{ik}\,\sqrt{2g}\; \text{sign}\; x_{ik}\sqrt{|x_{ik}|} \;,$$

$$x_{ik} = H_D - (Z_k + \frac{V_k}{A_k}) \;,$$

$$A_{ik} = L_{ik} H_{ik} \ ,$$

$$H_{ik} = \frac{1}{2} \left(\frac{V_i}{A_i} + \frac{V_k}{A_k} \right) \ ,$$

$$Q_{rk} = \rho_{rk} A_{rk} \sqrt{2g} \ \text{sign} \ x_{rk} \sqrt{|x_{rk}|}$$

$$x_{rk} = H_D - \left(z_k + \frac{V_k}{A_k} \right) \ ,$$

$$Q_{ke} = \rho_{ke} A_{ke} \sqrt{2g} \ \text{sign} \ x_{ke} \sqrt{|x_{ke}|} \ ,$$

$$x_{ke} = z_k + \frac{V_k}{A_k} - \left(z_e + \frac{V_e}{A_e} \right) \ ,$$

$$A_{ke} = L_{ke} H_{ke} \ ,$$

$$H_{ke} = \frac{1}{2} \left(\frac{V_k}{A_k} + \frac{V_e}{A_e} \right) \ ,$$

$$k = 1, 2, \ldots, n \quad (\text{in our application } n = 6).$$

where: $V_i/V_k/V_e$ is the volume of water in the lake $i/k/e$, p and e – the precipitation and the evaporation, $Q_{ik}/Q_{rk}/Q_{ek}$ – the inflow (the outflow) of water entered (exit) in (from) the lake k by canals or channels/under the floating reed islet/by canals or channels between two lakes, $\rho_{ik}/\rho_{rk}/\rho_{ke}$ – the rugosity coefficient, $A_i/A_k/A_e$ – the surfaces of the lakes $i/k/e$, $z_i/z_k/z_e$ – the level of the bottom of the lakes $i/k/e$, L_{ik}/L_{ke} – the width of the canals (or channels), H_{ik}/H_{ke} – the depth of water in the canals (channels), H_D – the level of the water of the Danube.

The lakes water's refreshment factor :

$$W_k = V_{ik}/(V_k + V_{ik})$$

The constraints (admissible intervals of variation) :

$$V_{k1} \leq V_k \leq V_{k2} \ ,$$
$$W_{k1} \leq W_k \leq W_{k2} \ ,$$
$$u_{k1} \leq u_k \leq u_{k2}.$$

4.2. The heuristical knowledge (heuristical rules).

Behavioural rules:

$$\langle (p < p_1) \wedge (e > e_2) \wedge ((Q_{ik}(t) < Q_{ik}(t-1)) \vee (Q_{rk}(t) < Q_{rk}(t-1) \vee (Q_{ke}(t) <$$
$$(Q_{ke}(t-1))) \longrightarrow (W_k(t) < W_k(t-1)) \rangle \ ;$$

$$\langle (p > p_2) \wedge (e < e_1) \wedge ((Q_{ik}(t) > Q_{ik}(t-1)) \vee (Q_{rk}(t) > Q_{rk}(t-1)) \vee (Q_{ke}(t)$$
$$Q_{ke}(t-1))) \longrightarrow (W_k(t) > (W_k(t-1)) \rangle \ , \qquad (k = 1, 2, \ldots, n)$$

Control rules:

$$\langle (p < p_1) \wedge (e > e_2) \wedge ((Q_{ik}(t) < Q_{ik}(t-1)) \vee (Q_{rk}(t) < Q_{rk}(t-1) \vee (Q_{ke}(t) <$$
$$Q_{ke}(t-1))) \wedge (W_k(t) < W_k(t-1)) \wedge (u_k(t+1) \longleftarrow u_k(t)) + \Delta u_k(t)) \longrightarrow$$
$$(W_k(t+1) > W_k(t)) \rangle \ ;$$

$$\langle (p > p_2) \wedge (e < e_1) \wedge ((Q_{ik}(t) > Q_{ik}(t-1)) \vee (Q_{rk}(t) > Q_{rk}(t-1) \vee (Q_{ke}(t) >$$
$$Q_{ke}(t-1))) \wedge (W_k(t) > W_k(t-1)) \wedge (u_k(t+1) \leftarrow\mid u_k(t) - \Delta u_k(t)) \longrightarrow$$
$$(W_k(t+1) < W_k(t)) \rangle \; . \qquad (k = 1,2,\ldots, n)$$

<u>Decision rules</u> :
$$\langle (W_k(t) < W_k(t-1)) \longrightarrow (u_k(t+1) \leftarrow\mid u_k(t) + \Delta u_k(t)) \rangle \; ;$$
$$\langle (W_k(t) > W_k(t-1)) \longrightarrow (u_k(t+1) \leftarrow\mid u_k(t) - \Delta u_k(t)) \rangle \; . \qquad (k = 1,2,\ldots,n)$$

In the rules above, $u_k(t)$ represents modifications of the inflow $Q_{ik}(t)$ and/or of the inflow (outflow) $Q_{ke}(t)$.

<u>Remark</u>: The same rules can be written for the simulation and control of water volumes $V_k(t)$.

The knowledge base, the structure of which has been described above, has been used to simulate the behaviour of the hydrological system, to forecast water volumes and the water refreshment factor, and to control these variables. The simulation results obtained on Romanian made mini-computer CORAL 4030 (PDP equivalent), are credible, and are used by the hydrologists, to statuate some decision concerning the lakes of the Danube Delta. Figure 4 shows the simulation results, concerning the water refreshment factor of two lakes of the Danube Delta, using the knowledge base.

5. CONCLUSIONS.

The knowledge-based simulation and control of a large-scale and complex system, is essentially based on a knowledge base. The key of the construction of a knowledge base is its structure and the knowledge representation. In this paper we present both the structure of the knowledge base and the knowledge representation. An application concerning the construction of a knowledge base, for a hydrological system composed of six interconnected lakes of the Danube Delta is presented, and we can conclude, that the simulation results obtained by using this knowledge base are credible and useful for the specialists.

REFERENCES
1. Davis, R., D.B.Lenat: Knowledge – Based System in Artificial Intelligence. Mc. Graw-Hill, Inc., New York, 1982.
2. Fedorowicz, Jane, G.D. Williams: Representing Modelling Knowledge in an Intelligent Decision Support System. In: DSS, 2 (1982), 1,3-14.
3. Kerckhoffs, E.J.H., G.C. Vansteenkiste: Advanced Simulation: Advanced Data/Knowledge Processing. 2nd Int. Symp. Systems Anal. Simul., Berlin, 1985.
4. Reddy, Y.V., M.S. Fox: An Artificial Intelligence Approach to Flexible Simulation. Res. Report, Carnegie – Mellon Univ., 1982.
5. Stănciulescu, F.: Sequential Logic and its Applications to the Synthesis of Finite Automata. In: IEEE Trans., vol. EC-14 (1965), 6, 786-791.
6. Stănciulescu, F.: Modelling and Simulation of Large Scale and Complex Systems Using a Knowledge Base and Applications. In: Proc. AMSE Confer. "Modelling & Simulation", Karlsruhe, 1987.
7. Sydow, A.: Hierarchical Concepts in Modelling and Simulation. In: Progress in Modelling and Simulation (F.E. Cellier, ed.), Academic Press, London, New York, 1982.
8. Wonham, W.M., P.J. Ramadge: Modular Feedback Logic for Discrete Event Systems. Systems Control Group Report no. 8614, University of Toronto, 1986.

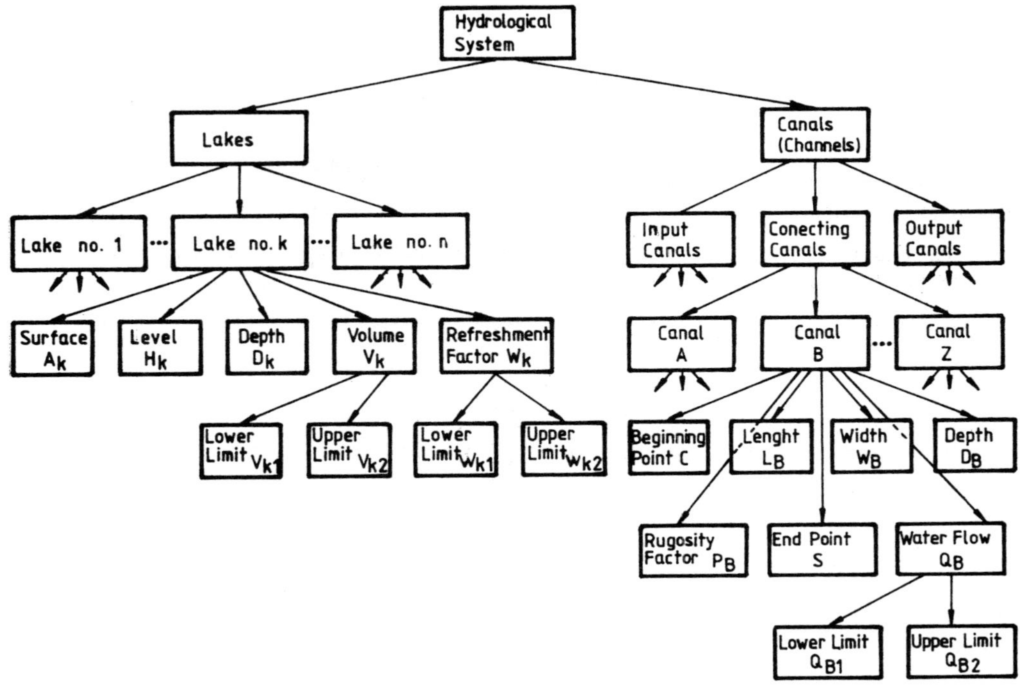

Figure 1. The Tree Structure of a Hydrological Network

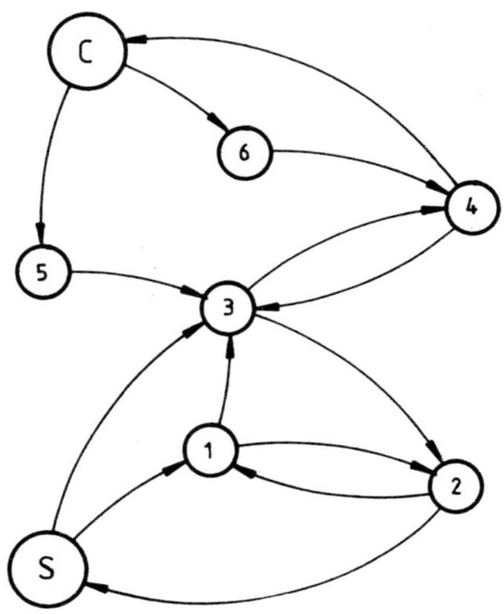

Figure 2. The Graph of a Hydrological System Composed of
Interconnected Lakes

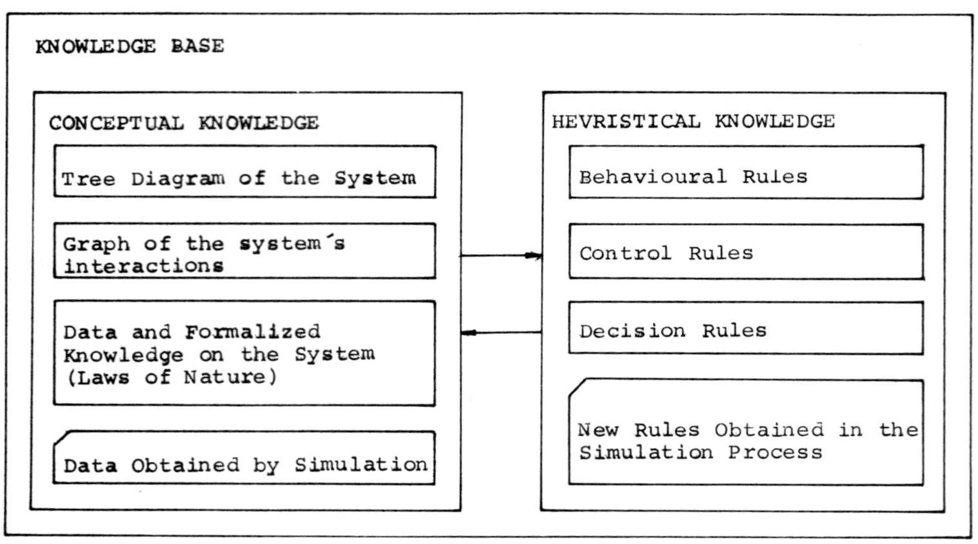

Figure 3. The Typological Structure of a Knowledge Base

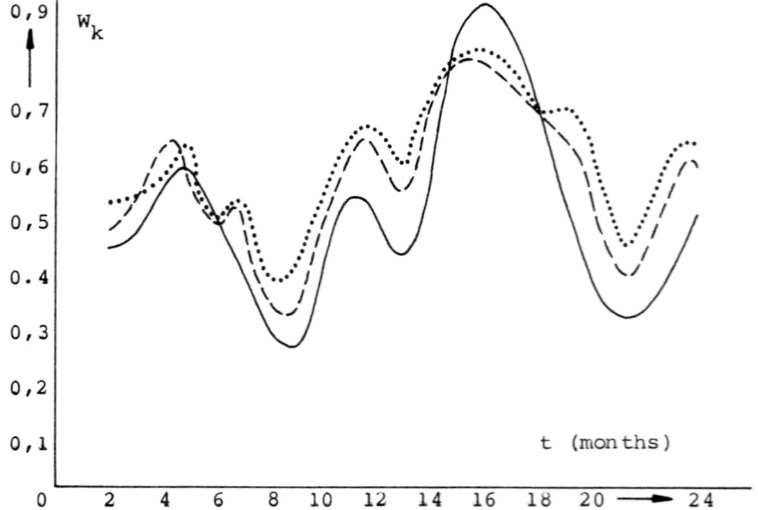

Figure 4. Simulation of the Water Refreshing Factor for two Lakes
of the Danube Delta

   ——————   simulation results for both lakes (real W_k)

   - - - - -   simulation results for the lake no. 3 (necessary W_{k3})

   · · · · · · · ·   simulation results for the lake no. 4 (necessary W_{k4})

Knowledge Based Inference Controlled Logic Simulation

András Jávor [1]

ABSTRACT

The possibilities provided by the methods and tools applied in expert systems, as knowledge bases and inference engines to control the simulation
experiments are outlined and their practical value shown on an implementation in investigating digital logic systems by simulation.

INTRODUCTION

Simulation as a tool for investigating and designing complex systems has
been used already for many years. The increasing demands posed by the increasing complexity of the systems to be modelled has been met by better,
more efficient and more sophisticated simulation tools. These simulators
have removed a large amount of the burden from the persons undertaking the
investigations.

It has however been realised that these traditional simulation tools could
solve only problems that where strictly related to conventional model
building and simulation. This area has been termed as *simulation in the
small* by Vansteenkiste [13]. In contrast *simulation in the large* has been
understood as comprising all the activities that were - possibly - more
loosely related to the process of simulation, but nevertheless have been
unavoidable to perform the necessary experimentations. These operations
included pre-, and postprocessing of information, evaluation and compression of the information gained from simulation.

As the aim of simulation has been to take certain tasks over from the investigator of the system the functions performed within *simulation in the
large* by means of auxiliary programs or program packages have helped to
achieve this end.

A further and more advanced possibility may be provided by the development
that took place in recent years resulting in practical application
of AI methods in the form of *expert system* methodologies. The application
of knowledge bases and inferencing based decision making and advising may
provide a large step towards fully automating the process of analysing
systems by means of simulation.

---

[1] Central Research Institute for Physics of the Hungarian Academy of
Sciences, H-1525 Budapest, P.O.Box 49., Hungary

In recent years the synergy of AI and simulation methods has become an essential trend [1][10][11][12][6][8]. In this paper the implementation of the principles that have been advocated is illustrated on a practical example to prove the close relation of the theories to practical problem solving.

THE CONTROLLING OF SIMULATION EXPERIMENTS

The general principle of the control of simulation experiments using a *knowledge base* and an *inference engine* is depicted in Fig. 1.

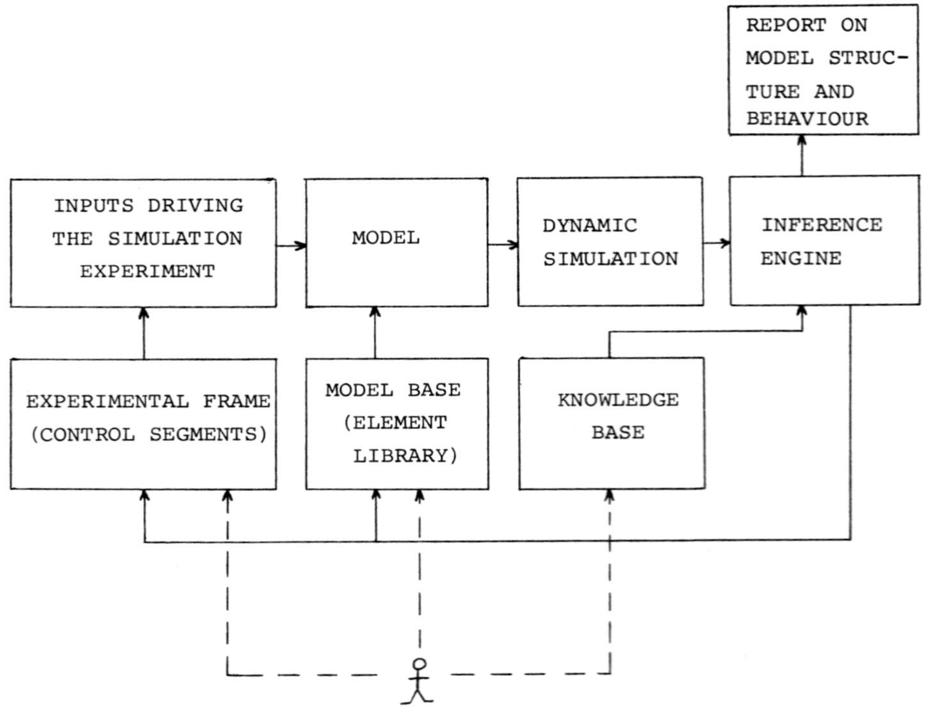

Fig. 1. Simulation experiment control using a knowledge base and inference engine

The structure shown is actually a synergy of a traditional simulator and an expert system. In contrast to conventional simulators the results of dynamic simulation are monitored and evaluated continuously by an *inference engine* with the aid of a *knowledge base* where the information with regard to expected results as well as possible deviations from them are stored. It is important to note that the information on which the actions are based are not of a "go - no go" nature i.e. in case of deviations, certain information on their direction and causes as well as on the possible actions to be taken has to be stored in the knowledge base.

The actions that are taken may be of two different kinds.

(a) The course of the control of the dynamic simulation experiment may be altered by means of changing the *control sequences* or *boundary values* determining the simulation run. This can be undertaken by using an *experimental frame* [14] in which the·*control sequences* have to be stored [6]. These control sequences consist of a set of *input sequences* and control conditions consisting of logic conditions together with the respective testing times and destinations (i.e. the control sequence to which control is transferred in case of fulfillment of the conditions) and finally reports that may be obtained on the fulfillment of the conditions tested.

The structure of the control sequences is the following.

$$C_i = (S_{i1}, S_{i2} \ldots S_{ik} \ldots S_{in})(t_{i1}; L_{11} = C_{m1}; R_1) \ldots$$
$$\ldots (t_{in}; L_{1n} = C_{m2}; R_n) \tag{1}$$

Where

C_i denotes the control sequence identifier,

S_{ik} the k-th input sequence of the i-th control sequence,

t_{in} the n-th testing time within the control sequence C_i,

L_{1n} the logic condition tested at the time value given in the same bracket,

C_{mi} the control sequence (given after the equation sign) the destination to which transfer is made on fulfillment of the corresponding logic condition at the prescribed time value,

R_i finally denote the reports that are generated on fulfillment of the logic condition at the prescribed time value.

With respect to the control structure the following have to be noted.

(i) All testing times as well as the time values describing the input sequences are in *relative values*. This means that they refer to the start of the control sequences as t=0. All these time values are updated at the time of their activation during dynamic simulation. The time parameters that are stored in the experimental frame are increased by the time of their activation. The control sequences may obviously be reactivated after their completion and may thus get other actual time values.

(ii) The reports generated are optional that may or may not be included. Nevertheless it is advisable to take advantage of this possibility since that way the person conducting the simulation experiment is automatically notified about the fact that the control of the simulation run is changed. This may give information on
- the reason that has been determined by the inference engine, and
- the further continuation of the control of the dynamic run.

It has to be noted that either the transfer of control or the report
may be omitted, but obviously not both because in that case the fulfillment of the logic condition would result in no action at all and
thus would be unnecessary.

In case the transfer of control is omitted, then we may obtain information only in certain situations that may be of value. On the other
hand if the reports are omitted the transfer of control of the experiment takes place without warning. The procedure is illustrated in
Fig. 2.

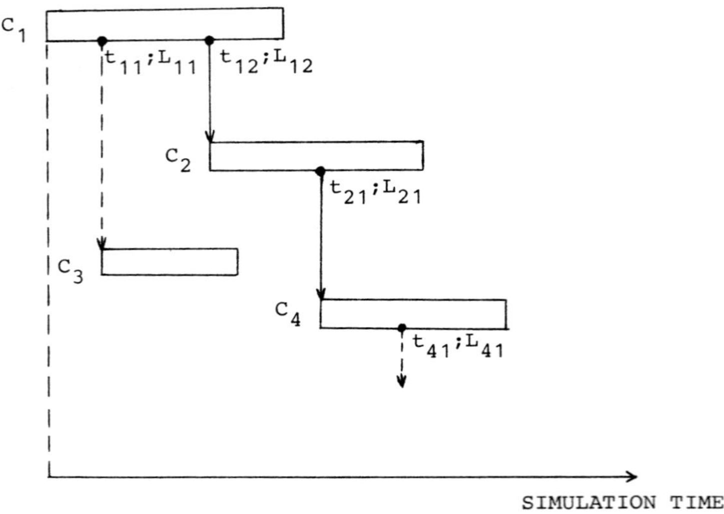

Fig. 2.   The process of simulation with conditional
transfers between control sequences

(b) The second type of action that may be evoked by the inference engine
is the changing of the model structure itself. This means that on the
fulfillment of certain conditions during dynamic simulation the model
structure itself is being modified and the simulation continued with the
modified model. The modification may be the removal, addition or exchange
of certain building elements of the model or a modification of its structure. Evidently the actions of modifying the experiment control and that
of the model itself may be combined.

STRUCTURAL QUESTIONS

To implement the structure described [4][6][3] there are a number of
problems that have to be solved.

The models ought to be built as a network of building elements having a

describing algorithm and communicating with each other via their inputs
and outputs. This non-procedural model structure [8] enables the convenient modification of the models as well as good correspondance to the respective parts in the knowledge base.

Another aspect that may contribute to system flexibility is the construction of the simulator in such a way that enables problem specific parts
to be inserted in a general purpose structure [6][3] that has its analogous counterpart is *shell* structures in expert systems.

CONTROLLING THE SIMULATION OF LOGIC

For the practical implementation of the principles of controlling simulation experiments by inferencing using a knowledge base, the simulation of
digital logic systems lends itself as an excellent area. The structure of
the models is essentially of non-procedural nature.

A realisation of the experiment control regarding the control loop including the experimental frame has been realized within the LOBSTER-MPC system [7][9]. This is the latest version of the LOBSTER (Logic Operational
Behaviour Simulator for Time and Effort Reduction) system where M stands
for Mixed mode and the letters PC indicate that it is running on IBM PC
XT/AT or 100 per cent compatible machines [5]. The main features of the
simulator are
a) Asynchronous time representation.
b) Quasideterministic state representation [2] providing for handling indeterminacies resulting from transient states and significant hazards
   caused by feedback loops.
c) Separate model building and dynamic simulation phases.
d) Element library containing elements with fixed and variable parameters.
e) Hierarchical model structures enabling nested subnetworks to be inserted repeatedly and at arbitrary depths.
f) Mixed mode simulation providing for the concurrent mixed simulation of
   logic and switch level networks. This possibility provides for bidirectional signal transfer, dynamic storage, etc.
g) Interactive graphic postprocessor for displaying the results.
h) Interactive conversational, user friendly input for model building and
   defining experimental conditions.
i) Experimental frame containing input sequences and sequence of transfer
   conditions in relative simulation time. The assignment of actual absolute time values being assigned at their activation.
j) The system is written in FORTRAN and runs under MS-DOS.

Finally it is mentioned that different versions of the simulator are currently being used in research, development and higher education in a number of places both in Hungary and abroad.

An example for solving a practical problem could be undertaken by determining the limiting speed of the operation of a digital logic circuit [9].
This is a well known problem in practical testing of such systems. Sequences have to be applied to the respective inputs of the circuit and the
outputs monitored for testing the proper operation of the system.

In case the system operates at the given speed adequately, the speed of
the test sequences has to be increased, if not it is decreased until the
limiting speed is found. This procedure has been undertaken with classical
hardware measurements, by tuning the frequencies of pulse generators. In
case of several inputs however several such input sequences had to be
varied.

The procedure in simulation can be undertaken in a way that the sets of
input sequences have to be grouped in *control sequences* together with
tests at properly chosen time instants and depending on the results of
testing accordingly to that in Eq. (1) transfer is made to a faster or
slower *control sequence*. This way similarly to a search in a binary tree
the proper speed may be determined to an arbitrary precision depending on
the number of control sequences applied. The search consisting of five
control sequences may illuminate this procedure. The transfers between
them could be described in the following way.

CONTROL SEQUENCE	FREQUENCY [Mc/s]	TRANSFER TO IF FASTER	IF SLOWER
1	5	[2]	–
2	50	–	[3]
3	10	[5]	4
4	6,6	–	–
5	20	[–]	–

During a given experiment – used for illustrating the problem – the
transfers of control have been realized at the conditions in rectangles.
This means that for the frequency f the following were true and reported
by the system

$$f > 5 \text{ Mc/s} \tag{2}$$

$$f < 50 \text{ Mc/s} \tag{3}$$

$$f > 10 \text{ Mc/s} \tag{4}$$

$$f > 20 \text{ Mc/s} \tag{5}$$

As a result it could be concluded that

$$20 \text{ Mc/s} < f < 50 \text{ Mc/s} \tag{6}$$

The "tuning" undertaken by the simulation system and displayed by its
graphic output on the screen can be seen in Fig. 3 where the increasing
and decreasing of the test sequences can be observed.

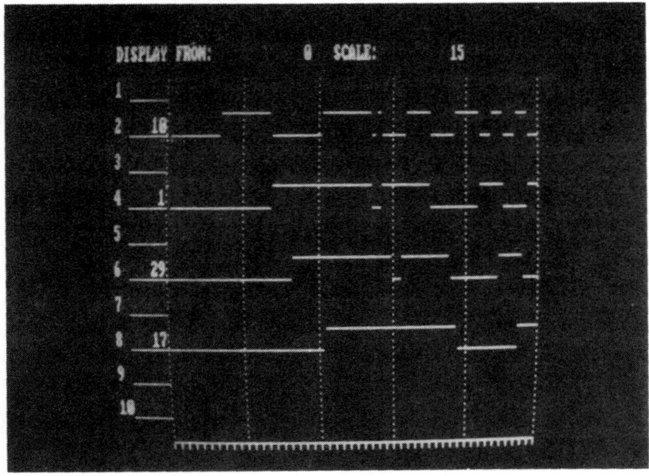

Fig. 3. Tuning of test sequence frequencies by
the simulator

CONCLUSIONS

As a rule in the field of research it can generally be accepted that "The
proof of the theory is in the practice" similarly to the popular proverb
about testing the pudding ... The application of AI based expert system
methodologies in simulation have seemed to be very promising. Nevertheless some of the scientists have been rather sceptical whether these
theories can be implemented for practical purposes. As the whole field of
simulation is intended to serve a manyfold of different disciplines actually as a heuristic tool of experimentation and analysis of complex systems the best way of deciding the practical value of these methodologies
seems to implement them for problem solving in some concrete area.

The field that we have chosen has been the simulation of digital logic.
Here the structure and descriptive style of the models have been well
adaptable for the purpose of the investigations. On the other hand the
complexity of the models and the extreme need to make the procedure of simulation more time efficient because of the significant economical impact
of its utilization have been important facts to try the methods.

The matching of state vectors of the models during simulation to expected
values stored in the rules of the knowledge base provided a state space
where the descriptions were easy to realize.

The investigations undertaken in simulating indeterminacies by handling
them with the paradism of the *Quasideterministic State Description* [2]
- and not simply by using pseudorandom generators - that has been unavoidable in the simulation of digital logic has been of value in describing
indeterminacies in knowledge representation and the synergy of the two [8].

The application of Expert Simulation System (ESS) principles [6] in analysing logic circuits can not only be applied in design verification but
also in simulation used in technological fault detection. It is hoped
that the utilization of the methods in a general purpose system may lead
to new and useful results.

ACKNOWLEDGEMENT

The author is indebted to M. Benkő and Á. Vigh whose valuable contribu-
tions in implementing the principles in the LOBSTER-MPC system has been
of high value.

REFERENCES

[1] Elzas, M.S., Ören, T.I., Zeigler, B.P. (eds.): Modelling and Simula-
    tion Methodology in the Artificial Intelligence Era, Elsevier, 1986.

[2] Jávor, A.: An Approach to the Modelling of Uncertainties in the Sim-
    ulation of Quasideterministic Discrete Event Systems, Problems of
    Control and Information Theory, $\underline{4}$ (1975) 3, 219-229.

[3] Jávor, A.: Proposals on the Structure of Simulation Systems, in Jávor,
    A. (ed.), Discrete Simulation and Related Fields, North-Holland, 1982,
    9-18.

[4] Jávor, A.: Applications of Expert Systems Concepts to Adaptive Experi-
    mentation with Models, in Elzas, M.S., Ören, T.I., Zeigler, B.P.
    (eds.), Modelling and Simulation Methodology in the Artificial Intel-
    ligence Era, Elsevier, 1986, ch.3. 153-163.

[5] Jávor, A.: LOBSTER-M: A Mixed Mode Simulator for CAD, in Trappl, R.
    (ed.), Cybernetics and Systems'86, D. Reidel Publishing Company, 1986,
    669-676.

[6] Jávor, A.: Proposals for the Architecture of Expert Simulation Sys-
    tems, in Proc. of the 2nd European Simulation Congress, 1986, 384-390.

[7] Jávor, A., Benkő, M.: Automatic Knowledge Based Decision Feedback
    Control of Simulation Experiments, in Proc. of IFAC/IMACS Internation-
    al Symposium on Simulation of Control Systems, 1986, 449-454.

[8] Jávor, A.: Declarative Programming in AI and Simulation, in Proc. of
    the 12th IMACS World Congress on Scientific Computation, Paris, 1988,
    (in publication)

[9] Jávor, A., Rőmer, M., Benkő, M.: Knowledge Base Controlled Simulation
    for Testing Digital Logic Circuits, in Proc. of the 12th IMACS World
    Congress on Scientific Computation, Paris, 1988, (in publication)

[10] Ören, T.I., Aytac, K.Z.: Architecture of MAGEST: A Knowledge-Based Modelling and Simulation System, in Jávor, A. (ed.), Simulation in Research and Development, North-Holland, 1985, 99-109.

[11] Ören, T.I.: Artificial Intelligence and Simulation: From Cognitive Simulation toward Cognizant Simulation, Simulation, $\underline{48}$ (1987) April, 129-130.

[12] Shannon, R.E., Mayer, R., Adelsberger, H.H.: Expert Systems and Simulation, Simulation, (1985) June, 275-284.

[13] Spriet, J.A., Vansteenkiste, G.C.: Trends in the Role of Modelling and Simulation, in Jávor, A. (ed.), Discrete Simulation and Related Fields, North-Holland, 1982, 3-8.

[14] Zeigler, B.P.: Multifacetted Modelling and Discrete Event Simulation, Academic Press, 1984.

Knowledge-Based Systems for Distributed Decision-Making[1]

Dj. B. Petkovski[2]

Summary. The paper presents three computer-based systems for (i) decentralized design of robust controllers, (ii) robustness analysis of large-scale economic systems and (iii) modeling the food production and distribution sector. The systems combine algorithmic and expert systems techniques. An important feature of the considered expert systems is the modularization of the software packages which allows a distributed problem solving approach.

Keywords: expert systems, distributed problem solving, large-scale systems, robustness, decentralized control, economic systems, production systems.

1. Introduction

The rapid advances in modern system and control theory will not find their potential application in modeling, analysis and design of large-scale systems unless they are pouched in computer-aided packages which are flexible, higher interactive, user friendly and easily accessible to the ultimate users. Existing computer-aided packages are primarily focused on certain isolated parts of the overall design process (e.g. system identification, model building, dynamic simulation, model reduction, optimization, etc.) without unifying them in a comprehensive modeling, analyze and design setting. Another criticism of the existing packages nowdays is that they require considerable skill and system and control theory background for their proper use. Thus, there is a real need for the development of knowledge based techniques in those packages to make them more user-friendly and be able to offer "intelligent" help at each step of the modeling, analysis and design process. Many researchers have already started investigation of expert systems for control systems work. The field of Artifical Intelligence is producing many tools and techniques applicable to the modeling, analysis, design and implementation of complex control systems (e.g. [3]). Much attention has been drawn to expert systems (e.g. [4]).

1) This work was supported in part by the U.S. - Yugoslav Joint Fund for Scientific and Technological Cooperation, in cooperation with NSF under Grant PP-736 and DOE under Grant PP-727.
2) Centre for Large-Scale Control and Decision Systems, Faculty of Technical Sciences, University of Novi Sad, Veljka Vlahovića 3, 21000 Novi Sad, Yugoslavia

In this paper we present three computer-aided systems for (i) decentralized design of robust controllers, (ii) robustness analysis of large--scale economic systems and (iii) modeling the food production and distribution sector. The systems combines algorithmic and expert system techniques. The knowledge based systems are programmed in PROLOG and the algorithmic parts in FORTRAN. In fact, many of the quantitative programs have been under development for many years, and with some modifications they were directly incorporated. An important feature of the presented knowledge-based systems is the modularization of the software packages.

The paper is organized as follows. The conceptual framework of the presented knowledge-based systems is given in Section 2. A knowledge-based system for stability robustness analysis and robust decentralized control systems design is presented in Section 3. In Section 4, a knowledge-based system for robustness analysis of large-scale economic systems is given. Finally, in Section 5 two knowledge-based systems are presented for modeling the food production and distribution sector.

2. Conceptual Framework

To design a control law i.e. to determine the decision strategy for complex systems such that a set of performance requirements are satisfied, a user first creates the model(s) of the system. Together with the model(s) is a set of performance specifications that constraint the final design. Then, experts from different fields are required to determine the control law. Therefore, typically, an design environment consists of a set of model building and control design experts that cooperate with each other. In addition to the experts, there is a large body of experience gained through a number of complex systems that have already been designed.

Time-critical and numeric in nature, many algorithms capable for modeling, identification and optimization of complex systems exhibit computational characteristics which are radically different from those exibited by existing expert systems. This is one of the most critical factor that complicate the use of expert system ideas in modeling and design oriented computer-aided packages for complex systems. As known the strength of expert systems comes from a symbolic processing capabilities, i.e., their ability to reason with non-numerical data. Therefore, the challenge is to integrate symbolic and numeric computation.

Therefore an important problem associated with the application of knowledge-based systems to large-scale systems lies in their computational complexity. Complex problems require large knowledge base. As a result, search space for problem solving algorithms tend to be large. The corresponding knowledge-based system may thus require a fair amount of time to reason before reaching conclusions and, thus, fail to provide timely solutions.

Real-life complex problems often require distributed problem solving approaches, that is approaches that involve the collaborative efforts of several problem solving agents with different field of expertise. Collaboration is necessary when no single agent can solve the entire problem. The agents must often work in parallel for reasons of speed and feasibility.

In order to reduce the complexity involved in modeling, analysis and design of large-scale systems, the decision-making process has been divided into a number of sub-processes. The advantages of organizing the modeling, analysis and design knowledge into different levels of complexity are as follows:

- This provides the necessary basis for the application of parallel processing algorithms.

- The performances of the closed-loop system, particularly those whose evaluation is very time consuming (e.g. robustness analysis) can be evaluated in parallel. If an acceptable level of the overall system behaviour is estimated by numericaly simpler algorithms, the use of more complex algorithm can be avoided.

- The lower level of design algorithms and techniques can be tried first before a more complex algorithm is used to tackle the design problem. This approach will give solutions which are no more complicated than necessary.

A variety of radically new architectures, such as parallel, fully or partially distributed (coupled or uncoupled) are now available to the decision-makers. Earlier restriction on memory, and the amount of time that can be allocated to a given process, may be overcome with these new architectures. With a new multiprocessor parallel and distributed software architecture many formally intractable problems associated with the modeling, analysis and design of large-scale economic systems can be sucsessfully solved.

In order to provide tractability, to facilitate maintenance of large bodies of knowledge, and to ease users from memorizing every detail of the system, one effective possibility is to apply a deductive knowledge base [5] approach. The essential idea of this approach is to bridge the modeling, analysis and design system and the user(s) of the system with a deductive knowledge base, which is a map, or a summary, of the system.

3. Decentralized Design of Robust Controllers

In order to reduce the complexity involved in large-scale control systems analysis and design the design problem could be divided into a number of sub-problems. In our case it was done by breaking the analysis and design stages down into three basic levels:

(1) data handling
(2) control law design
(3) analysis

These levels are organized in such a way that they can be represented and structured in computer(s). In other words, each level is organized as object-oriented system. Each level consists of various modules to accomplish the task of that particular level. This has the significant advantage that the associated knowledge base is extremely modular.

* Data handling level: accepts process data and specifications (e.g. modeling data, dynamic performance specifications, etc.) The Modeling Module, which is part of this level, manipulates the data performing linearization, model reduction, transformations between state-space and transfer functions, etc. The Pre-design Analysis Module has a function to analyze the model of the plant by checking the open-loop stability of the system, controllability, observability, calculating the uncertainties bounds, etc.

* Control law design level consists of various controller synthesis techniques in time and frequency domain. The systematic and modular nature of these design techniques facilitates the development of the corresponding expert system.

* Analysis level consists of various time-domain and frequency-domain techniques and provides additional information so that the designer can determine whether the current design is acceptable or not, i.e. it advances the user whether or not the selected control algorithm was in fact the one most appropriate to the given plant condition.

In this context, particular attention was given to: the stability robustness analysis module and the module for decentralized design of robust controllers.
The first module covers the following three categories of approaches for stability robustness analysis

(a) frequency domain approach
(b) time domain approach
(c) frequency domain approach which uses a state space representation of the system.

In the emerging field of interactive computing, the increasing speed of computers allows a designer to modify his design rapidly and observe the results of his different designs. It is well known that the design is a trade-off activity as well as an experimental and exploratory process. The use of expert systems techniques is invaluable in this aspect because an expert system can guide a designer to go through the design process and remind him of the various options he or she can take at each stage of the design. In developing our expert system for a decentralized design of

robust controller we closely followed the results presented in [11]. It
is assumed that the decentralized control system can be represented in
the form:

$$\dot{x}(t)=Ax(t)+\sum_{i=1}^{N} B_i u_i(t), \qquad x(t_0)=x(0) \tag{1}$$

where $x \in R^n$ is a state vector and $u \varepsilon_i R^{m_i}$ is a control vector, $\sum_{i=1}^{N} m_i = m$. The
information available to the local controller is assumed to be

$$y_i(t)=C_i x(t), \qquad i=1,2,\ldots,N \tag{2}$$

where $y_i \varepsilon R^{r_i}$ is a local output vector $\sum_{i=1}^{N} r_i = r$. The local control u_i is
assumed to be a direct feedback form the local output y_i namely

$$u_i(t)=E_i y_i(t) \tag{3}$$

where E_i is a time-invariant gain matrix. For more details see reference [13].

4. Robustness Analysis of Large-Scale Economic Systems

The problem of controlling large-scale economic systems has lead to
the introduction of the state space forms as an alternative representation of traditional model forms in various theoretical and empirical studies of dynamic economic systems, especially in the application of optimal
decision rules for macroeconomic planning and policy models. The application of optimal control techniques to macroeconomics has demonstrated the
potential of optimal control theory for macroeconomic growth theory, development and stabilization (e.g. [6,20]). To open up the field of econometric modeling to the techniques of optimal control, econometric models,
in either structural, reduced or final form, have usually been translated
into state space forms (e.g. [19]).

In order to simplify the notation, the focus of our attention is on
a large-scale discrete-time system controlled by a set of k agents - each
having different information and control variables:

$$x(n+1)=Ax(n)+\sum_{i=1}^{k} B_i u_i(n) \tag{4}$$

$$u_i(n)=F_i y_i(n), \qquad i=1,2,\ldots,k \tag{5}$$

$$u_i(n)=C_i x_i(n), \qquad i=1,2,\ldots,k \tag{6}$$

where $x(n)$ is the state of economy, $u_i(n)$ is the vector of exogenous variables (inputs) and $y_i(n)$ is the vector of endogenous variables (outputs).

The mathematical model of any economic system involves a number of
approximations brought about by the simplification of the theory, reduction of order, elimination of nonlinearities and the assumption of parametric invariance. All these approximations introduce a degree of uncertainty into any prediction of performance of the actual closed loop system.
While the economic processes of interest may vary greatly and performance
objectives may differ from application to application, most control strategies share the common requirement that stability be maintained in the
face of significant system uncertainties and perturbations.

Following some of the results presented in the previous section an
expert system is under development [17] for stability robustness analysis
of large-scale economic systems which mathematical models are given in
the form (4)-(6). Both unstructural perturbations (when only a bound on
the norm at the perturbation matrices is given) and structural perturbations (when the structure of perturbation matrices is specified and the
bounds on such structured perturbations are given) are included in the
robustness analysis. Additional modules are being developed so as to cover
some other system-theoretic properties such as suboptimality degree, controllability, observability, etc.

5. <u>Modelling the Food Production and Distribution Sector</u>

In the recent years, the development of new production technologies
for an increasing range of food products, incorporating a wide variety of
raw materials, has multiplied the problem of efficient strategic management and production planning for agricultural enterprises. This problem is
getting more difficult because the rate of technological inovation is
accelerating. Furthermore, because of the complexity of the environment in
which agroindustrial enterprises operate, it would be misleading to focus
on only a few elements when addressing the question of defining the optimal growth path. It became clear that it is neccessary to develop new,
computer-aided planning methodologies which will incorporate all relevant
parameters needed to clearly identify enterprise's operational parameters,
constraints and issues in its operations, and broadly outline possibilities for further growth.

The need to provide the capability of achieving higher efficiency of
resource usage, including the use of energy, raw materials, and human resources in complex engineering-economic systems, and in particular agroindustrial systems, has provided part of the motivation for the development of COSEFIM (Complex Systems Efficiency Improvement)[18], a dynamic
network model in which system variables described the system operation
while preserving structure. The development of COSEFIM was based on ALINET
(Alimentary Industry Network), a model for assessing energy conservation
opportunities in the food processing industry [7,8,14-16]. COSEFIM, a computer based modeling system, has been designed to bridge the gap between
detailed process and plant models (engineering models) on the one hand
and economic models on the other. The corresponding knowledge-based system, which is still under development, could help a user to develop and
use large-scale models for production planning and monitoring. The system
automates many of the tedious processes associated with large-scale modeling and provides an environment with a number of different forms of problem representation. As known, no single representation scheme can hope to
be universally most powerful accross all applications. Furthermore, addition work has been done in order to build intelligence into other aspects
of the problem solving process: model testing and validation, automatic
generation of alternative scenarious and aids for analyzing the results of
simulation. The software environment requires enhanced capabilities that
aid a decision-makers in validating the results produced from a modeling
effort.

One of the expert systems under development is "Expert System for
Small Farms". The primary objective of this research is to develop an expert system to help small farm operators improve their profitability by
enhancing their access to new technology and managerial information. With
the help of the expert system, small farmer operators will have equal access to new technology and managerial information in making production,
marketing, and farm management decisions. They can become efficient managers and complete more effectively with large farm operators and survive
and prosper in today's high technology and highly competative environment.
The expert system consists of two parts:

- The first part will be used to advice small farm operators the most
profitable combination of crops and livestock for their farms, taking into
consideration the farm size, i.e., acres of land by soil type, availability of labor, both own and hired labor, availability of cash and credit,
cost of production, such as seeds, fertilizers, chemicals, etc.

- The second part will advice farmers how to operate and manage each
crop or livestock selected.

Due to the lack of space, the refferences are omitted, but available
on request.

Knowledge-Based Modelling and Simulation: Restrictions, Alternatives and Applications

Axel Lehmann

Universität der
Bundeswehr München
Institut für Techn. Informatik
D-8014 Neubiberg, F.R. Germany

Universität Karlsruhe
Institut für Rechnerentwurf
und Fehlertoleranz
D-7500 Karlsruhe, F.R. Germany

ABSTRACT

This paper gives a survey of different categories of knowledge-based systems or expert systems, respectively, for supporting modelling and simulation of system dynamics in discrete systems. At first, actual requirements concerning support of users in model construction and goal-directed, efficient application of models are summarized. Regarding similarities existing between expert systems and simulation models with respect to their intended purpose, principal approaches for the combined and adjusted application of expert system and simulation techniques are discussed in this paper. As a result, knowledge-based modelling and simulation environments are classified regarding their taxonomy and their intended applicability. Some examples for practical applications of expert systems as classification and advisory systems for user supporting tools in the modelling process are finally presented.

1. ACTUAL REQUIREMENTS IN SIMULATION

Current trends in modelling and simulation can be characterized essentially by:

* a permanently enlarging application range of simulation, in general,
* an increasing number and diversity of modelling methods, solution techniques, simulation languages and tools (most of them are only applicable with respect to specific analysis goals and under certain restrictions) and
* the necessity for the construction and solution of extremely complex simulation models or a hierarchy of simulation models.

Regarding these trends, major requirements in modelling and simulation concern especially:

* improved support of model construction and application
  (concerning e.g. selection of modelling and simulation techniques, construction of an executable model, model verification and validation, interpretation of experimental results)
* availability of enhanced modelling and simulation tools
  (e.g. the development of efficient analytical modelling techniques applicable for hierarchical modelling, adequate application of uncertain or fragmentary data about the systems behaviour, goal driven control of simulation experiments, causal and qualitative modelling techniques).
* offer of powerful computer system architectures, configurations and networks for parallel or distributed modelling and simulation.

As demonstrated by recent publications, some innovative approaches are coming up to capture these requirements [Kit88], [ZeD86], [Wil88]. One basic approach is coupling of simulation and artificial intelligence techniques by the realization of knowledge-based simulation environments. This paper summarizes different categories, taxonomies and applications of coupling expert systems and simulation techniques.

2. ASPECTS FOR COUPLING EXPERT SYSTEMS AND SIMULATION MODELS

When comparing the intended purpose of knowledge-based systems or of expert systems with simulation models, some characteristic similarities and differences can be observed. Similarities exist in that sense that both approaches can be used to capture and to apply knowledge about the structure, organization and application of discrete or continuous changing systems. The difference is the way of knowledge representation, as well as the way of reasoning about this knowledge by applying numerical, procedural computing techniques in simulation versus symbolic, declarative techniques in knowledge-based systems.

Simulation models, as well as knowledge-based systems represent <u>know-ledge and expertise</u> in a modular form about a system and its behaviour in specific domains. For example, in discrete event simulation models, this knowledge is implemented by events, processes or activities in procedural form. In knowledge-based systems, frames, rules, semantic nets etc. are used to represent the knowledge about a system mostly in a declarative form. The <u>control flow</u> in simulation and in knowledge-based systems depends on logical decisions in contrast to numerical calculated values in other programs. The control flow for an event-driven simulation can be based on a next-event-time-scheduling algorithm in contrast to an inference mechanism like foreward or backward chaining in a knowledge-based system environment. Another difference between simulation and knowledge-based systems concerns the representation and processing of <u>fragmentary, uncertain or heuristic knowledge.</u> In stochastic simulation, uncertain knowledge is expressed by probabilities. In knowledge-based systems, uncertainty can be expressed by certainty factors, fuzzy-logic or fuzzy-sets [HaK85].

This comparison indicates that simulation and knowledge-based systems offer different methods, techniques and tools for knowledge acquisition and representation, as well as for inferencing and processing schemes. The similar scope of application of simulation models and of knowledge-based systems, on one hand side, and the different techniques for realization, on the other hand side, offer various possibilities for useful and efficient interrelations and integrations of knowledge-based systems, expert systems and simulation models.

3. APPLICATION OF EXPERT SYSTEMS IN SIMULATION

As indicated by Fig. 1, goal-directed modelling of system dynamics has to be seen as a multi-phase process. The solution, validation and experimental application of a model requires the availability of domain knowledge, modelling knowledge, tool knowledge and knowledge about statistical analysis. In general, this knowledge is distributed in documentations and among various experts in form of well-known facts, as well as in form of heuristics and experiences. Expert systems or knowledge-based systems are programs that emulate human expertise in a specific domain by applying techniques of logical inferences to a knowledge base. The knowledge base stores information about how to carry a task including facts, uncertain and heuristic knowledge, as well as non-algorithmic inference procedures. In contrast to conventional numerical, procedural software, expert systems (realized by symbolic programming languages) offer a higher degree of flexibility regarding knowledge acquisition and adaptation, explanation facilities of its line of reasoning and the contents of the knowledge bases [Joh84], [Hak85].

Applying the high potential of these features of declarative, symbolic and object-oriented programming environments, expert systems can be used more efficient than conventional software techniques to enhance the acceptance and effectiveness of modelling and to support users in the different modelling phases. Regarding their specific task, taxonomy and accessibility by the user as analyst, at least two categories have to be distinguished [OKe86], [Leh87]:

- **Expert systems as integral part of (simulation) models**
  The global goal of an expert-system-integrated modelling environment can be seen as an approach to represent and to reason about fragmentary, uncertain or frequently changeable domain knowledge in a model by means of an expert system technique, e.g. concerning the non-deterministic job-flow or user behaviour in a system. Regarding the degree of integration of expert system and coded model, a distinction can be made between:
  * an **interactive cooperation** of an expert system and an executable model (Fig. 2a), both systems mostly realized in different programming environments, and
  * an **expert system, embedded in the coded model** (Fig. 2b), mostly realized in the same programming environment as the simulation model, e.g. in PROLOG/TPROLOG.
- **Expert systems as supporting framework for a goal-directed application of modelling tools**
  This category of knowledge-based systems could be applied by unexperienced users in different stages of a modelling process, like for the selection of a problem-adapted modelling method or for the construction and experimental application of a simulation model with respect to cost-benefit considerations. In contrast to the first category, these expert systems are directly accessible by the user. Regarding the task of these expert systems, we have to distinguish between:
  * **decision support systems** (Fig. 2c) applicable as **classification systems** (giving the user final, weighted recommendations) or as **advisory systems** (leading the analyst in a step-by-step consultation, indicating alternative solutions weighted e.g. by certainty factors) [OKe86b], [Leh87];
  * **intelligent front-ends** (Fig. 2d), which offer the user several application-dependent interfaces to a single modelling tool or to a simulation language; this class of expert systems is used to bridge over the gap between problem domains and a modelling tool by inter-

nal mapping of the meaning and attributes of objects, items etc. of a specific application domain in terms used by the modelling tool.

4. EXPERT SYSTEMS AS DECISION SUPPORTING SYSTEMS IN INT^3

The goal of our research and tutorial project INT^3 concerns the implementation of a highly _interactive_, _intelligent_ (knowledge-based) and _integrated_ modelling facility on a standard PC. Regarding the different phases of a modelling process (see Fig. 1), we have fixed the general concept of an INT^3-environment by the following intentions [LKK86b], [LKK86c]:

- to provide the user with a comfortable interface in all phases of a modelling process by means of an interactive, advise-giving dialog component and by means of computer graphics,
- to offer on an INT^3-PC different modelling and simulation tools, which can be processed not only on PC but also on a host system connected to the PC,
- to enhance the efficiency of modelling supporting users by expert systems as decision supporting systems or intelligent front-ends during model construction and experimental application [LRS88],
- to collect and to manage all the information concerning a specific problem and the models belonging to it in a single data base; most of these data collected once are accessible for several tools in different modelling phases.

A prototype version of INT^3 is available for students for application and testing on IBM PC/XT or AT.

5. CONCLUSIONS

This paper summarizes some new perspectives for the development and application of simulation models by coupling knowledge-based or expert system computing techniques, respectively, and conventional simulation techniques for several reasons:

- to enhance the flexibility and effectiveness of modelling for analysis of system dynamics, in general;
- to support an increasing user community - confronted with a permanently growing number of simulation techniques and tools - in the different phases of a modelling process;
- to improve the efficiency of simulation, especially regarding the representation and adaptation of changable knowledge, the model optimization and the understandability of the way of reasoning.

First prototypes and practical applications with knowledge-based simulation indicate their usefulness and importance. Other remarkable research on that topic includes:

* qualitative modelling and causal reasoning, as well as
* inductive modelling.

<u>REFERENCES</u>

[Hak85]   Harmon, P.; King, D.
          "Expertensysteme in der Praxis;
          Oldenbourg-Verlag, 1987

[Joh84]   Johnson, T.
          "The commercial application of expert systems technology";
          Ovum Ltd., London/England, 1984

[Kit88]   Kitzmüller, C.T.
          "Simulation and AI: Coupling symbolic and numeric computing";
          in: AI and Simulation, Society for Computer Simulation, 1988

[Leh87]   Lehmann, A.
          "Expert Systems for Interactive Simulation of Computer System
          Dynamics"; in: Simulation Series Vol.18, No.3: AI and Simulation, Society for Computer Simulation, San Diego, 1987

[LKK86b] Lehmann, A.; Knödler, B.; Kwee, E.; Szczerbicka, H.
          "INT^3: Interactive, Intelligent and Integrated Modelling in a
          Typical PC Environment"; Application Brief 3, IBM Germany,
          September 1986

[LKK86c] Lehmann, A.; Knödler, B.; Kwee, E.; Szczerbicka, H.
          "Interactive Modelling and Simulation in an Intelligent PC-
          Environment"; in: Proceedings of the European Simulation Congress, Antwerpen/Belgium; September 1986

[LRS88]   Lehmann, A.; Roll, G.; Szczerbicka, H.
          "Application of expert systems in INT^3"; in: AI and Simulation; Society for Computer Simulation, 1988

[OKe86a] O'Keefe, R.
          "Simulation and expert systems - A taxonomy and some examples"; in: Simulation, 46/1; Jan. 1986

[OKe86b] O'Keefe, R.
          "Advisory systems in simulation"; in: AI Applied to Simulation; Kerckhoffs, E.J., Vansteenkiste, G.C., Zeigler, B.P.
          (Eds.); Simulation series, Vol.18, No.1; Febr. 1986

[Wil88]   Wildberger, A. M.
          "Integrating an Expert System Component into a Simulation"
          General Physics Corp., Columbia, MD, 21044, 1988

[ZeD86]   Zeigler, B. P.; De Wall, L.
          "Towards a Knowledge-Based Implementation of Multifacetted
          Modeling Methodology";
          in: AI Applied to Simulation, SCS Simulation Series, Vol. 18,
          No. 1, 1986

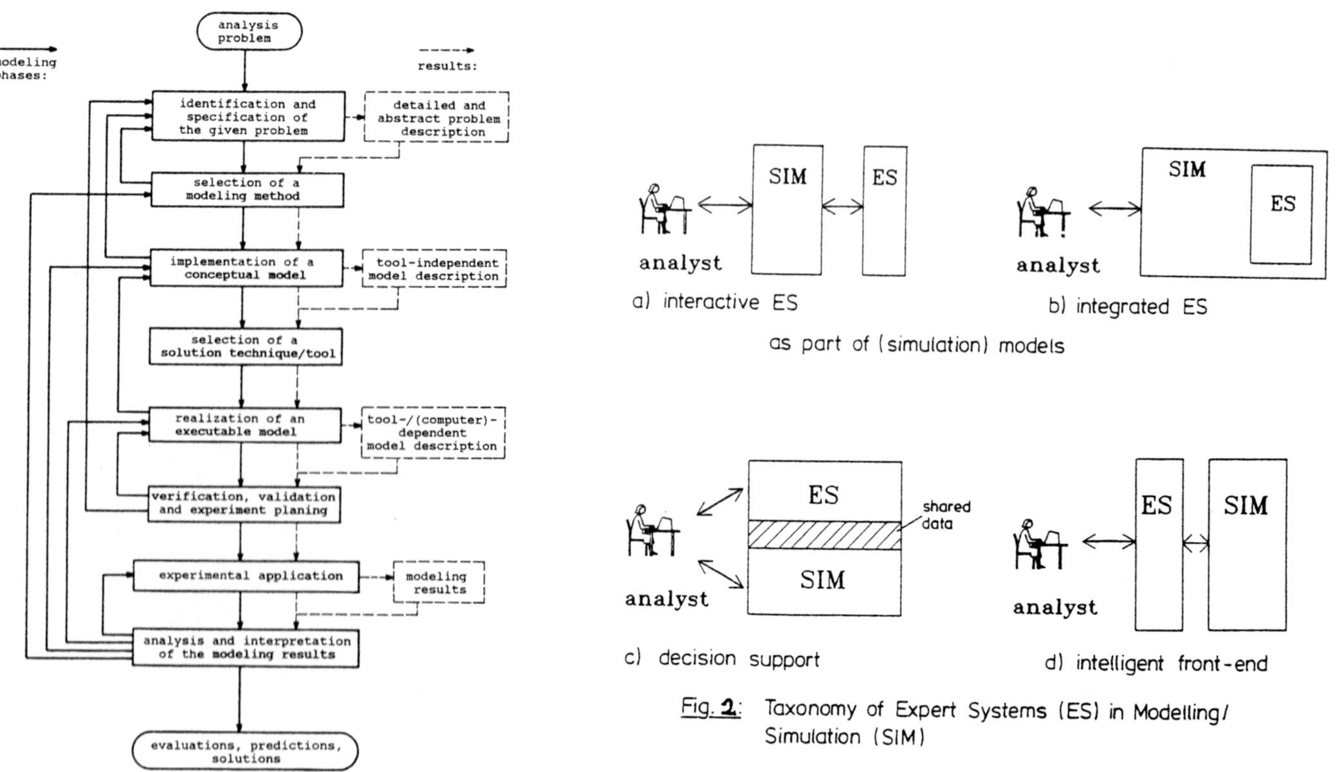

Fig. 1: Phases and results of a modelling process

Fig. 2: Taxonomy of Expert Systems (ES) in Modelling/ Simulation (SIM)

Knowledge Based Process Control

Berndt Böhme, Ralf Wieland, Uwe Starke 1)

1. INTRODUCTION

The development in microprocessor technology and computer science in
the last ten years has caused substantial changes in process automation
technology. Nowadays the new generation of decentralized process control systems which are sufficiently reliable and satisfy the requirements for high performance allows to an increasing extent the implementation of the so-called "higher automation functions".
The application of these automation tasks to complex systems faces us
usually with a lot of features making substantial trouble.
Such features are
- markedly incomplete information on the system and the running processes
- fuzzy and/or confusing information
- multi-objective decision making
- disturbances are characterized by high amplitudes
- numerous data
- the set of admissible control actions cannot be overlooked

Due to these characteristics and the usually high complexity of the
systems under consideration classical methods of mathematical modelling
and control fail, mainly because of the enormous expenditure spent on
the development and the strong real-time conditions.
There is no chance to overcome all these difficulties alone by revising
conventional methods or creating and improving nonconventional methods
of process automation and their application to automatic control systems. Past experience has shown that the application of the majority of
higher automation functions is not reasonable without the decision maker (DM) and his central position in a process control system even in
future.

Owing to the high responsibility of the DM, the features of the system
to be controlled and the generally hectic conditions during decision
making the DM needs a suitable decision support system for determining
an effective control. Using expert knowledge it would be desirable to
create an expert system the roots of which lie in the field of artificial intelligence.
Today the development in AI has reached a stage which to an ever increasing extent allows serious applications /1/, /2/, /3/, /4/.

---

1) TH Leipzig, Sektion Automatisierungsanlagen, Karl-Liebknecht-Str.
132, Leipzig, 7030

2. FUNCTIONAL STRUCTURE OF PROCESS CONTROL SYSTEMS WITH INTEGRATED EXPERT SYSTEMS

Modern process automation systems are characterized by a hierarchical structure. Usually, we can distinguish three functional levels
- field level: decentralized process data acquisition and primary data processing
- control level: process stabilization, -optimization, -malfunction diagnosis, -emergency control
- top level: process/production management

The operator or DM in both the process control level and the top level should be supported by different expert systems according to the given automation tasks to be solved. Even though the above-mentioned tasks widely differ, in most cases the following general conception for designing the intelligent control system seems to be favourable.
To satisfy the demands of real-time mode algorithms for performing an automatic situation recognition should be separated from the knowledge base of the expert system and implemented in the basic process automation system.
Situation recognition as a nonconventional method of real-time identification can be characterized by the following two partial tasks:
- assignment of the process input and/or state vectors in a limited but not empty set of natural or semantic classes
- detection and interpretation of beginning changes in the controlled process which lead to unwanted or dangerous states.

If there exists only one admissible control action for the situation detected (e.g. emergency control) then this elementary situation control will be performed automatically. In all other cases the results of situation recognition and, if necessary, further information given by the DM form the input information of the expert system.
The expert system is used now to solve two tasks:

- DIAGNOSIS
  Determination and localization of the reasons causing the current or prognosticated system state
- DECISION MAKING
  Generation of an effective control corresponding to the diagnosis achieved.

The interactive diagnosis and decision making is based on the knowledge represented in the expert system as well as on a continual interpretation and assessment of the external world by the DM. Thus the DM can verify control actions proposed by the expert system concerning their consistency with further information that is not acqired by the process automation system and, if necessary, demand alternative solutions.

3. KNOWLEDGE REPRESENTATION IN PROCESS AUTOMATION

With regard to its semantic background the knowledge base (KB) includes:

1. A copy of the objective reality - that is a mathematical model of the technological system to be supervised or controlled on the basis of nonconventional forms of description (declarative part of KB).
2. Strategies and procedures for solving the desired automation tasks. This knowledge is based in a high degree on human experience and notformalized information about the process to be controlled. This expert knowledge represents the procedural part of the KB. It includes also instructions for applying the mathematical model stored in the declarative part of the KB.

A typical feature of the procedural knowledge base consists in its limited truth. This situation can only partly be improved by an efficient knowledge acquisition component of the expert system. In general the fuzzyness of the expert knowledge and data should be taken into consideration for decision making.
The trustworthiness of expert knowledge (and also unreliable data) can be valued by
- Bayesian probabilities
- certainty factors
- using fuzzy sets
Without doubt the creation of a fuzzy knowledge base leads to a markedly reduced error risk and may be regarded as the most elegant solution. However, when applying the fuzzy set theory we are always faced with additional expenditure in the fields of knowledge acquisition and knowledge processing.
Furthermore should be mentioned that the derivation rules for logical deduction applied in the inference engine (e.g. modus ponens) are not valid for any realization of fuzzy variables. It can be shown that these deduction rules lose their tautological character in the case of fuzzy logic. This is a very interesting matter of fact which, however, has been investigated insufficiently so far.
In satisfying the demands of real-time operation expert systems for process automation require the integration of conventional control software that could also be denoted as deep knowledge. The utilization of algorithmic procedures representing mathematical models as well as diagnosis and control algorithms on the basis of conventional forms of description is necessary or desirable for the following reasons:
1. Knowledge implemented in this way is sharper and mostly safer than the formalized expert knowledge.
2. This knowledge will be processed faster, thus improving the ability of the system to operate under real-time conditions.
3. In most of the industrial branches is a lot of well-tried conventional software that can be applied in expert systems without substan-

tial alterations. By utilizing this software the expenditure for
development could markedly be reduced, above all, in the phase of
knowledge acquisition.

Summerizing our experience in the field of knowledge representation we
come to the conclusion that it would be desirable and useful to create
hybrid knowledge representation schemas combining frames with other representation forms. This leads to a special frame taxonomy, which describes typical events, states, situations, plants, processing units up
to elementary devices.

In such a schema the slots represent the knowledge of facts or point to
subframes whereas the expert knowledge for solving the automation task
is integrated into the frames by production rules, semantic networks or
other forms (Fig.).

FRAME NAME	: name of the given frame
SLOTS	: sequence of pairs - variable / value
SUBCLASSES	: name of the frames linked directly
DERIVATION RULES	: rules for selecting preferential frames in per- forming the search
PROOF RULES	: rules for verifying the selected subframes

Fig. Structure of the diagnosis frame in PROCON

In that way we obtain a well structured knowledge base leading e.g. in
the case of production rules to a drastically reduced set of these
rules to be matched by the interpreter performing the inference process. Furthermore solutions for proving the KB concerning its consistency as well as components for knowledge acquisition and learning become simpler because the expert knowledge is arranged semanticly.

The idea discussed was applied in the shell PROCON /5/. An application
(production management in a chemical engineering plant) in which the
shell PROCON is used will be demonstrated.

4. REFERENCES

/1/ KOMMTECH'87. 4. Europ. Kongreßmesse für techn. Automation. 12.-15.
Mai 1987, Essen. Dokumentation Kongreß VI "Wissensbasierte Systeme
in der Praxis".

/2/ 10th IFAC World Congress on Automatic Control. Munich, July 27-31,
1987. Preprints, subject area 15.1 (vol. 6).

/3/ GI-Kongreß Wissensbasierte Systeme. München 20./21. Oktober 1987.
Informatik Fachberichte 155. Springer Verlag Berlin, Heidelberg,
New York, Tokyo.

/4/ INFO'88. 4. Kongreß der Informatiker der DDR. Dresden, 22.-27.
Februar 1988.

/5/ Böhme,B;Balzer,D.;Wieland,R.;May,V. PROCON I - Ein Expertensystem-
Shell für die Prozeßsteuerung. In /4/, Proceedings S. 195-197.

A Symbolic-Numerical Support for Computer-Aided Modelling[1]

Antoni Ligęza and Maciej Szymkat[2]

1. Introduction

Recently a great attention has been paid to Computer Aided Control Systems Design (CACSD) (see [3,6]). This new area covers such trends as specialized control-oriented Computer Aided Design (CAD), knowledge-based support of the creative processes, symbolic computer calculations and interactive man-machine dialog based on a user-friendly communication system (interface). A number of papers have been devoted to the design and implementation problems arising during the development of CACSD systems (see [2]).

In this paper we present a programming approach to symbolic-numerical manipulation inside a specific CACSD system. The work reported here constitutes a part of a software project called DS including also graphically supported modelling and interactive simulation of such systems for both engineering and educational purposes. The DS package is destinated for the implementation on an IBM PC XT/AT or compatible computers. The DS modules responsible for resources management, carrying out numerical tasks and supervising the communication with the user are written in C language, while the other modules performing symbolic calculations are implemented in PROLOG. The knowledge base containing the current information about the system under analysis is organised with the use of both C and PROLOG data structures. The problems of the interaction of the above modules is solved through linking and procedure parameter passing.

2. Overview of the DS system

The following theses constitute the basic assumptions concerning the development and operation principles of the DS system:
- the system is meant to be user-friendly, easy to communicate with and manipulate for a control engineer rather than a programist,
- the system operates fast enough to provide the possibility of real-time interaction with the user, and,

---

[1] Supported by the Polish Ministry of Higher Education and Technology under the contract PR I.02 ASO 2.1/1987.

[2] Institute of Automatics, Academy of Mining and Metallurgy, al.Mickiewicza 30, 30-059 Kraków, Poland

- the communication with the user is organized with extensive use of the control engineer's natural language, i.e. block diagrams, symbolic transfer functions, etc.

The system is to be used throughout the complete design or analysis process, including initial structural design, mathematical model setting, numerical simulation and experiments, symbolic calculations, stability and performance analysis, etc., while modifications can be introduced at any stage of analysis.

The main modules of the system include:
- a full-screen graphical editor for creating, inspecting and modifying the block diagrams of control systems,
- a block of numerical routines for interactive simulation support,
- a symbolic calculations component for automatic handling of symbolic transformations and determining the desired transfer functions,
- a graphical display module for on-line visual presentation of the simulation results.

The data concerning the designed/analysed control system are specified in a graphical, symbolic and numerical form, depending on their detailed character. The structure of the control system is defined with the use of the full-screen graphical editor. All the defined blocks, inputs and outputs are marked with their proper names (defined by the user or system-generated default names). At this level the transfer functions are also represented by symbolic names. Any block can be redefined so as to constitute a system (of blocks) itself, i.e. the specification of the control system structure can be performed at several levels of hierarchy. The environment of a subsystem is accessible via special block represented by a block diagram frame. A number of the editing functions is performed automatically and the editor possesses some degree of intelligence. In principle there are no predefined quantitative constraints for any data categories (the only limitation is the amount of the computer memory available).

The modules responsible for graphical editing/representation and numerical calculations are being written in C language, both for its conciseness and speed. The modules destinated to perform symbolic calculations and model tramsformations are written in PROLOG language (see [1]) which is considered to be a modern and convenient tool for higher-order symbolic manipulation on data. A crucial problem affecting the efficiency concerns the inter-module data exchange. The traditional solutions, i.e. linking the modules together into a single stand-alone executable program turned out to be unsatisfactory. On the other hand the implementations of the object-oriented languages on PC-s seem to be premature. Finally, we decided to organize the communication between PROLOG and C through common predicates and standard facilities and to use the dynamic linking approach in the interactive simulation module.

<u>3. Symbolic-numerical calculations</u>

The DS package provides the user with two ways of model parameter handling. A declared parameter may be interactively accessed, redefined and assigned a value; it can be also treated in a purely symbolic way, i.e. identified with its name only. Such an approach is applied in the standard software for symbolic manipulations (see [4]). The symbolic-numerical operations within the DS package are being implemented in PROLOG employing termal and list structures. The main features of symbolic computation include cascade (path) finding, loop detection and symbolic calculation of rational transfer functions.

Let us briefly discuss, for example, the problem of determining the symbolic transfer functions between given two points in a block diagram of a linear system. The first approach is the diagram reduction, including such operations as node shifting, replacement of serial connections by a single resulting connection, signal summation and subtraction and loops elimination. The second approach consists in the direct application of Mason rule in order to find the transmittances between nodes of interest without the middle steps of the block diagram modification. The first approach results in a chain expanded fraction form of the transfer function while the second one gives a simple fraction form. The representation of the systems structure is transferred to PROLOG via specially defined predicates for communication with C language. The basic format of facts representing the structure of the investigated system is as follows:

```
connected(Node1,Node2,TransferFunction),
```

where Node1 and Node2 stand for an input or output nodes of a certain block and TransferFunction is the symbolic inter-node transmittance. Note that the proposed format of the connected predicate subsumes all the possible connections in the block diagram. In fact, there are three possibilities:

- Node1 is an input node of a block and Node2 is an output node of the same block with TransferFunction being the symbolic transmittance between the nodes (inside the block),
- Node1 is an output of some block and Node2 is an input of some other block with the TransferFunction equal to 1; in this case the predicate represents a connection between two blocks (this case includes the possibility of signal branching),
- there are two (or more) connected predicates having the same symbol in the place of Node2 with the TransferFunction being equal to 1 for all of them; in this case Node2 is a summation node.

A set of PROLOG clauses defining the rules of calculations of symbolic transfer functions is to be defined. The symbolic names of the transfer functions are represented as lists in order to ensure an easy manipulation. The following, simplified code excerpt is an example of cascade finding in a previously defined block diagram:

```
cascade(X,Y,[connected(X,Y,G)]) :-
    connected(X,Y,G).
cascade(X,Y,[connected(X,Z,G)|H]) :-
    connected(X,Z,G),
    cascade(Z,Y,H).
```

(Note: the predicates for loop checking were removed).

The above example of PROLOG code is a simplified recursive definition of a cascade, i.e. there exists a cascade between nodes X and Y if they are connected directly with the transfer function G or if node X is connected directly with node Z by G and there is a cascade between node Z and node Y. The results of the operations are kept in a symbolic form and can be used for the analysis of the connections and signal flows. Once the structure is (at least temporarily) fixed, the transfer functions can be defined as rational functions and "substituted" into the previously calculated symbolic expressions. Then specialized routines for symbolic transformation of the polynomials structures are to be applied. The resulting transfer functions can be used then for stability tests, simulation models construction and frequency analysis.

4. Conclusions and further work

The approach to building a CACSD system described in this paper was primarily aimed at highlighting some of the most important problems arising during the construction of the DS system. The main assumptions constituting the guidelines for the project were given and some details concerning the implementation were outlined. The presented symbolic-numerical approach to the control design problems constitutes a useful tool for engineering applications. The area of future work should cover the development of a conceptual framework for the formalization of the control system design procedures (compare [5]). This should result in the construction of the knowledge-based support system enhancing the user's creative thinking during all the stages of the design process.

References

[1] Clocksin, W.F. and C.S.Mellish: Programming in PROLOG (second edition). Springer, Berlin 1984.
[2] Isermann, R. (ed.): Preprints for the 10th World Congress on Automatic Control, Vol.7, Subject Area 12.4. IFAC, Munich 1987, pp.235-294.
[3] Jamshidi, M. and C.J.Herget,(eds.): Computer-Aided Control Systems Engineering. North-Holland, Amsterdam 1985.
[4] Microsoft muMATH-83 Symbolic Mathematics Package. Honolulu 1983.
[5] Pang, G.K.H. and A.G.J.MacFarlane: An Expert Systems Approach to Computer-Aided Design of Multivariable Systems. Lect. Not. in Contr. and Inf. Sci., Vol. 89, Springer, Berlin 1987.
[6] Rimvall, C.M.: Man-Machine Interfaces and Implementational Issues in Computer-Aided Control System Design. Diss. ETH No 8200, Zurich 1986.